The Complete
HOME ELECTRICAL WIRING HANDBOOK

The Complete
HOME ELECTRICAL WIRING HANDBOOK

J. T. Adams

ARCO PUBLISHING, INC.
NEW YORK

Published by Arco Publishing, Inc.
219 Park Avenue South, New York, N.Y. 10003

Library of Congress Cataloging in Publication Data
Adams, J. T.
The complete home electrical wiring handbook.
Includes index.
1. Electric wiring—Amateurs' manuals. I. Title.
TK9901.A29 621.319'24 78-21969
ISBN 0-668-04525-6 (Library Binding)

Printed in the United States of America

Contents

SECTION I: FUNDAMENTALS AND PROCEDURES OF ELECTRICAL WIRING

Acknowledgments

THE AUTHOR desires to acknowledge with thanks the assistance of the following firms and national organizations and branches of the government that have cooperated in the production of this book:

American Home Lighting Institute, American Institute of Electrical Engineers, American Society of Agricultural Engineers, American Standards Association, American Wire Gage Standards, Brown & Sharpe Mfg. Co., Edison Electric Institute, Electric Energy Association, Electronics Laboratories, Illuminating Engineering Society, Industry Committee on Interior Wiring Design, International Association of Electrical Inspectors, Minneapolis Honeywell Regulator Co., National Adequate Wiring Bureau, National Board of Fire Underwriters, National Electrical Code, National Electrical Contractors Association, National Electrical Manufacturers' Association, National Electrical Wholesalers Association, Radio Manufacturers Association, Underwriters' Laboratories, South Dakota State University, U.S. Department of the Air Force, U.S. Department of the Army, U.S. Naval Bureau, Westinghouse Electric Corporation.

Preface

THIS BOOK provides practical information in the design, layout installation, and maintenance of electrical wiring systems. Design of interior wiring systems, construction methods, and tools and material used are discussed in detail, and profusely illustrated.

In addition, *The Complete Home Electrical Wiring Handbook* includes information relating to methods of repairing, remodeling, and extending domestic and foreign systems; troubleshooting; necessary safety precautions in electrical installations; and chapters on batteries and transformers.

This book has been produced to help not only apprentices and beginners, but homeowners, experienced electricians, teachers, architects, and others to understand the principles of interior wiring.

In the Appendices you will find Electrical Terms; Electric Data; Design Procedures for Electrical Wiring; Electrical Abbreviations; Electrical Symbols of American Standards Association; Formulas; Wire and Motor Data; Component Color Code; and forty-two reference tables.

A comprehensive index adds to the usefulness of this handbook.

It is the wish of the author that the material in *The Complete Home Electrical Wiring Handbook* be interesting, helpful, and beneficial to those who realize the importance of interior wiring and the tremendous possibilities it offers.

J.T.A.

The Complete
HOME ELECTRICAL WIRING HANDBOOK

Section I

Fundamentals and Procedures of Electrical Wiring

Never compromise on the quality or safety of your electric wiring.

Wire for your future needs as well as for your present needs.

A home wiring system must be designed with attention to all details—service entrance; circuits; outlets; symptoms of inadequacy; planning modernization; additional wiring; sizes of conductors; types of circuits.

Chapter 1

Fundamentals of Electricity

Throughout this book, emphasis is placed on the constructional aspects of electric wiring. The term "phase" is used when referring to the angular displacement between two or more like quantities, either alternating electromotive force (EMF) or alternating currents. It is likewise used in distinguishing the different types of alternating current generators. For example, a machine designed to generate a single EMF wave is called a single-phase alternator, and one designed to generate two or more EMF waves is called a polyphase alternator.

Power Generators

Power generators will produce single- or three-phase voltages that may be used for electrical power systems at generated voltage or through transformer systems.

Single-phase generators. Single-phase generators are normally used only for small lighting and single-phase motor loads. If the generated voltage is 120 volts, a two-wire system is used. (*See* Table 7(A), Appendix 2.) One of the conductors is grounded and the other is ungrounded, or hot. The generated single-phase voltage may be 240 volts. This voltage is normally used for larger single-phase motors. In order to provide power for lighting loads, the 240-volt phase is center-tapped to provide a three-wire single-phase system. (*See* Table 7(B), Appendix 2.) The center tap is grounded neutral conductor. The voltage from this grounded conductor to either of the two ungrounded (hot) conductors is 120 volts. This is half of the total phase value. The voltage between the two ungrounded conductors is 240 volts.

This system provides power for both lighting and single-phase 240-volt motors.

Three-phase system. The most common electrical system is the three-phase system. The generated EMF's are 120 degrees apart in phase. As shown in Table 7 (C, D, E), Appendix 2, three-phase systems may be carried by three or four wires. If connected in a delta (Δ), the common phase voltage is 240 volts. Some systems generate 480 or 600 volts. If the delta has a grounded center tap neutral, a voltage equal to one-half the phase voltage is available. If the phases are wye (γ) connected, the phase voltage is equal to $\sqrt{3}$ (1.73) times the phase-to-neutral voltage.

Single-phase three-wire and *three-phase four-wire systems* provide voltages for both lighting and power loads. If the load between each of the three phases or between the two ungrounded conductors and their grounded center tapped neutral are equal, a balanced circuit exists. When this occurs there is no current flowing in the neutral conductor. Because of this, two ungrounded conductors and one grounded neutral may be used to feed two circuits. Thus, three conductors may be used where four are normally required.

The preceding discussion leads logically into one-, two-, and three-phase electric light circuits. Electric lamps for indoor lighting in the United States are generally operated at 110 to 120 volts from constant-potential circuits. Two- and three-wire distribution systems, either direct current or single-phase alternating current, are widely used for lighting installations.

These systems of distribution are capable of handling both lamp and motor loads connected in parallel between the constant-potential lines. The three-wire system provides twice the potential difference between the outside wires than it does between either of the outside wires and the central or neutral wire. This system makes it possible to operate the larger motors at 240 volts while the lamps and smaller motors operate at 120 volts. When the load is unbalanced, a current in the neutral wire will correspond to the difference in current taken by the two sides. A balance of load is sought in laying out the wiring for lighting installations.

Drawing Symbols and Schematic Wiring Diagrams

Symbols. The more common symbols and line conventions used in wiring plans are shown in Fig. 1. These symbols allow precise location of any electrical equipment in a building from the study of a drawing.

Schematic wiring diagrams. Electrical plans show what items are to be installed, their approximate locations, and the circuits to which they are to be connected. Single lines indicate wires connecting the fixtures and equipment. Two conductors are indicated in a schematic diagram by a single line. If there are more than two wires together, short parallel lines through the line indicate the number of wires represented by the line. Connecting wires are indicated by placing a dot at the point of intersection. No dot is used where wires cross without connecting. The electrician may encounter drawings in which the lines indicating the wiring have been omitted. In this type of drawing only the fixture and equipment symbols are shown; the location of the actual wiring is to be determined by the electrician. No actual dimensions or dimension lines are shown in electrical drawings. Location dimensions and spacing requirements are given in the form of notes or follow the standard installation principles. (*See* Fig. 1.)

Drawing Notes

A list of drawing notes is ordinarily provided on a schematic wiring diagram to specify wiring requirements and indicate building conditions that alter standard installation methods.

Color Coding

The National Electrical Code requires that a grounded or neutral conductor be identified by an outer color of white or natural gray for Number 6 or smaller wire. For larger conductors the outer identification of white or natural gray may be used, or they should be identified by white markings at the terminals. The ungrounded conductors of a circuit should be identified with insulation colored black, red, and blue, used in that order, in two-, three-, or four-wire circuits, respectively. All circuit

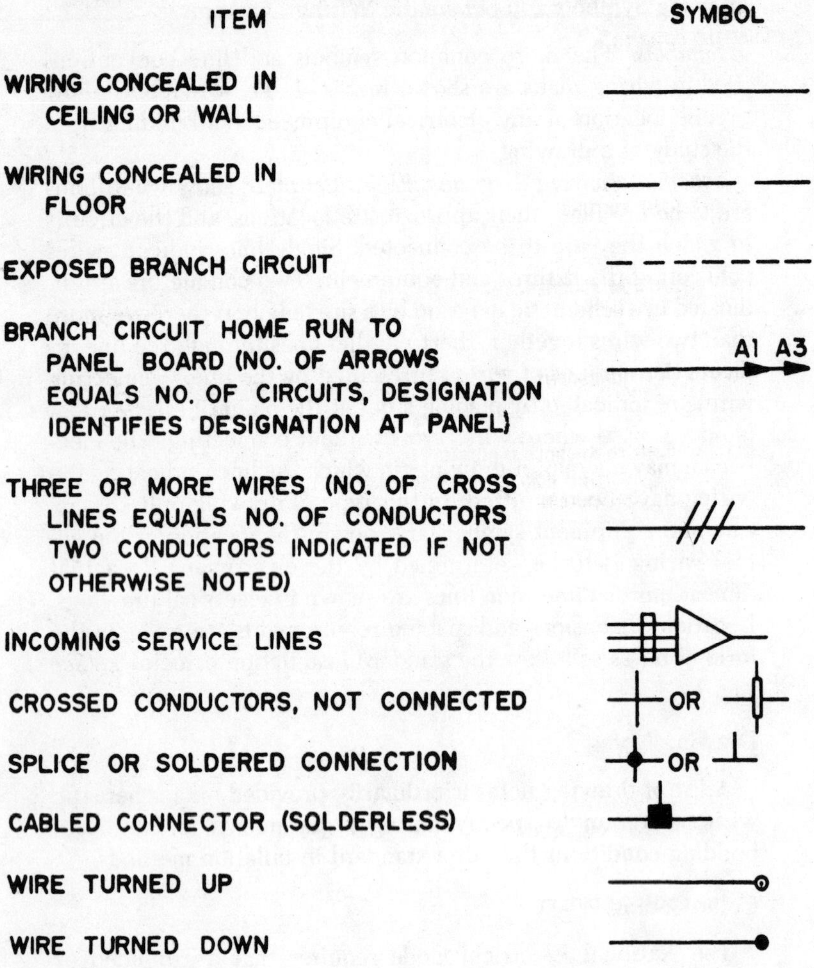

ITEM	SYMBOL
WIRING CONCEALED IN CEILING OR WALL	
WIRING CONCEALED IN FLOOR	
EXPOSED BRANCH CIRCUIT	
BRANCH CIRCUIT HOME RUN TO PANEL BOARD (NO. OF ARROWS EQUALS NO. OF CIRCUITS, DESIGNATION IDENTIFIES DESIGNATION AT PANEL)	A1 A3
THREE OR MORE WIRES (NO. OF CROSS LINES EQUALS NO. OF CONDUCTORS TWO CONDUCTORS INDICATED IF NOT OTHERWISE NOTED)	
INCOMING SERVICE LINES	
CROSSED CONDUCTORS, NOT CONNECTED	OR
SPLICE OR SOLDERED CONNECTION	OR
CABLED CONNECTOR (SOLDERLESS)	
WIRE TURNED UP	
WIRE TURNED DOWN	

Fig. 1. Standard electrical symbols.

ITEM	SYMBOL	ILLUSTRATION
LIGHTING OUTLETS*-		
CEILING		
WALL		
FLUORESCENT FIXTURE		
CONTINUOUS ROW FLUORESCENT FIXTURE		
BARE LAMP FLUORESCENT STRIP		

*LETTERS ADDED TO SYMBOLS INDICATE SPECIAL TYPE OR USAGE

J- JUNCTION BOX R- RECESSED
L- LOW VOLTAGE X- EXIT LIGHT

ITEM	SYMBOL	ILLUSTRATION
RECEPTACLE OUTLETS**-		
SINGLE OUTLET	OR 1	
DUPLEX OUTLET		
QUADRUPLEX OUTLET	OR 4	
SPECIAL PURPOSE OUTLET	OR	
20-AMP, 250-VOLT OUTLET		
SINGLE FLOOR OUTLET (BOX AROUND ANY OF ABOVE INDICATES FLOOR OUTLET OF SAME TYPE)	OR	

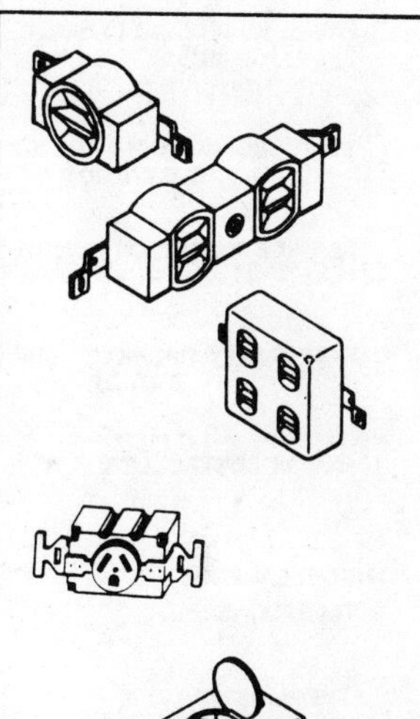

**LETTER G NEXT TO SYMBOL INDICATES GROUNDING TYPE

Fig. 1. (continued)

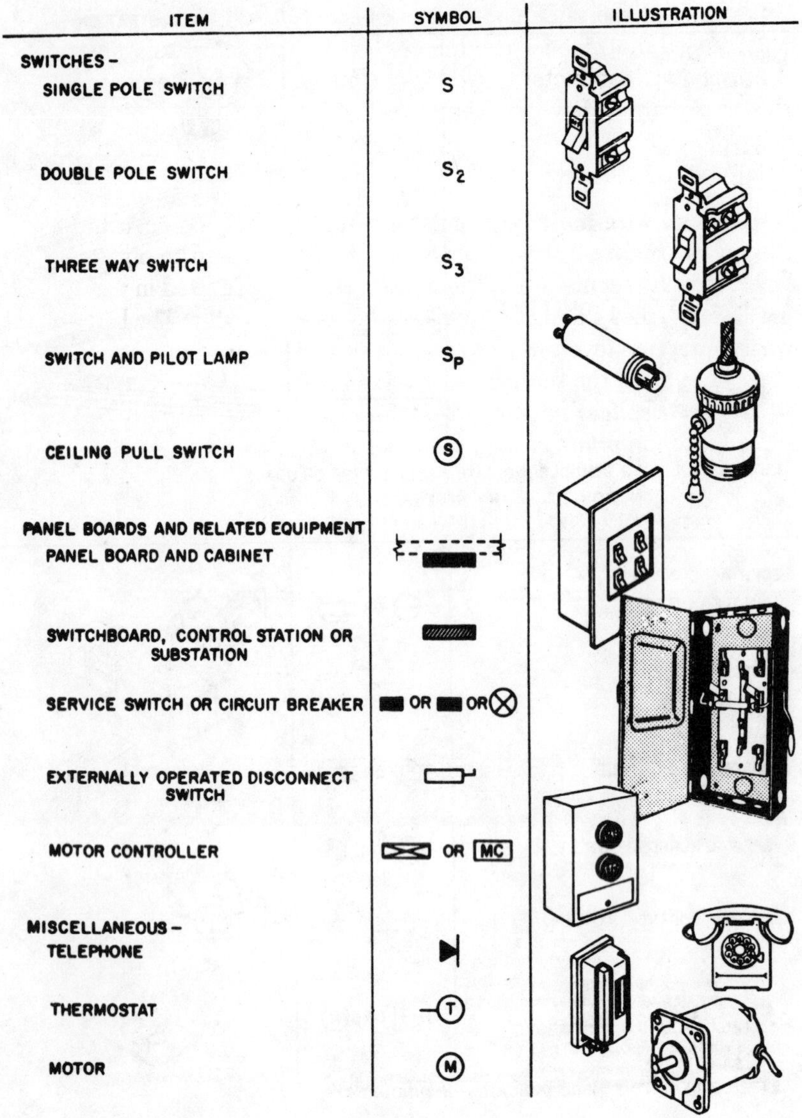

ITEM	SYMBOL	ILLUSTRATION
SWITCHES –		
SINGLE POLE SWITCH	S	
DOUBLE POLE SWITCH	S_2	
THREE WAY SWITCH	S_3	
SWITCH AND PILOT LAMP	S_P	
CEILING PULL SWITCH	Ⓢ	
PANEL BOARDS AND RELATED EQUIPMENT		
PANEL BOARD AND CABINET		
SWITCHBOARD, CONTROL STATION OR SUBSTATION		
SERVICE SWITCH OR CIRCUIT BREAKER	OR OR ⊗	
EXTERNALLY OPERATED DISCONNECT SWITCH		
MOTOR CONTROLLER	OR MC	
MISCELLANEOUS –		
TELEPHONE		
THERMOSTAT	Ⓣ	
MOTOR	Ⓜ	

Fig. 1. (continued)

conductors of the same color should be connected to the same ungrounded (hot) feeder conductor throughout the installation. A grounding conductor, used solely for grounding purposes, should be bare or have a green covering.

Splices

A spliced wire must be as good a conductor as a continuous conductor. Figure 2 shows variations of splicing used to obtain an electrically secure joint. Though splices are permitted in wiring systems, they should be avoided whenever possible. The best wiring practice (even in open wiring systems) is to run continuous wires from the service box to the outlets. *Under no conditions should splices be pulled through conduit. Splices must be placed in appropriate electrical boxes.*

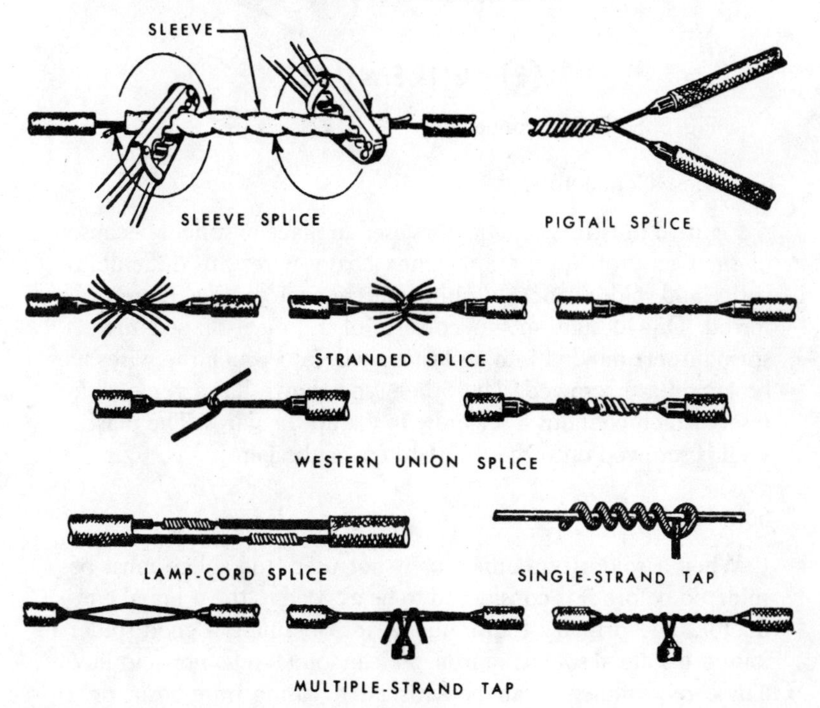

Fig. 2. Typical wire splices and taps.

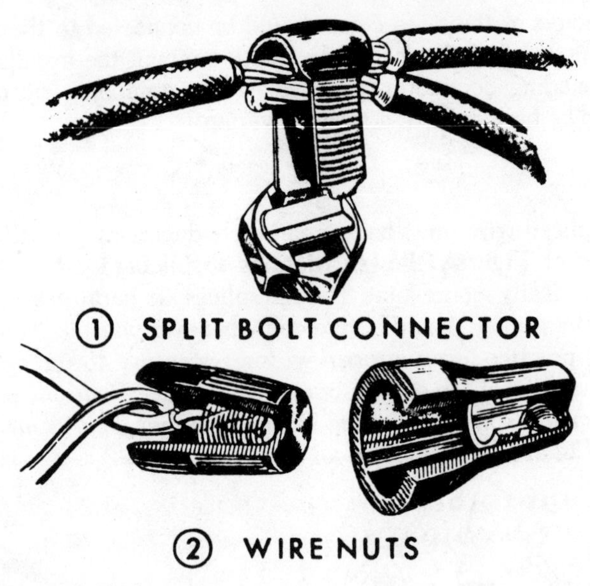

Fig. 3. Solderless connectors.

Solderless Connectors

Figure 3 illustrates connectors used in place of splices because of their ease of installation. Since heavy wires are difficult to splice and solder properly, split-bolt connectors (Fig. 3) are preferred. One design shown consists of a funnel-shaped metal-spring insert molded into a plastic shell, into which the wires to be joined are screwed. The other type shown has a removable insert which contains a setscrew to clamp the wires. The plastic shell is screwed onto the insert to cover the joint.

Soldering

When a solderless connector is not used, the splice must be soldered before it is considered to be as good as the original conductor. The primary requirements for obtaining a good solder joint are a clean soldering iron, a clean joint, and a nonacid flux. These requirements can be satisfied by using pure rosin or a rosin core solder on the joint.

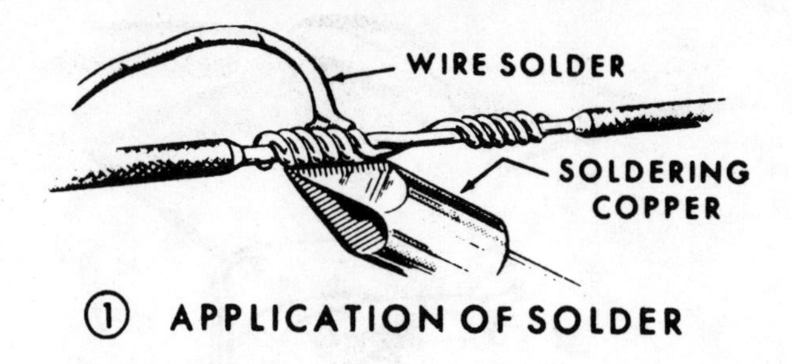

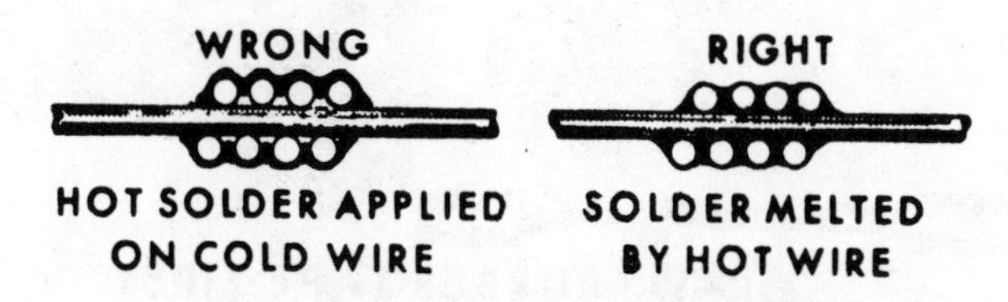

Fig. 4. Soldering and solder joints.

To insure a good solder joint, the electric heated or copper soldering iron should be applied to the joint (Fig. 4) until the joint melts the solder by its own heat. Figure 4 shows the difference between good and bad solder joints. The bad joint has a weak crystalline structure.

Figure 5 illustrates dip soldering. This method of soldering is frequently used by electricians because of its convenience and relative speed for soldering pigtail splices. (*See* Chapter 4, Electrical Conductors and Wiring Techniques, sections on Soldering.)

Taping Joints

Every soldered joint must be covered with a coating of rubber or varnished cambric and friction tape to replace the wire insulation of the conductor. In taping a spliced solder joint (Fig. 5),

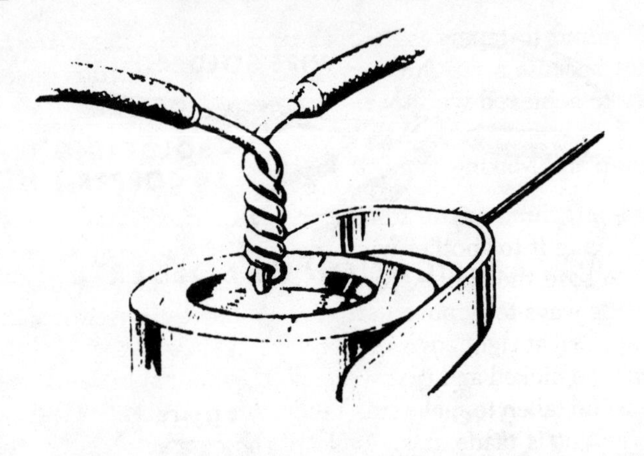

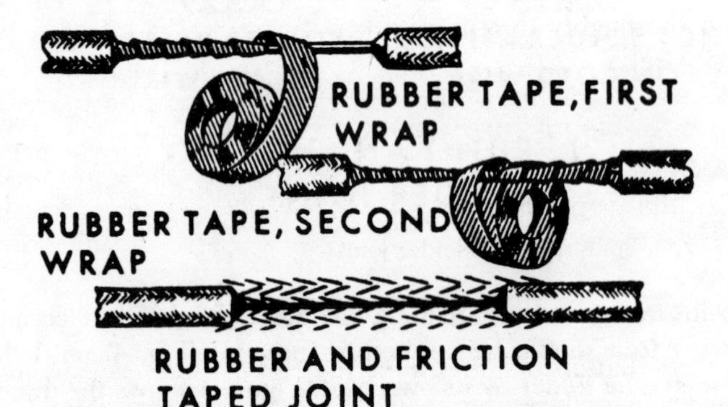

Fig. 5. Rubber- and friction-tape insulating.

the rubber or cambric tape is started on the tapered end of the wire insulation and advances toward the other end, with each succeeding wrap, by overlapping the windings. This procedure is repeated from one end of the splice to the other until the original thickness has been restored. The joint is then covered with several layers of friction tape.

Though the method for taping joints described in the previous paragraph is still considered standard, the plastic electrical tape, which serves as an insulation and a protective covering, should be used whenever available. This tape materially reduces the time required to tape a joint and reduces the space needed by the joint because a satisfactory protective and insulation covering can be achieved with three-layer taping.

Insulation and Making Wire Connections

When attaching a wire to a switch or an electrical device or when splicing it to another wire, the wire insulation must be removed to bare the copper conductor. Figure 6 shows the right and wrong ways to remove insulation. When the wire-stripping tool is applied at right angles to the wire, there is danger that the wire may be nicked and thus weakened. Therefore extreme caution must be taken to make sure the wire is not nicked. To avoid nicks, the cut is made at an angle to the conductor. After the protective insulation is removed, the conductor is scraped or sanded thoroughly to remove all traces of insulation and oxide on the wire.

The correct method of attaching the trimmed wire to terminals is shown in Fig. 6. The wire loop is always inserted under the terminal screw, as shown, so that tightening the screw tends to close the loop. The loop is made so that the wire insulation ends close to the terminal.

Job Sequence

The installation of interior wiring is generally divided into two major procedures: roughing-in and finishing. *Roughing-in* is the installation of the outlet boxes, cable, wire, and conduit. *Finishing* is the installation of the switches, receptacles, covers, and fixtures, and the completion of the service. The interval between these two work periods is used by other trades for plastering, enclosing walls, finishing floors, and trimming.

The procedure for *roughing-in* is as follows:

1. The first step in the roughing-in phase of a wiring job is the mounting of outlet boxes. The mounting can be expedited if the locations of all boxes are first marked on the studs and joists of the building.

2. All the boxes are mounted on the building members on their own or by special brackets. For concealed installation, all

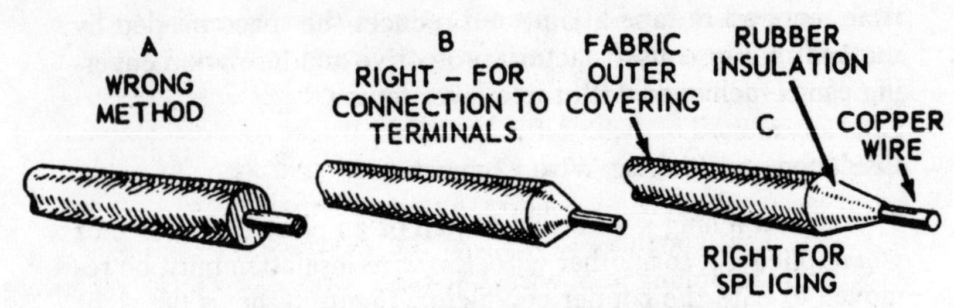

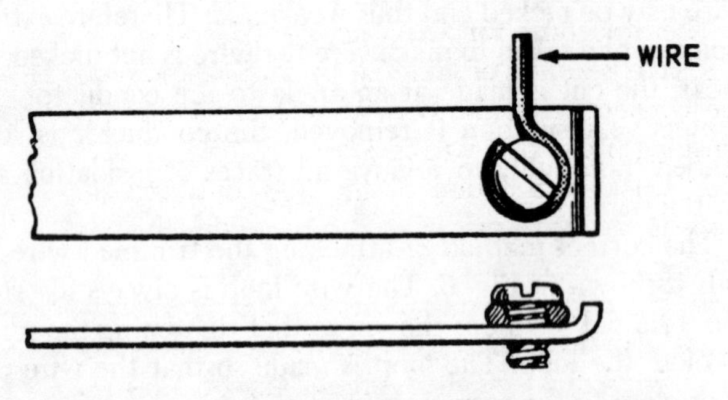

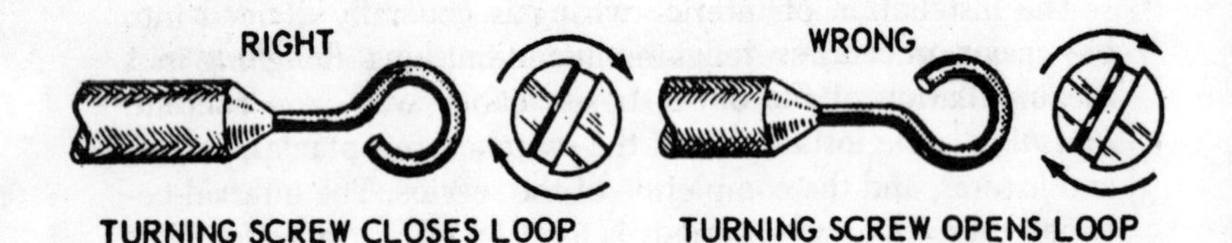

Fig. 6. Removing insulation and attaching wire to terminals.

boxes must be installed with the forward edge or plaster ring of the boxes flush with the finished walls.

3. The circuiting and installation of wire for open wiring,

cable, or conduit should be the next step. This involves the drilling and cutting-out of the building members to allow for the passage of the conductor or its protective covering. The production-line method of drilling the holes for all runs (as the installations between boxes are called) at one time, and then installing all of the wire, cable, or conduit, will expedite the job.

4. The final roughing-in step in the installation of conduit systems is the pulling in of wires between boxes. This can also be the first step in the finishing phase and requires care in the handling of the wires to prevent the marring of finished wall or floor surfaces.

The procedure for *finishing* is as follows:

1. The splicing of joints in the outlet and junction boxes and the connection of the bonding circuit is the initial step in the finishing phase of a wiring job.

2. Upon completion of the first finishing step, the proper leads to the terminals of switches, ceiling and wall outlets, and fixtures are installed.

3. The devices and their cover plates are then attached to the boxes. The fixtures are generally supported by the use of special mounting brackets called fixture studs or hickeys.

4. The service-entrance cable and fusing or circuit breaker panels are then connected and the circuits fused.

5. The final step in the wiring of any building requires the testing of all outlets by the insertion of a test prod or test lamp, the operation of all switches in the building, and the loading of all circuits to insure that proper circuiting has been installed.

Chapter 2

Electrician's Tools and Equipment

The electrical apparatus and materials that an electrician is required to install and maintain are different from other building materials. Their installation and maintenance require the use of special hand tools. This chapter describes and illustrates the tools normally used by an electrician in interior wiring.

Pliers

Pliers are furnished with either uninsulated or insulated handles. Although the insulated-handle pliers are always used when working on or near "hot" wires, they must not be considered sufficient protection alone and other precautions must be taken. Long-nose pliers are used for close work in panels or boxes. *Wire clippers* are used to cut wire to size. One type of wire clippers shown in Fig. 1 has a plastic cushion in the cutting head which grips the clipped wire end and prevents the clipped piece from flying about and injuring personnel. The slip-joint pliers are used to tighten locknuts or small nuts on devices.

Fuse Puller

The *fuse puller* shown in Fig. 2 is designed to eliminate the danger of pulling and replacing cartridge fuses by hand. It is also used for bending fuse clips, adjusting loose cutout clips, and handling live electrical parts. The second type of fuse puller, although having the same general configuration, is made of molded plastic. Encased in the handle is an electrical circuit similar to a voltmeter except that the indicating device is a neon glow tube. Test probes are attached to the handle of this fuse puller and may be used to determine if voltage is present in a circuit.

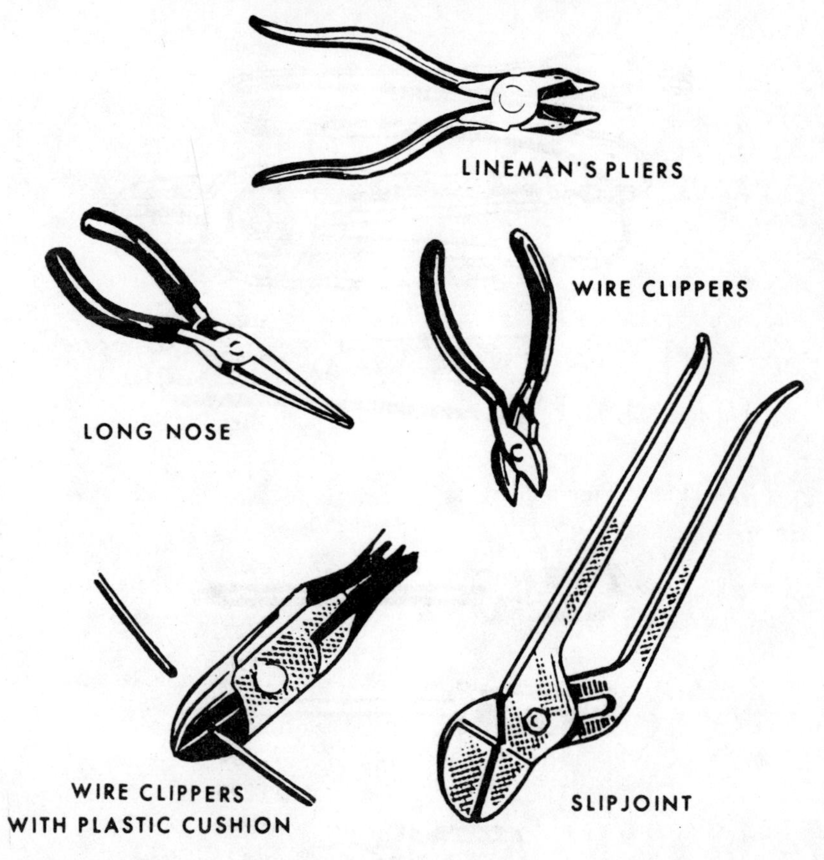

Fig. 1. Pliers.

Screwdrivers

Screwdrivers (Fig. 3) are made in many sizes and tip shapes. Those used by electricians should have insulated handles. Generally the electrician uses screwdrivers in attaching electrical devices to boxes and attaching wires to terminals. One variation of the screwdriver is the screwdriver bit, which is held in a brace and used for heavy-duty work. For safe and efficient application, screwdriver tips should be kept square and properly tapered and should be selected to match the screw slot.

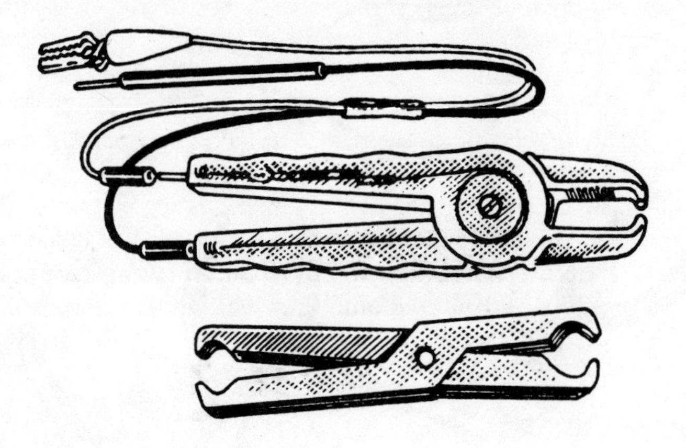

Fig. 2. Fuse pullers.

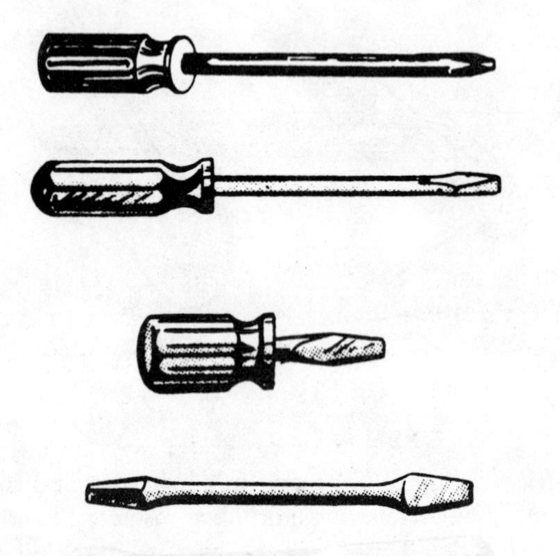

Fig. 3. Screwdrivers.

Wrenches

Figure 4 shows four types of *wrenches* used by electricians. Adjustable open-end wrenches—commonly called crescent wrenches—open end, closed end, and socket wrenches are used on hexagonal and square fittings such as machine bolts, hexagon nuts, or conduit unions. Pipe wrenches are used for pipe and conduit work and should not be used where crescent, open end, closed end, or socket wrenches can be used. Their construction will not permit the application of heavy pressure on square or hexagonal material, and the continued misuse of the tool in this manner will deform the teeth on the jaw faces and mar the surfaces of the material being worked.

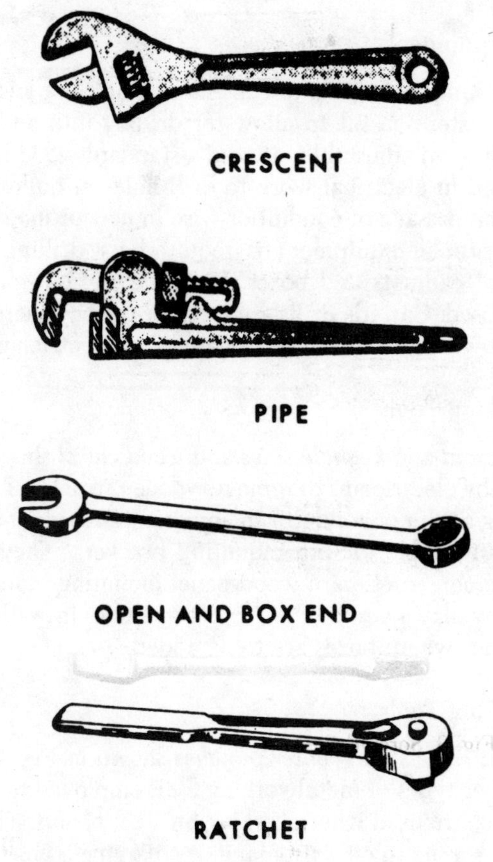

Fig. 4. Wrenches.

Soldering Equipment

A standard *soldering kit* (Fig. 5) used by electricians consists of nonelectric or electric soldering irons or both, a blowtorch (for heating a nonelectric soldering iron and pipe or wire joints), a spool of solid tin-lead wire solder or flux core solder, and soldering paste. An alcohol or propane torch may be used in place of the blowtorch. Acid core solder should never be used in electrical wiring. (*See* Chapter 4, Electrical Conductors and Wiring Techniques, sections on soldering.)

Drilling Equipment

Drilling equipment (Fig. 6) consists of a brace, a joist-drilling fixture, an extension bit to allow for drilling into and through deep cavities, an adjustable bit, and a standard wood bit. These are required in electrical work to drill holes in building structures for the passage of conduit or wire in new or modified construction. Similar equipment is required for drilling holes in sheet metal cabinets and boxes. In this case high-speed drills should be used. Carbide drills are used for tile or concrete work. Electric power drills aid in this phase of an electrician's work.

Woodworking Tools

The *crosscut* and *keyhole saws* and *wood chisel* shown in Fig. 7 are used by electricians to remove wooden structural members obstructing a wire or conduit run and to notch studs and joists to take conduit, cable, or box-mounting brackets. They are also used in the construction of wood-panel mounting brackets. The keyhole saw may again be used to cut openings in walls of existing buildings where boxes are to be added.

Metalworking Tools

The *cold chisels* and *center punches* shown in Fig. 8, besides several other types of metalworking tools employed by the electrical trade, are used when working on steel panels. The *knockout punch* is used in either making or enlarging a hole in a steel cabinet or outlet box. The *hacksaw* is usually used by an electri-

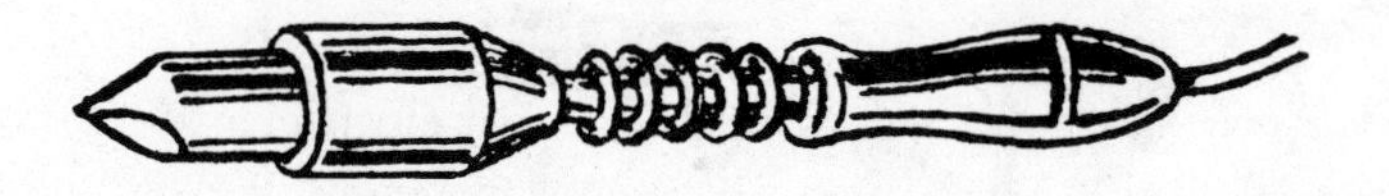

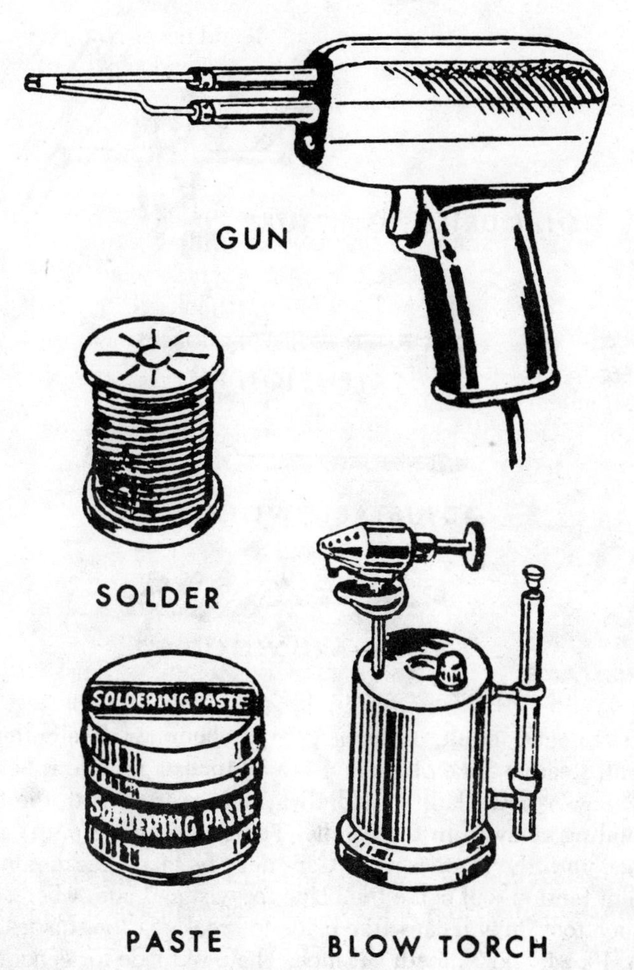

Fig. 5. Soldering equipment.

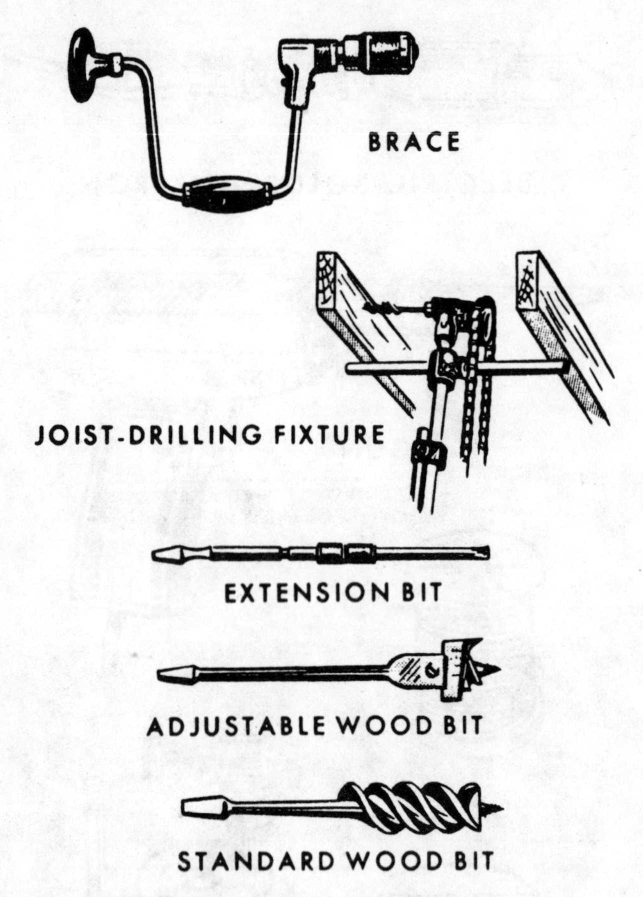

Fig. 6. Drilling equipment.

cian to cut conduit, cable, or wire too large for wire cutters. A light, steady stroke of about 40 to 50 times a minute is best for sawing. A new blade should always be inserted with the teeth pointing away from the handle. The tension wing nut is tightened until the blade is rigid. Care must be taken because insufficient tension will cause the blade to twist and jam, whereas too much tension will cause the blade to break. Various blades have 14, 18, 24, and 32 teeth per inch. The best blade for general use is one having 18 teeth per inch. A blade with 32 teeth per inch is

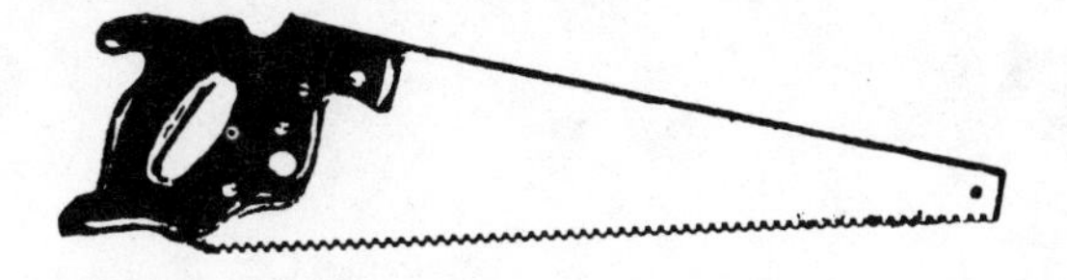

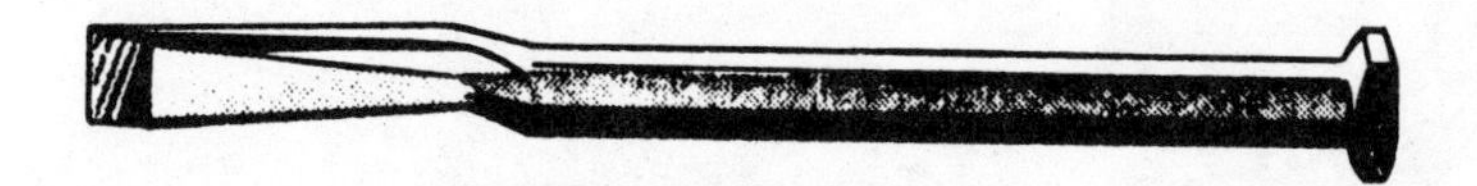

Fig. 7. Woodworking tools.

best for cutting thin material. The *mill file* shown in Fig. 8 is used in filing the sharp ends of cutoffs as a precaution against short circuits.

Masonry Tools

An electrician should have several sizes of masonry drills in his tool kit. These normally are carbide-tipped and are used to drill holes in brick or concrete walls, either for anchoring apparatus with expansion screws or for the passage of conduit or cable. Figure 9 shows the carbide-tipped bit used with a power drill and a hand-operated masonry drill.

Conduit Threaders and Dies

Rigid conduit is normally threaded for installation. Figure 10 illustrates one type of conduit threader and dies used in cutting pipe threads on conduit. The tapered pipe reamer is used to

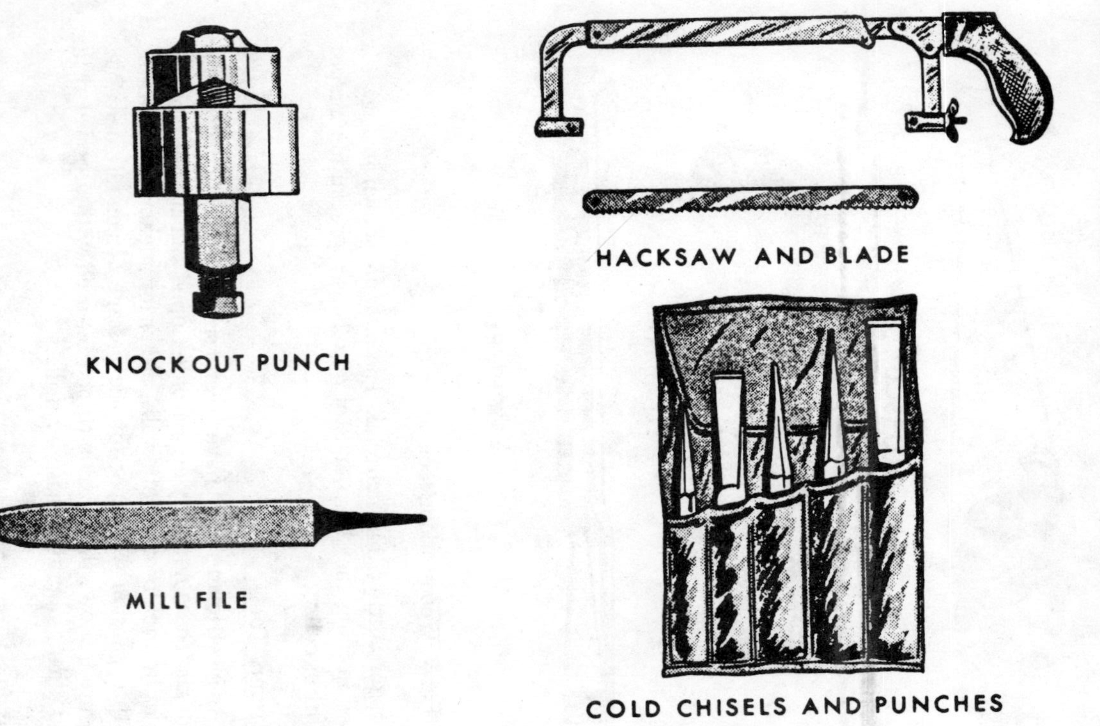

Fig. 8. Metalworking tools.

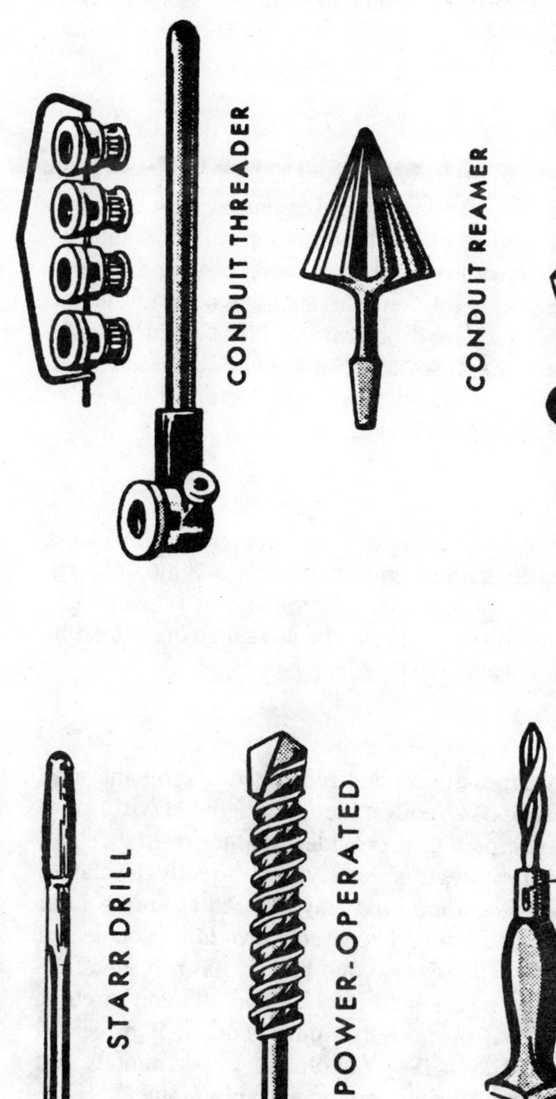

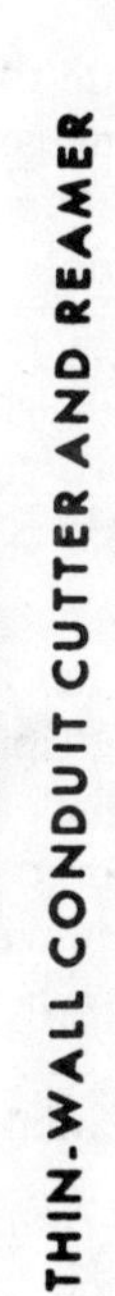

Fig. 10. Conduit threader, reamers, and cutter.

Fig. 9. Masonry drills.

ream the inside edge of the conduit as a precaution against wire damage. The conduit cutter is used when cutting thin-wall conduit and has a tapered blade attachment for reaming the conduit ends.

Knives and Other Insulation Stripping Tools

Wire and cable insulation is stripped or removed with the tools shown in Fig. 11. The knives and patented wire strippers are used to bare the wire of insulation before making connections. The scissors shown are used to cut insulation and tape. A multipurpose tool designed to cut and skin wires, attach terminals, gage wire, and cut small bolts may also be used. The armored cable cutter may be used instead of a hacksaw in removing the armor from the electrical conductors at box entry or when cutting the cable to length.

Hammers

Hammers are used either in combination with other tools, such as chisels, or in nailing equipment to building supports. Figure 12 shows a carpenter's claw hammer and a machinist's ball peen hammer, both of which can be advantageously used by electricians in their work.

Tape

Various kinds of tapes are used to replace insulation and wire coverings. *Friction tape* is a cotton tape impregnated with an insulating adhesive compound. It provides weather resistance and limited mechanical protection to a splice already insulated. *Rubber* or *varnished cambric tape* may be used as an insulator when replacing wire covering. *Plastic electrical tape* is made of a plastic material with adhesive on one face. It has replaced friction and rubber tape in the field for 120- and 208-volt circuits, and as it serves a dual purpose in taping joints, it is preferred over the former methods. (*See* Chapter 1, Fundamentals and Procedures of Electric Wiring, section on taping joints.)

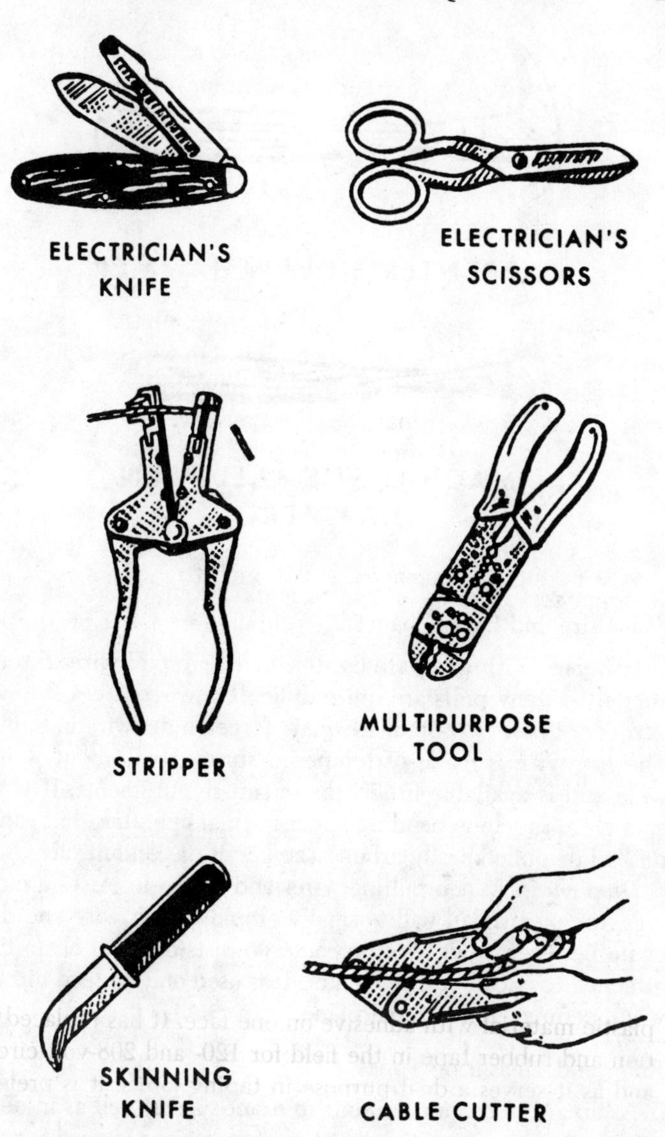

Fig. 11. Insulation-stripping tools.

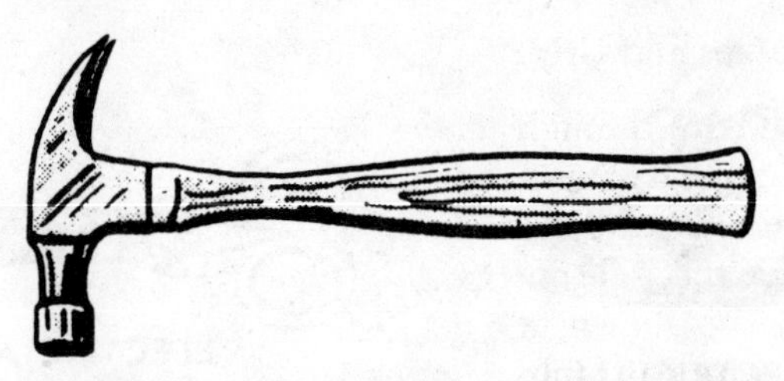

Fig. 12. Hammers.

Fish Wire and Drop Chain

Fish wire. Fish wires are used primarily to pull wires through conduits. Many pulls are quite difficult and require a fish-wire "grip" or "pull" to obtain adequate force on the wire in pulling. The fish wire is made of tempered spring steel about ¼ inch wide and is available in lengths to suit requirements. It is stiff enough to preclude bending under normal operation but can be pushed or pulled easily around the bends or conduit elbows.

Drop chain. When pulling wires and cables in existing buildings, the electrician will normally employ a fish wire and drop chain between studs. A *drop chain* consists of small chain links attached to a lead or iron weight. It is used only to feed through wall openings in a vertical plane.

Ruler and Measuring Tape

As an aid in cutting conduit to exact size as well as in determining the approximate quantities of material required for each job, the electrician should be equipped with a folding rule and a steel tape.

Wire Clamps and Grips

To pull wire through conduit and to pull open wire installations tight, the *wire grip* shown in Fig. 13 is an invaluable aid. As seen in the illustration, the wire grip has been designed so that the harder the pull on the wire, the tighter the wire will be gripped. Also shown in Fig. 13 is the *splicing clamp,* used to twist the wire pairs into a uniform and tight joint when making splices.

Extension Cord and Light

The *extension light* shown in Fig. 14 normally is supplied with a long extension cord and is used by the electrician when normal building lighting has not been installed or is not functioning.

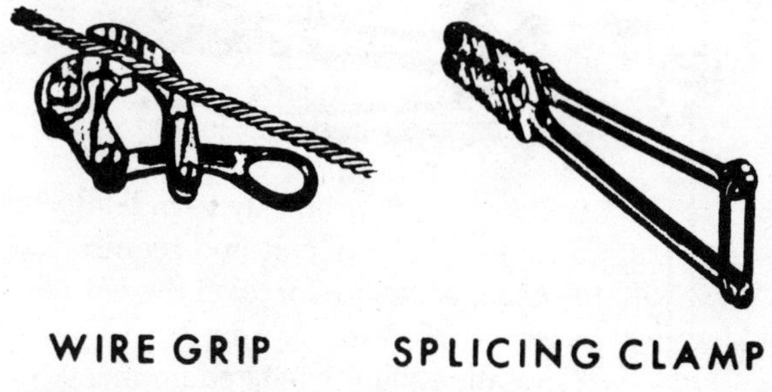

Fig. 13. Wire grip and splicing clamp.

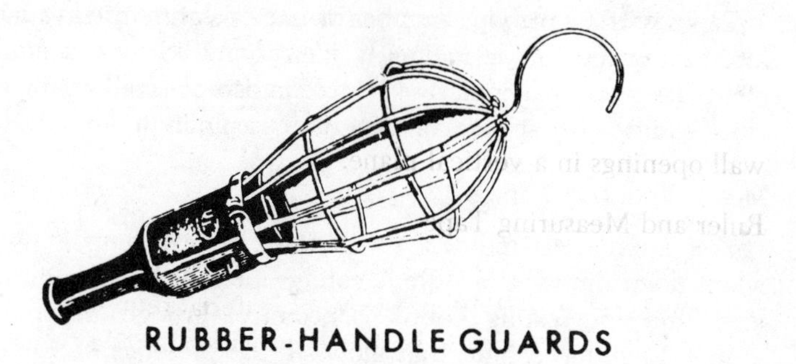

Fig. 14. Extension light (without bulb).

Thin-Wall Conduit Impinger

When the electrician uses indenter-type couplings and connectors with thin-wall conduit, an *indenter tool* (a thin-wall conduit impinger shown in Fig. 15) must be used to attach these fittings permanently to the conduit. This tool has points which, when pressed against the fitting, form indentations that press into the wall of the tubing to hold the fitting on the conduit. The use of these slip-on fittings and the impinger materially reduces the installation time required in electrical installations and thus reduces the cost of thin-wall conduit installations considerably.

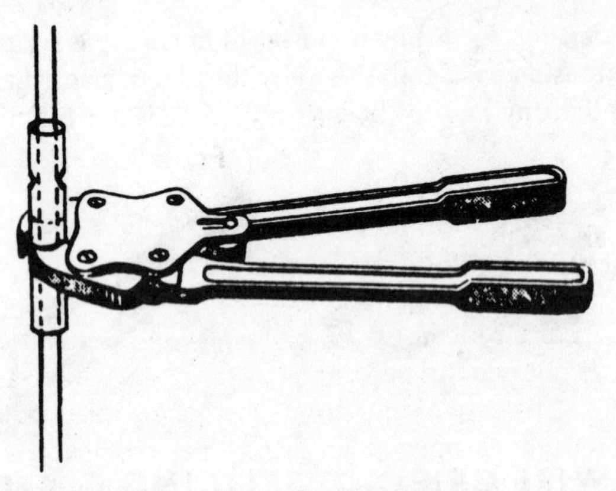

Fig. 15. Thin-wall conduit impinger.

Wire Code Markers

Tapes with identifying number or nomenclature are available for the purpose of permanently identifying wires and equipment. These are particularly valuable in complicated wiring circuits, in fuse circuit breaker panels, or in junction boxes.

Meters and Test Lamps

Test Lamps. An indicating voltmeter or test lamp is useful when determining the system voltage or locating the ground lead, and for testing circuit continuity through the power source. Each has a light that glows in the presence of voltage. Figure 16 shows a test lamp used as a voltage indicator.

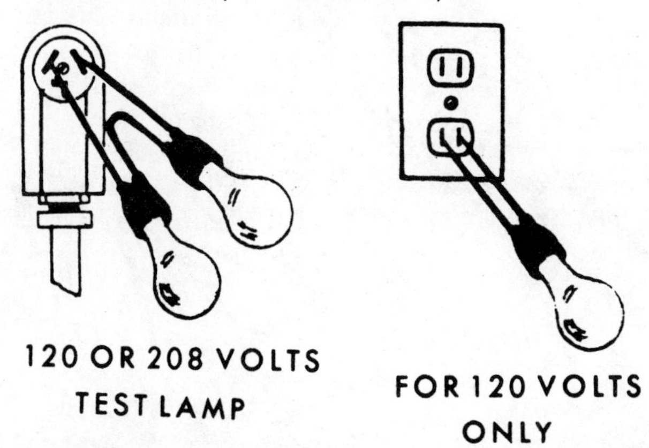

Fig. 16. Test lamps.

Hook-On Volt-Ammeter. A modern method of measuring current flow in a circuit uses the hook-on volt-ammeter (Fig. 17), which does not need to be hooked into the circuit, and the illustration shows its ease of operation. To make a measurement, the hook-on section is opened by hand and the meter is placed against the conductor. A slight push on the handle snaps the section shut; a pull springs the hook on the C-shaped current transformer open and releases the conductor. Applications of this meter are shown in Fig. 17, where voltage is being measured using the hook-on section. With three coils around the meter (Fig. 17), the current reading will be three times the actual current flowing through the wire. To obtain the true current, therefore, this reading is divided by three. The hook-on volt-ammeter can be used only on alternating-current circuits and can measure current only in a single conductor.

Wattmeter. The basic unit of measurement for electric power is the watt. In the power ratings of electric devices used by domestic consumers of electricity, the term watts signifies that, when energized at the normal line voltage, the apparatus will use electricity at the specified rate. In alternating-current cir-

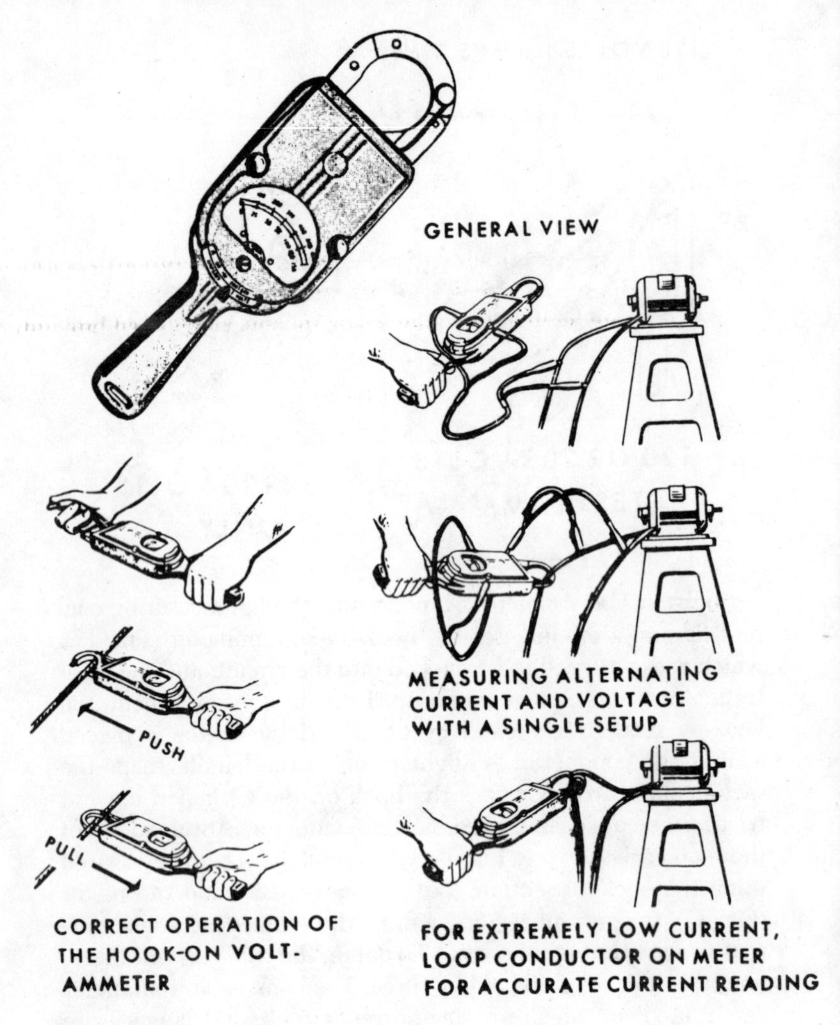

Fig. 17. Hook-on volt-ammeter.

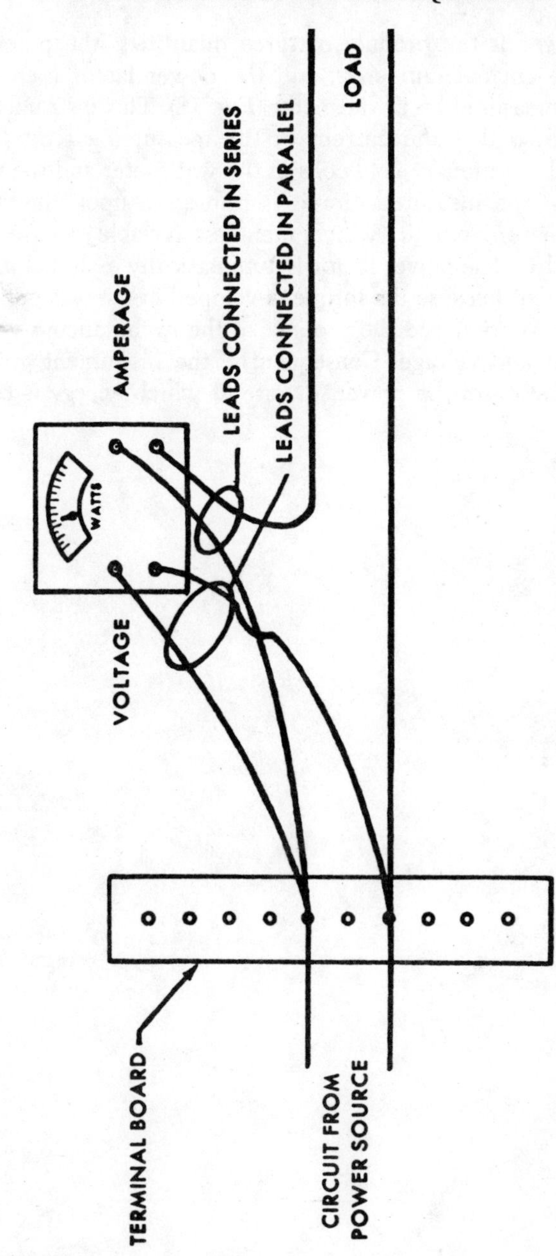

Fig. 18. Wattmeter connection.

cuits, power is the product of three quantities: the potential (volt), the current (amperes), and the power factor (percent). *Power* is measured by a wattmeter (Fig. 18). This instrument is connected so that the current in the measured circuit flows through the stationary field coils in the wattmeter and the voltage across the measured circuit is impressed upon the wattmeter-armature circuit, which includes movable coils and a fixed resistor. The power factor is automatically included in the measurement because the torque developed in the wattmeter is always proportional to the product of the instantaneous values of current and voltage. Consequently, the instrument gives a true indication of the power, or rate at which energy is being utilized.

Chapter 3

Wiring Materials

There are many different wiring systems currently in use which vary in complexity from the simple-to-install open wiring to the more complex conduit systems. These various systems contain common components. This chapter describes these common or general-use materials.

Electrical Conductors

Single conductors. Electrical conductors that provide the paths for the flow of electric current generally consist of copper or aluminum wire or cable over which an insulating material is formed. The insulating material insures that the path of current flow is through the conductor rather than through extraneous paths, such as conduits, water pipes, and so on. The wires or conductors are initially classified by type of insulation applied and wire gage. The various types of insulation are in turn subdivided according to their maximum operating temperatures and nature of use. Figure 1 illustrates the more common single conductors used in interior wiring systems. Table 8, in Appendix 2, lists the common trade classification of wires and compares them with regard to type, temperature rating, and recommended use.

Wire sizes. The wire sizes are denoted by the use of the American Wire Gage (AWG) standards. The largest gage size is No. 0000. Wires larger than this are classified in size by their circular mil cross-sectional area. One circular mil is the area of a circle with a diameter of 1/1,000 of an inch. Thus, if a wire has a diameter of 0.10 inch or 100 mil, the cross-sectional area is 100 × 100, or 10,000 circular mils. The most common wire sizes used in interior wiring are 14, 12, and 10, and they are usually of solid

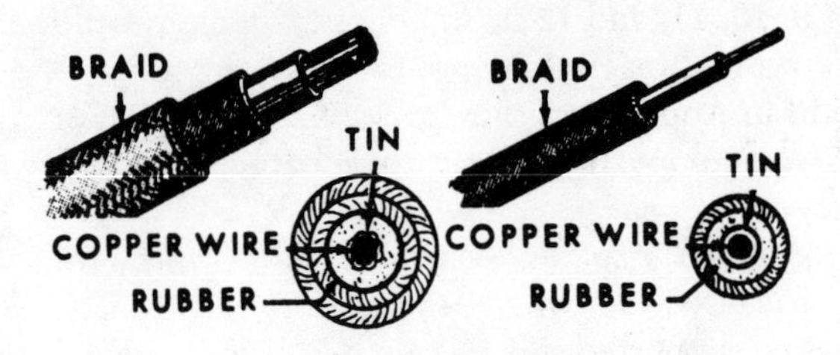

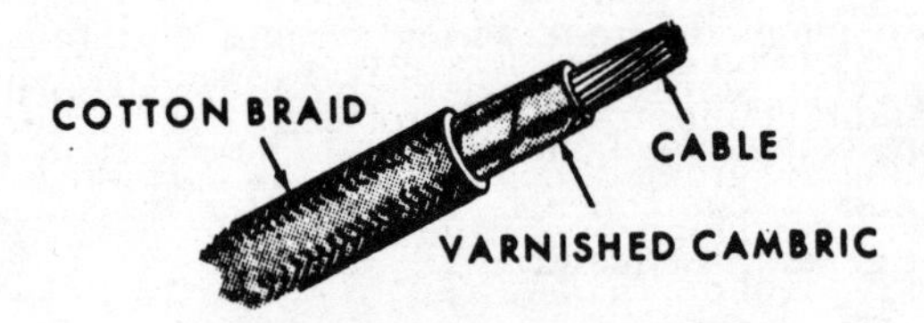

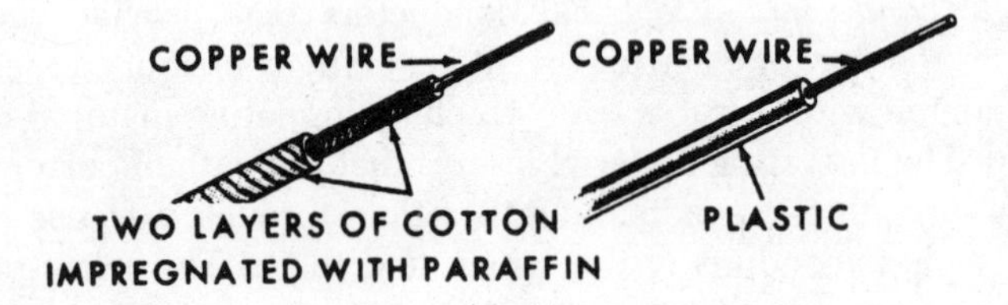

Fig. 1. Single conductors.

construction. Some characteristics of the numbering system are as follows.

1. As the numbers become larger, the size of the wire decreases.

2. The sizes normally used have even numbers, such as 14, 12, and 10.

3. Numbers 8 and 6 wires, which are furnished either solid or stranded, are normally used for heavy-power circuits or as service-entrance leads to buildings. Wire sizes larger than these are used for extremely heavy loads and for poleline distributions.

Tables 9, 10, 11, and 12 in Appendix 2 show the allowable current-carrying capacity for copper and aluminum conductors. Table 13 in Appendix 2 shows the percent reduction in current capacity if more than three conductors are in a cable or raceway.

Multiconductor cables. There are many types of installations of electrical wiring where the use of individual conductors spaced and supported side by side becomes an inefficient as well as hazardous practice. For these installations, multiconductor cables have been designed and manufactured. Multiconductor cables consist of the individual conductors, as outlined previously in the section on wire sizes, arranged in groups of two or more. An additional insulating or protective shield is formed or wound around the group of conductors. The individual conductors are color coded for proper identification. Figure 2 illustrates some of the types of multiconductors. The description and use of each type are as follows.

1. *Armored cable,* commonly referred to as BX, can be supplied in either two- or three-wire types and with or without a lead sheath. The wires in BX, matched with a bare equipment ground wire, are initially twisted together. This grouping, totaling three or four wires with the ground, is then wrapped in coated paper and a formed self-locking steel armor. The cable without a lead sheath is widely used for interior wiring under dry conditions. The lead sheath is required for installation in wet locations and through masonry or concrete building partitions where added protection for the copper conductor wires is required.

2. *Nonmetallic-sheathed cable* consists of two or three rubber- or thermoplastic-insulated wires, each covered with a jute type of filler material which acts as a protective insulation against mishandling. This in turn is covered with an impregnated cotton braid. The cable is light in weight, simple to install, and comparatively low-priced. It is used quite extensively in interior wiring, but is not approved for use in wet locations. A dual-purpose plastic-sheathed cable with solid copper conductors can be used underground outdoors or indoors. It needs no conduit, and its flat shape and gray or ivory color make it ideal for surface wir-

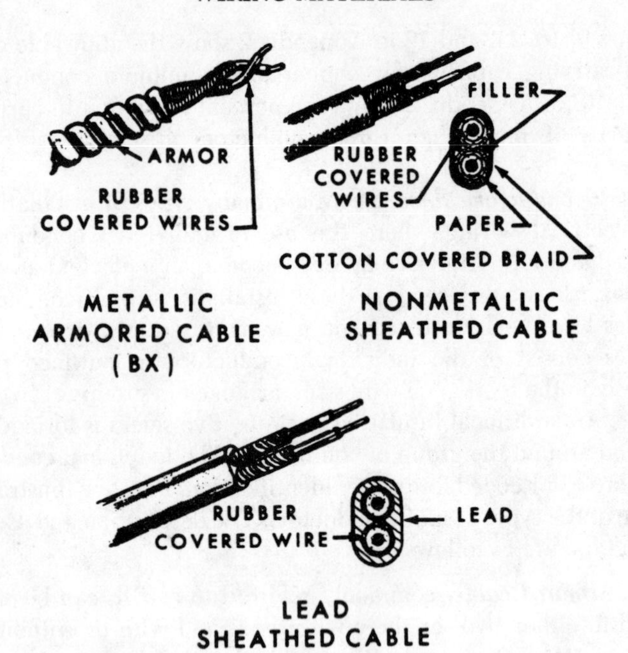

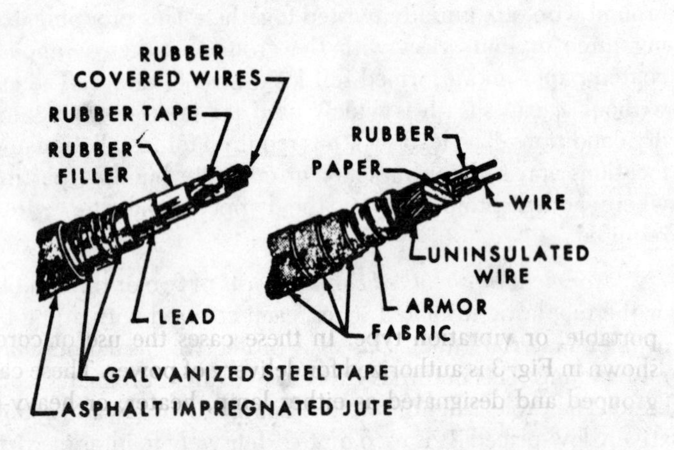

Fig. 2. Multiconductor cables.

ing. It resists moisture, acid, and corrosion, and can be run through masonry or between studding.

3. *Lead-sheathed cable* consists of two or more rubber-covered conductors surrounded by a lead sheathing which has been extruded around it to permit its installation in wet and underground locations. Lead-covered cable can also be immersed in water or installed in areas where the presence of liquid or gaseous vapors would attack the insulation on other types.

4. *Parkway cable* provides its own protection from mechanical injury and therefore can be used for underground services, buried in the ground without any protecting conduit. It normally consists of rubber-insulated conductors enclosed in a lead sheath and covered with a double spiral of galvanized steel tape which acts as a mechanical protection for the lead. On top of the tape, a heavy braid of jute saturated with a waterproofing compound is applied for additional weather protection.

5. *Service-entrance cable* normally has three wires with two insulated and braided conductors laid parallel and wound with a bare conductor. Protection against damage for this assembly is obtained by encasing the wires in heavy tape or armor, which serves as an inner cushion, and covering the whole assembly with braid. Though the cable normally serves as a power carrier from the exterior service drop to the service equipment of a building, it may also be used in interior-wiring circuits to supply power to electric ranges and water heaters at voltages not exceeding 150 volts to ground, provided the outer covering is armor. It may also be used as a feeder to other buildings on the same premises under the same conditions, if the bare conductor is used as an equipment grounding conductor from a main distribution center located near the main service switch.

Cords. Many items using electrical power are of the pendant, portable, or vibration type. In these cases the use of cords as shown in Fig. 3 is authorized for delivery of power. These can be grouped and designated as either lamp, heater, or heavy-duty power cords. *Lamp cords* are supplied in many forms. The most common types are the single-paired rubber-insulated and twisted-paired cords. The twisted-paired cords consist of two cotton-wound conductors which have been covered with rubber

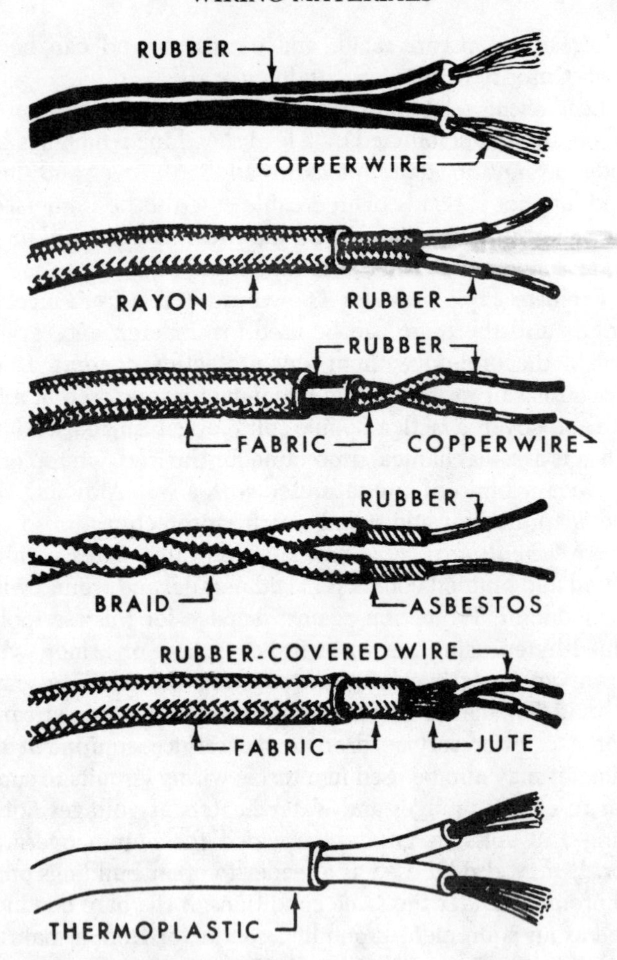

Fig. 3. Types of flexible cords.

and rewound with cotton braid. *Heater cords* are similar to this latter type except that the first winding is replaced by heat-resistant asbestos. *Heavy-duty* or *hard-service cords* are normally supplied with two or more conductors surrounded by cotton and rubber insulation. In manufacture, these are first twisted or stranded. The voids created in the twisting process are then filled with jute and the whole assembly covered with rubber. All cords, whether of this type or of the heater or lamp variety, have the conductors color coded for ease of identification. (*See* Appendix 12, Component Color Code.) Table 14 in Appendix 2 groups by common trade terms the cords found in general use and illustrates some of their characteristics.

Electrical Boxes

Design. Outlet boxes bind together the elements of a conduit or armored cable system in a continuous grounded system. They provide a means of holding the conduit in position, a space for mounting such devices as switches and receptacles, protection for the device, and space for making splices and connections. Outlet boxes are manufactured in either sheet steel, porcelain, bakelite, or cast iron, and are either round, square, octagonal, or rectangular. The fabricated steel box is available in a number of different designs. For example, some boxes are of the sectional or "gang" variety, while others have integral brackets for mounting on studs and joists. Moreover, some boxes have been designed to receive special cover plates so that switches, receptacles, or lighting fixtures may be more easily installed. Other designs facilitate installation in plastered surfaces. Regardless of the design or material, they all should have sufficient interior volume to allow for the splicing of conductors or the making of connections. For this reason the allowable minimum depth of outlet boxes is limited to 1½ inches in all cases except where building-supporting members would have to be cut. In this case the minimum depth can be reduced to ½ inch.

Selection. The selection of boxes in an electrical system should be made in accordance with Tables 15 and 16 in Appendix 2, which list the maximum allowable conductor capacity for each type of box. In these tables a conductor running through the box is counted along with each conductor terminating in the box. For example, one conductor running through a box and two terminating in the box would equal three conductors in the box. Consequently, any of the boxes listed would be satisfactory. The tables apply for boxes that do not contain receptacles, switches, or similar devices. Each of these mounted in a box will reduce by one the maximum number of conductors allowable as shown in the tables.

Outlet boxes for rigid and thin-wall circuit and armored cable. Steel or cast-iron outlet boxes are generally used with rigid and thin-wall conduit or armored cable. The steel boxes are either zinc- or enamel-coated, the zinc coating being preferred when installing conduit in wet locations. All steel boxes have "knockouts." These knockouts are indentations in the side, top, or back of an outlet box, sized to fit the standard diameters of conduit fittings or cable connectors. They usually can be removed with a small cold chisel or punch to facilitate entry into the box of the conduit or cable. Boxes designed specifically for armored-cable use also have integral screw clamps located in the space immediately inside the knockouts, and thus eliminate the need for cable connectors. This reduces the cost and labor of installation. Box covers are normally required when it is necessary to reduce the box openings, provide mounting lugs for electrical devices, or to cover the box when it is to be used as a junction. Figure 4 illustrates several types of cable connectors and also a cable clamp for use in clamping armored cable in an outlet box. The anti-short bushing shown in Fig. 4 is inserted between the wires and the armor to protect the wire from the sharp edges of the cut with a hacksaw or cable cutter.

Outlet boxes for nonmetallic sheathed cable and open wiring.

1. *Steel.* Steel boxes are sometimes used for nonmetallic cable and open wiring. However, the methods of box entry are different from those for conduit and armored-cable wiring because the electrical conductor wires are not protected by a hard surface. The connectors and interior box clamps used in nonmetallic and open wiring are formed to provide a smooth surface for securing the cable rather than being the sharp-edged type of closure normally used.

2. *Nonmetallic.* Nonmetallic outlet boxes made of either porcelain or bakelite may also be used with open or nonmetallic sheathed wiring. Cable or wire entry is generally made by removing the knockouts of preformed weakened blanks in the boxes.

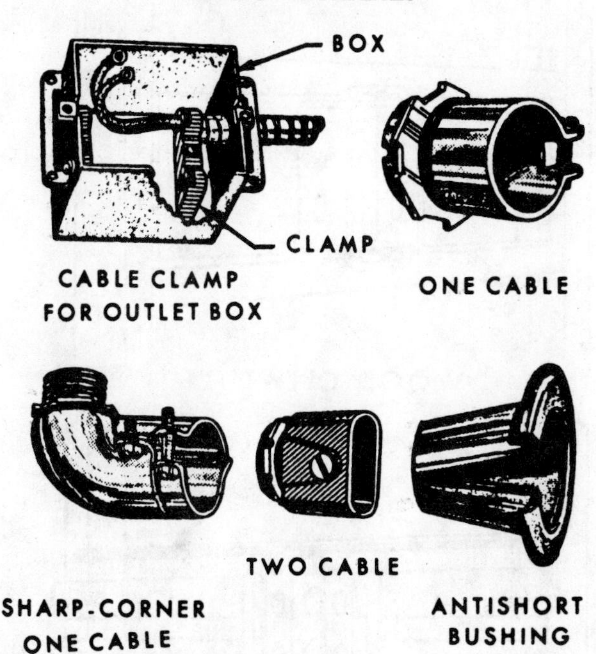

Fig. 4. Armored-cable fittings.

3. *Special.* In open wiring, conductors should normally be installed in a loom from the last support to the outlet box. Although all of the boxes described in 1 and 2 are permissible for open wiring, a special loom box is available which has its back corners "sliced off" and allows for loom and wire entry at this sliced-off position.

Attachment devices for outlet boxes. Outlet boxes which do not have brackets are supported by wooden cleats or bar hangers, as illustrated in Fig. 5.

1. *Wooden cleats.* Wooden cleats are first cut to size and nailed between two wooden members. The boxes are then either nailed or screwed to these cleats through holes provided in their back plates.

2. *Strap hangers.* If the outlet box is to be mounted between studs, mounting straps are necessary. The ready-made straps are

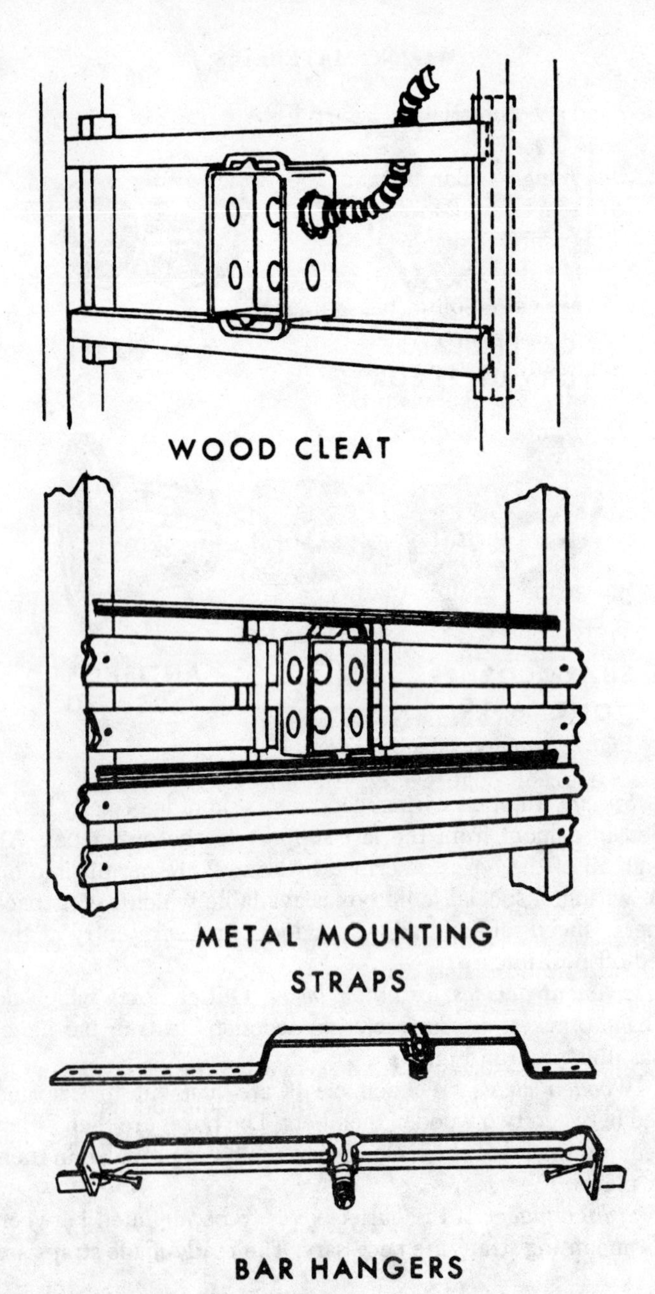

Fig. 5. Typical box mountings.

handy and accommodate not only a single box, but a 2, 3, 4, or 5 gang box.

3. *Bar hangers.* Bar hangers are prefabricated to span the normal 16-inch and 24-inch joist and stud spacings and are obtainable for surface or recessed box installation. They are nailed to the joist or stud exposed faces. The supports for recessed boxes normally are called offset bar hangers.

4. *Patented supports.* When boxes have to be installed in walls that are already plastered, several patented supports can be used for mounting. These obviate the need for installing the boxes on wooden members and thus eliminate extensive chipping and replastering.

Knobs, Tubes, Cleats, Loom, and Special Connectors

Open wiring requires the use of special insulating supports and tubing to insure a safe installation. These supports, called knobs and cleats, are smooth-surfaced and made of porcelain. Knobs and cleats support the wires which are run singly or in pairs on the surface of the joists or studs in the buildings. Tubing or tubes, as they are called, protect the wires from abrasion when passing through wooden members. Insulation of loom of the "slip-on" type is used to cover the wires on box entry and at wire-crossover points. The term "loom" is applied to a continuous flexible tube woven of cambric material impregnated with varnish. At points where the type of wiring may change and where boxes are not specifically required, special open wiring to cable or conduit wiring connectors should be used. These connectors are threaded on one side to facilitate connection to a conduit and have holes on the other side to accommodate wire splices, but are designed only to carry the wire to the next junction box. The specific methods of installation and use of these items are covered in Chapter 6, Open Wiring, Knobs, and Tubes.

Cable and Wire Connectors

Code requirements state that "Conductors shall be spliced or joined with splicing devices approved for the use or by brazing, welding, or soldering with a fusible metal or alloy. Soldered splices shall first be so spliced as to be mechanically and electrically secure without solder and then soldered." Soldering or splicing devices are used as added protection because of the ease of wiring and the high quality of connection of these devices. Assurance of high quality is the responsibility of the electrician, who selects the proper size of connector relative to the number and size of wires. Figure 6 shows some of the many types of cable and wire connectors in common use.

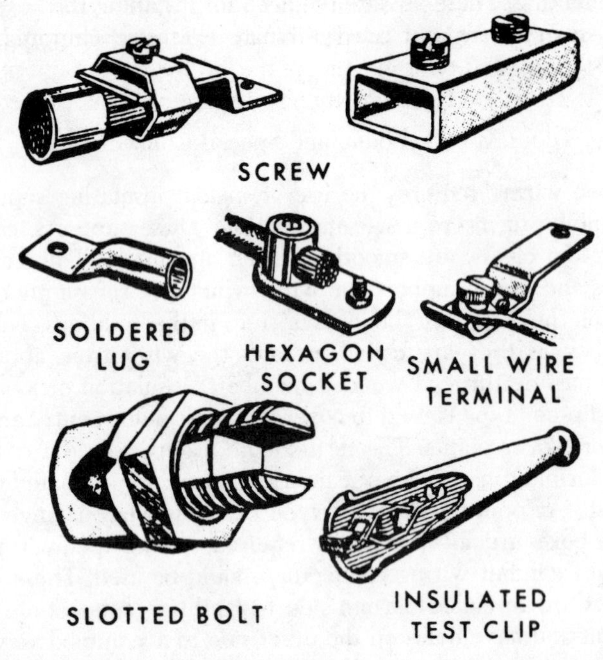

Fig. 6. Cable and wire connectors.

Straps and Staples

All conduits and cables must be attached to the structural members of a building in a manner that will preclude sagging. The cables must be supported at least every 4½ feet for either a vertical or a horizontal run and must have a support in the form of a strap or staple within 12 inches of every outlet box. Conduit-support spacings vary with the size and rigidity of the conduit. (*See* Table 17, Appendix 2, for support requirements of rigid nonmetallic conduit, and Chapter 9, Conduit Wiring, section on conduit.)

Cable staples. A very simple and effective method of supporting BX cables on wooden members is by the use of cable staples (Fig. 7).

Insulating staples. Bell or signal wires are normally installed in pairs in signal systems. The operating voltage and energy potential is so low in these installations (12 to 24 volts) that protective coverings such as conduit or loom are not required. To avoid any possibility of shorting in the circuit, they are normally supported on wood joists or studs by insulated staples of the type shown in Fig. 7.

Fig. 7. Straps and staples.

Straps. Conduit and cable straps (Fig. 7) are supplied as either one-hole or two-hole supports and are formed to fit the contour of the specific material for which they are designed. The conduit and cable straps are attached to building materials by "anchors" designed to suit each type of supporting material. For example, a strap is attached to a wood stud or joist by nails or screws. Expanding anchors are used for box or strap attachment to cement or brick structures and also to plaster or plaster-substitute surfaces. Toggle and "molly" bolts are used where the surface wall is thin and has a concealed air space which will allow for the release of the toggle or expanding sleeve.

Receptacles, Fixtures, and Receptacle Covers

Portable appliances and devices are readily connected to an electrical supply circuit by means of an outlet called a receptacle. For interior wiring these outlets are installed as either single or duplex receptacles. Receptacles previously installed, and their replacements in the same box, may be two-wire receptacles. All others must be the three-wire type. The third wire on the three-wire receptacle is used to provide a ground lead to the equipment which receives power from the receptacle. This guards against dangers from current leakage due to faulty insulation or exposed wiring and helps prevent accidental shock. The receptacles are constructed to receive plug prongs either by a straight push action or by a twist-and-turn push action. Fixtures are similar to receptacles but are used to connect the electrical supply circuit directly to lamps inserted in their sockets.

Knob-and-tube wiring. Receptacles with their entire enclosures made of some insulating material, such as bakelite, may be used without metal outlet boxes for exposed, open wiring or nonmetallic sheathed cable.

Conduit and cable. The receptacles (Fig. 8) commonly used with conduit and cable installations are constructed with yokes to facilitate their installation in outlet boxes. In this case they are attached to the boxes by metal screws through the yokes, threaded into the box. Wire connections are made at the receptacle terminals by screws which are an integral part of the outlet. Receptacle covers made of either brass, steel, or nonmetallic

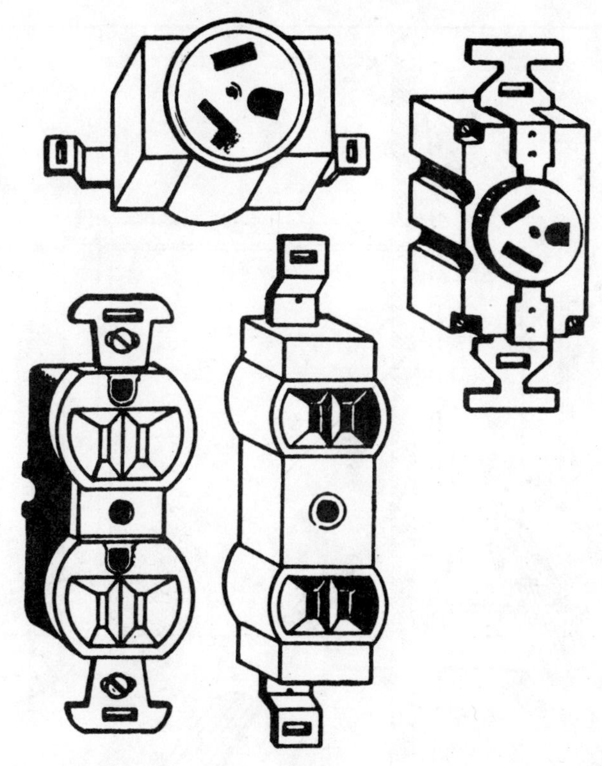

Fig. 8. Types of wall receptacles.

materials are then attached to box and receptacle installations to afford complete closure at the outlets.

Surface metal raceways. These raceways provide a quick, inexpensive electrical wiring installation method, since they are installed on the wall surface instead of inside the wall (Fig. 9).

Surface metal raceways are basically of two types—one-piece construction or two-piece construction. With the *one-piece construction type*, the metal raceway is installed like conduit, and then the wires are "pulled" to make the necessary electrical connections. With the *two-piece construction type*, the base piece is installed along the wiring run. Wiring is then laid in the base piece and held in place with clamps. After the wires are laid, the capping is snapped on and the job is complete.

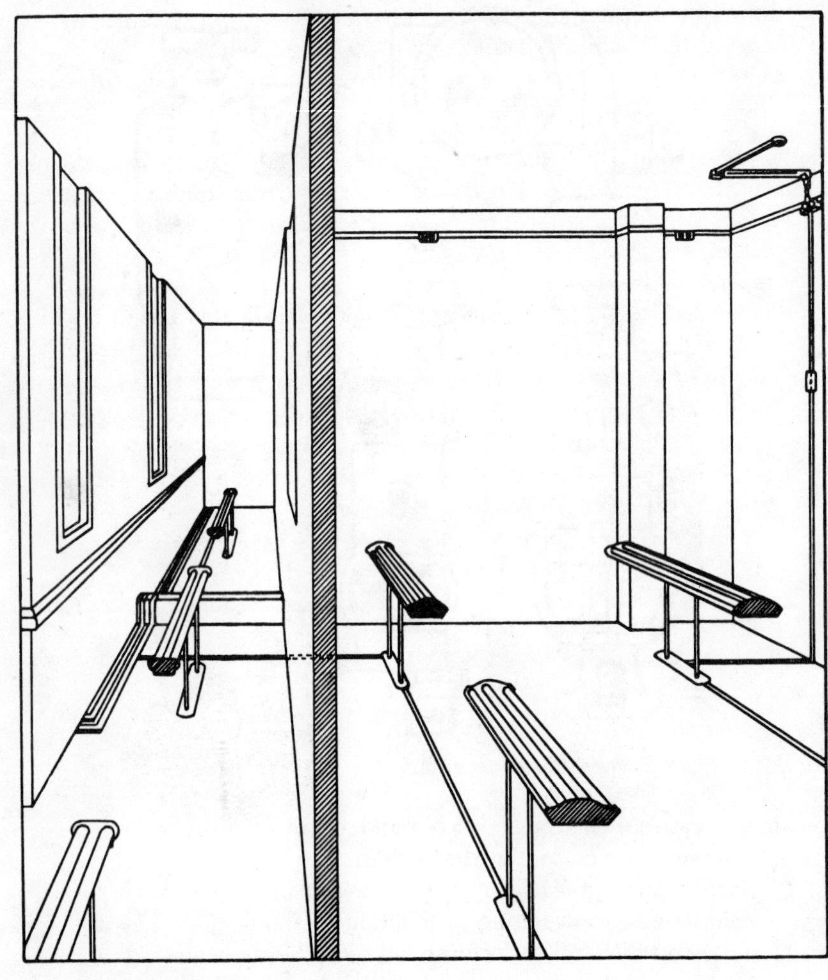

Fig. 9. Surface metal raceways.

A *multioutlet system*, with grounding inserts if desired, has outlets spaced every few inches so that several tools or pieces of equipment can be used simultaneously. An overfloor metal raceway system handles telephone and signal or power and light wiring where the circuits must be brought to locations in the middle of the floor area. These systems are all designed so that they can be installed independently of other wiring systems, or may be economically connected to existing systems.

Plugs and Cord Connectors

Plugs. Portable appliances and devices that are to be connected to receptacles have their electrical cords equipped with plugs (Fig. 10) that have prongs which mate with the slots in the outlet receptacles. A three-prong plug can fit into a two-prong receptacle by the use of an *adapter.* If the electrical conductors connected to the outlet have a ground system, the lug on the lead wire of the adapter is connected to the center screw holding the receptacle cover to the box. Many of these plugs are permanently molded to the attached cords. There are other types of cord-grips that hold the cord firmly to the plug. *Twist-lock plugs* have patented prongs that catch and are firmly held to a mating receptacle when the plugs are inserted into the receptacle slots and twisted. Where the plugs do not have cord-grips, the cords should be tied with an Underwriters knot (Fig.

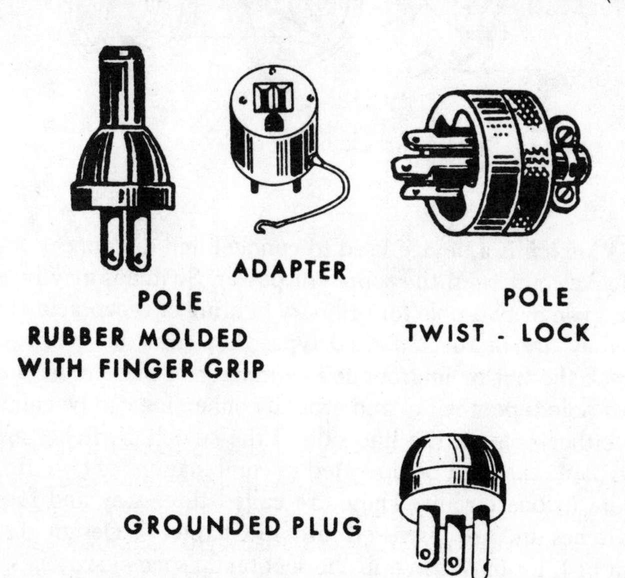

Fig. 10. Attachment plugs.

11) at plug entry to eliminate tension on the terminal connections when the cord is connected to and disconnected from the outlet receptacle. Figure 11 shows the steps to be used in tying this type of knot.

Cord connectors. There are some operating conditions where a cord must be connected to a portable receptacle. This type of receptacle, called a *cord connector body* or a *female plug,* is attached to the cord in a manner similar to the attachment of the male plug outlined previously in the section on Plugs.

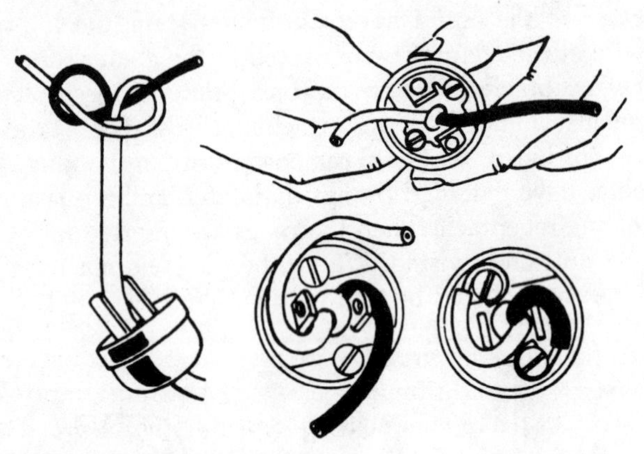

Fig. 11. Underwriters knot.

Switches and Covers

A *switch* is a device used to connect and disconnect an electrical circuit from the source of power. Switches may be either one-pole or two-pole for ordinary lighting or receptacle circuits. If they are of the one-pole type, they must be connected to break the hot or ungrounded conductor of the circuit. For the two-pole type, the hot and ground connection can be connected to either pole on the line side of the switch. Switches are also available that can be operated in combinations of two, three, or more in one circuit. These are called three-way and four-way switches and are discussed fully in Chapter 5, Design and Layout of Interior Wiring, in the section on special switches.

Open and nonmetallic sheathed wiring. Switches used for exposed open wiring and nonmetallic sheathed cable wiring are usually of the *tumbler type* with the switch and cover in one piece. Other less common ones are the *rotary-snap* and *pushbutton types.* These switches are generally nonmetallic in composition (Fig. 12).

Conduit and cable installations. The tumbler switch and cover plate (Fig. 12) normally used for outlet-box installation are mounted in a manner similar to that for box type receptacles and covers, and are in two pieces. Foreign installations may still use pushbutton switches, as shown in Fig. 12.

Entrance installations. At every powerline entry to a building, a switch and fuse combination or circuit breaker switch of a type similar to that shown in Fig. 13 must be installed at the service entrance. This switch must be rated to disconnect the building load while in use at the system voltage. Entrance or service switches, as they are commonly called, consist of one "knife" switch blade for every hot wire of the power supplied. The switch is generally enclosed and sealed in a sheet-steel cabinet. When the building circuit is connected or disconnected, the blades are operated simultaneously through an exterior handle by the rotation of a common shaft holding the blades. The neutral or grounded conductor is not switched but is connected at a neutral terminal within the box. Many entrance switches are equipped with integral fuse blocks or circuit breakers which protect the building load. The *circuit breaker type* of entrance switch is preferred, particularly in field installations, because of its ease of resetting after the overload condition in the circuit has been cleared.

Fuses and Fuse Boxes

Fuses. The device for automatically opening a circuit when the current rises beyond the safety limit is technically called a *cutout,* but more commonly is called a *fuse.* All circuits and electrical apparatus must be protected from short circuits or dangerous overcurrent conditions through correctly rated fuses.

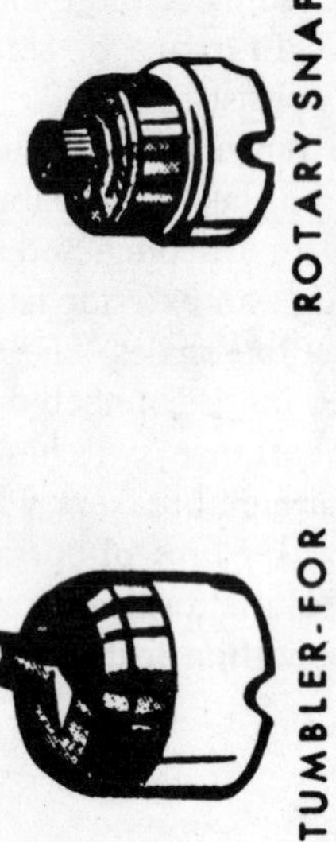

Fig. 12. Switches and covers.

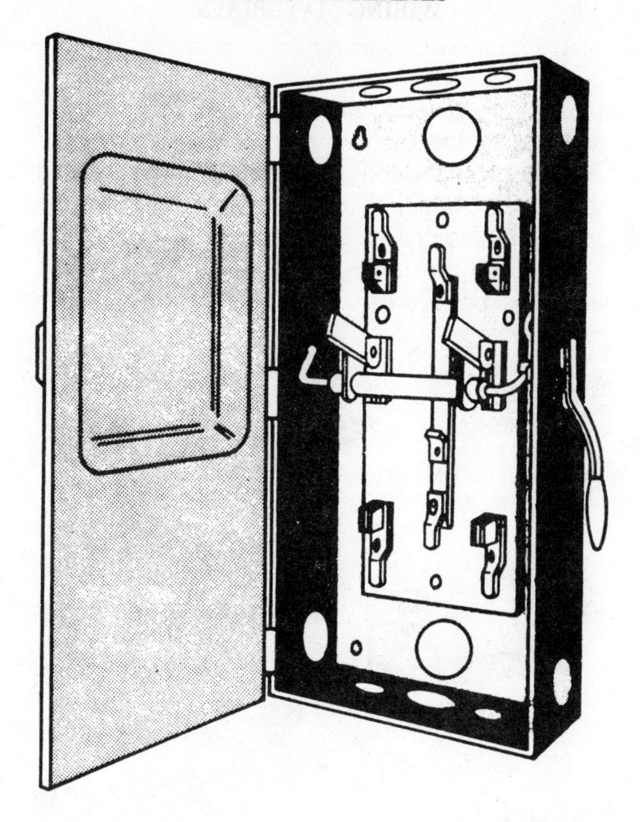

Fig. 13. Service switch box.

1. *Standard.* The *cartridge-type fuse* is used for current above 30 amperes in interior wiring systems. The ordinary *plug* or *screw-type fuse* is satisfactory for incandescent lighting or heating appliance circuits.

2. *Special.* On branch circuits, wherever motors are connected, *time-lag fuses* should be used instead of the standard plug or cartridge-type fuse. These fuses have self-compensating elements which maintain and hold the circuit in line during a momentary heavy ampere drain, yet cut out the circuit under normal short-circuit conditions. The heavy ampere demand normally occurs in motor circuits when the motor is started. Exam-

ples of such circuits are the ones used to power oil burners or air conditioners.

Fuse boxes. As a general rule, the fusing of circuits is concentrated at centrally located fusing or distribution panels. These panels are normally located at the service-entrance switch in small buildings or installed in several power centers in large buildings. The number of service centers or fuse boxes in the latter case would be determined by the connected power load. Fuses and a fuse box are shown in Fig. 14.

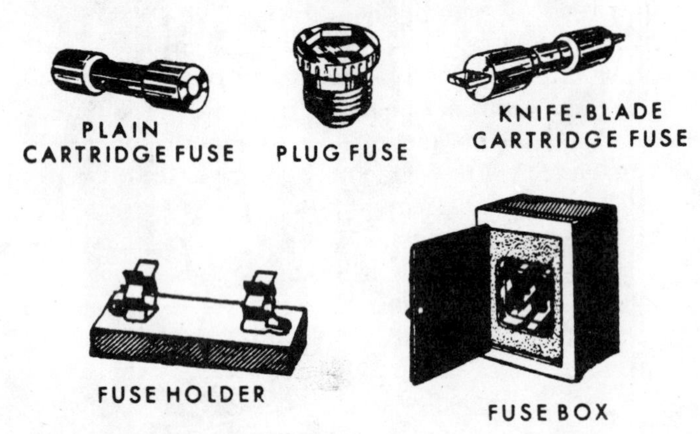

Fig. 14. Typical fuses and fuse box.

Circuit Breaker Panels

Circuit breakers are devices resembling switches that trip or cut out the circuit in case of overamperage. They perform the same function as fuses and can be obtained with time-lag opening features similar to the special fuses outlined in the section on fuses and fuse boxes. Based on their operation, they may be classified as thermal, magnetic, or combination thermal-magnetic reaction types. A *thermal-type* circuit breaker has a bimetallic element built within the breaker that responds only to fluctuations in temperature within the circuit. The element is made by bonding together two strips of dissimilar metal, each of which has a different coefficient of expansion. When a current is flowing in the circuit, the heat created by the resistance of the bimetallic element will expand each metal at a different rate, causing the strip to bend. The element acts as a latch in the circuit as the breaker mechanism is adjusted, so that the element bends just far enough under a specified current to trip the breaker and open the circuit. A *magnetic* circuit breaker responds to changes in the magnitude of current flow. In operation an increased current flow will create enough magnetic force to "pull up" an armature, opening the circuit. The magnetic circuit breaker is usually used in motor circuits for closer adjustment to motor rating, while the circuit conductors are protected, as usual, by another circuit breaker. The *thermal-magnetic* breaker, as the name implies, combines the features of the thermal and the magnetic types. Practically all of the molded case circuit breakers used in lighting panelboards are of this type. The thermal element protects against overcurrents in the lower range and the magnetic element protects against the higher range usually occurring from short circuits.

During the last decade, circuit breakers have been used to a greater extent than fuses because they can be manually reset after tripping, whereas fuses require replacement. Fuses may easily be replaced with higher-capacity ones that do not protect the circuit. This is difficult to do with circuit breakers. In addition, they combine the functions of fuse and switch, and when tripped by overloads or short circuits, all of the ungrounded conductors of a circuit are opened simultaneously. Each branch circuit must have a fuse or circuit breaker protecting each ungrounded conductor. Some installations may or may not have a main breaker that disconnects everything. As a guide during installation, if more than six movements of the hand are not required to open all the branch circuit breakers, a main breaker or switch is not required ahead of the branch-circuit breaker. However, if more than six movements of the hand are required, a separate disconnecting main circuit breaker is required ahead of the branch-circuit breaker. Each 208-volt circuit requires a single-pole (one-pole) breaker which has its own handle. Each 208-volt circuit requires a double-pole (two-pole) breaker to protect both ungrounded conductors. You can, however, place two single-pole breakers side by side, and tie the two handles together mechanically to give double-pole protection. Both handles can then be moved by a single movement of the hand. A two-pole breaker may have one handle or two handles which are mechanically tied together, but either one requires only one movement of the hand to break the circuit. Figure 15 illustrates a typical circuit breaker panel.

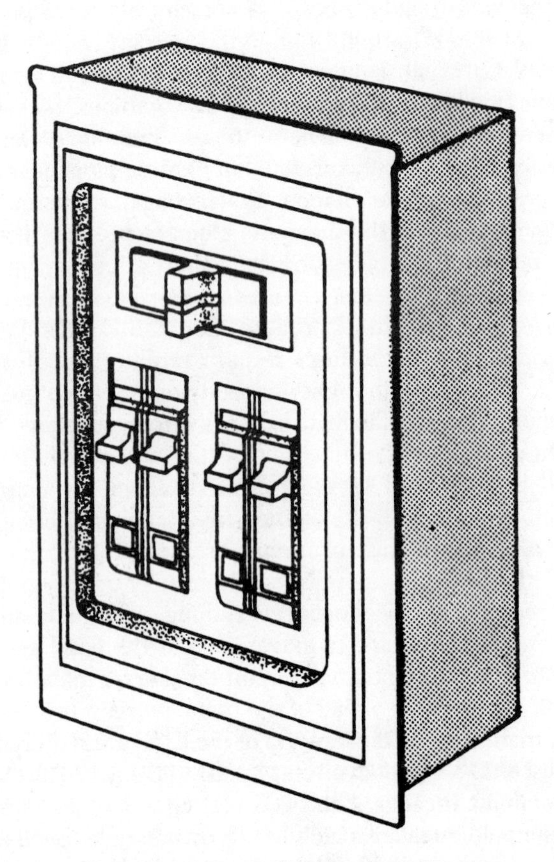

Fig. 15. Typical circuit breaker-box.

Lampholders and Sockets

Lamp sockets as shown in Fig. 16 are generally screw-base units placed in circuits as holders for incandescent lamps. A special type of lampholder has contacts, rather than a screw base, which engage and hold the prongs of fluorescent lamps when they are rotated in the holder. The sockets can generally be attached to a hanging cord or mounted directly on a wall or ceiling in open wiring installations by using screws or nails in the mounting holes provided in the nonconducting material which is molded or formed around the lamp socket. The two mounting holes in a porcelain lamp socket are spaced so the sockets may also be attached to outlet box "ears" or a plaster ring with machine screws. The screw threads molded or rolled in the ends of the lampholder sockets also facilitate their ready integration in other types of lighting fixtures such as table lamps, floor lamps, or hanging fixtures which have reflectors or decorative shades. In an emergency, a socket may also be used as a receptable. The socket is converted to a receptacle by screwing in a female plug. One type of ceiling lampholder has a grounded outlet located on the side. Lamp sockets are produced in many different sizes and shapes. A few of the most common sizes are shown in Fig. 17.

Signal Equipment

Figure 18 illustrates the most common components in interior wiring signal systems. Their normal operating voltages are 6, 12, 18, or 24 volts, ac or dc. As a general rule they are connected by open-wiring methods and are used as interoffice or building-to-building signal systems.

Reflectors and Shades

Figure 19 shows several types of *reflectors* and *shades* which are used to focus the lighting effect of bulbs. Of these, some are used to flood an area with high intensity light and are called floodlights. Others, called spotlights, concentrate the useful light on a small area. Both floodlights and spotlights can come in two- or three-light clusters with swivel holders. They can be mounted

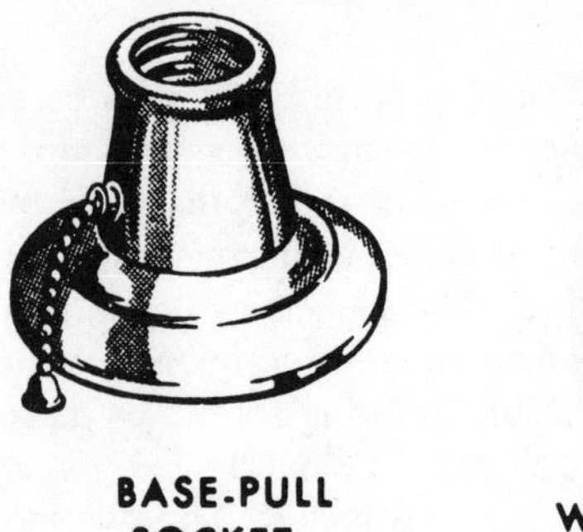

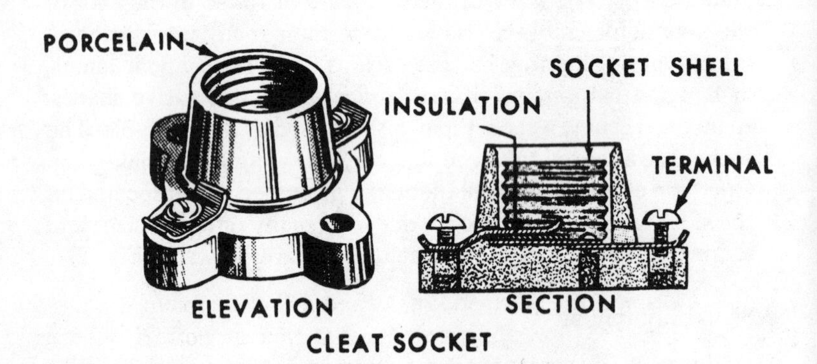

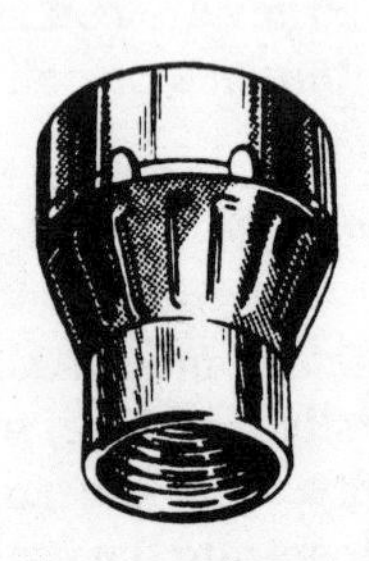

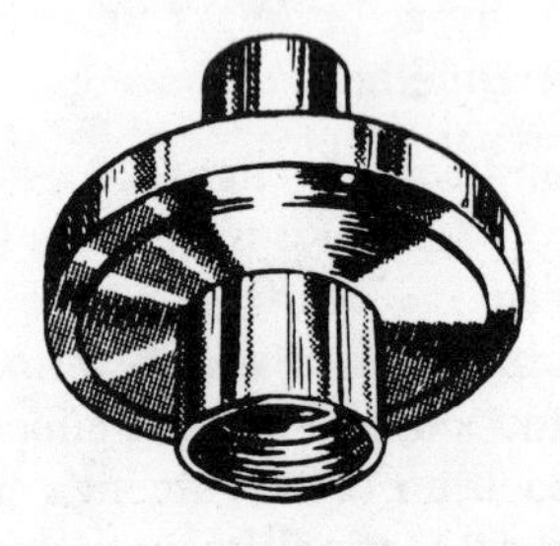

Fig. 16. Lampholders and sockets.

HANGING SOCKETS

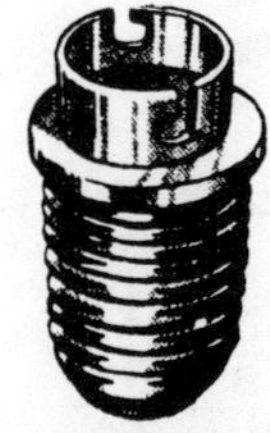

LAMP-SOCKET ADAPTER

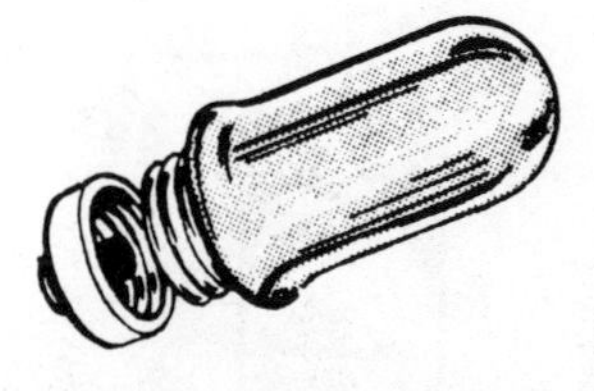

VAPORPROOF RECEPTACLE

Fig. 16. (continued)

on walls or posts or on spikes pushed into the ground. One and two, Fig. 19, illustrate reflectors that deliver normal building light of average intensity in a pattern similar to the floodlight shown in 3, Fig. 19.

Incandescent Lamps

The most common light source for general use is the *incandescent lamp.* Though it is the least efficient type of light, it is preferred over the fluorescent type because of its low initial cost, ease of maintenance, and convenience that is readily seen by the wide selection of wattage ratings that can be inserted in one type of socket. Further, since its emitted candlepower is directly proportional to the voltage, a lower voltage application will dim the light. A high-rated voltage application from a power source will increase its intensity. Although an incandescent light is eco-

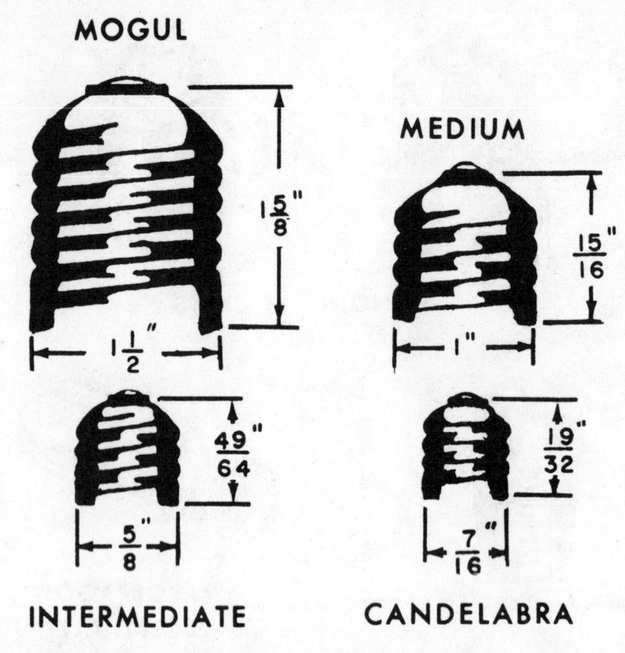

Fig. 17. General lamp-socket sizes.

nomical, it is also inefficient, because a large amount of the energy supplied to it is converted to heat rather than light. Moreover, it does not give a true light because the tungsten filament emits a great deal more red and yellow light than does the light of the sun. Incandescent lamps are shown in Fig. 20. Incandescent lights are normally built to last 1,000 hours when operating at their rated voltage.

Fluorescent lamps

Fluorescent lamps (Fig. 21) are of either the conventional "hot cathode" or "cold cathode" type. The "hot cathode" lamp has a coiled-wire type of electrode, which gives off electrons when heated. These electrons collide with mercury atoms, provided by mercury vapor in the tubes, which produces ultraviolet radiation. Fluorescent powder coatings on the inner walls of the

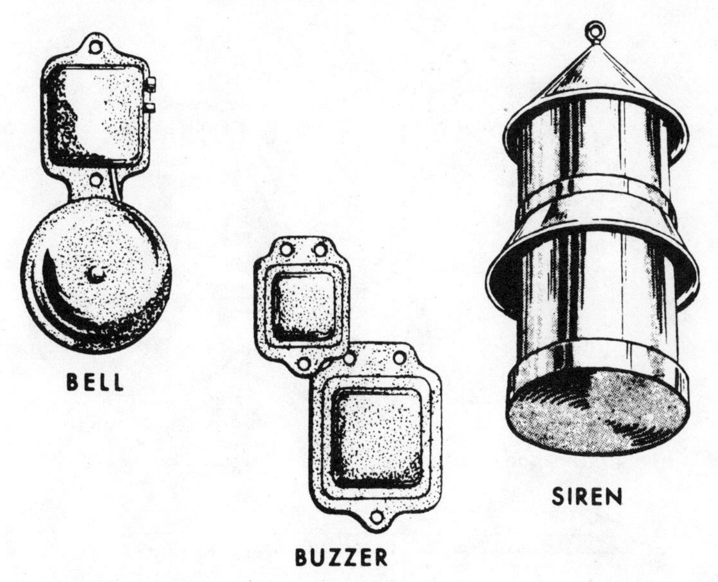

Fig. 18. Types of signal equipment.

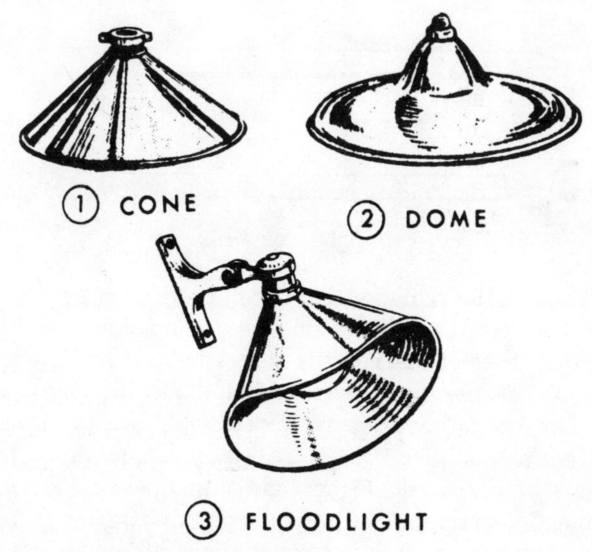

Fig. 19. Types of reflectors.

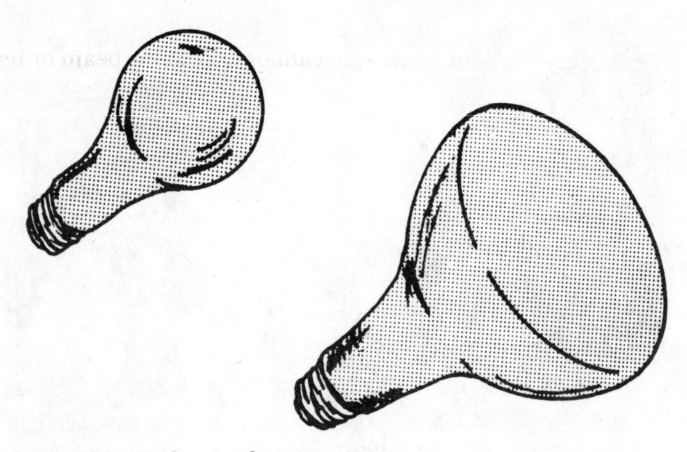

Fig. 20. Incandescent lamps.

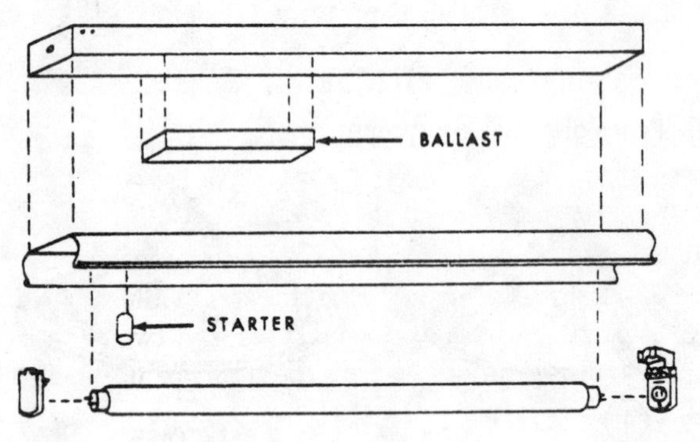

Fig. 21. Fluorescent light accessories.

tubes absorb this radiation and transform the energy into visible light. The "cold cathode" lamp operates in a similar manner except that its electrode consists of a cylindrical tube. It receives its name because the heat is generated over a larger area and, therefore, the cathode does not reach as high a temperature as in the "hot cathode" unit. It is used most frequently on flashing circuits. Because of the higher light output per watt input, more illumination and less heat is obtained per watt from fluorescent lamps than from incandescent ones. Light diffusion is also better, and surface brightness is lower. The life of fluorescent lamps is also longer compared to filament types. However, the fluorescent lamp, because of its design, cannot control its beam of light as well as the incandescent type and has a tendency to produce stroboscopic effects which are counteracted by phasing arrangements. Moreover, when voltage fluctuations are severe, the lamps may go out prematurely or start slowly. Finally, the higher initial cost in fluorescent lighting, which requires auxiliary equipment such as starters, ballasts, special lampholders, and fixtures (Fig. 21), is also a disadvantage compared with other types of illumination.

Construction. The fluorescent lamp is an electric discharge lamp that consists of an elongated tubular bulb with an oxide-coated filament sealed in each end to comprise two electrodes (Fig. 22). The bulb contains a drop of mercury and a small amount of argon gas. The inside surface of the bulb is coated with a fluorescent phosphor. The lamp produces invisible, short wave (ultraviolet) radiation by the discharge through the mercury vapor in the bulb. The phosphor absorbs the invisible radiant energy and reradiates it over a band of wavelengths that are sensitive to the eye.

1. Detail illumination is required where the intensity of general illumination is not sufficient, and in engineering spaces for examination of gages. The fixtures for detail illumination commonly use single fluorescent lamps. One and two, Fig. 22, illustrate the wiring arrangement for these single units, and three, Fig. 22, shows a multiple unit.

2. Because of greater cost and shorter life of 8-watt fluorescent lamps, as compared to 15-watt and 20-watt lamps, fixtures with 8-watt lamps are used only for detail illumination and general illumination within locations where space is restricted.

3. Although the fluorescent lamp is basically an ac lamp, it can be operated on dc with the proper auxiliary equipment. The current is controlled by an external resistance in series with the lamp (4, Fig. 22). Since there is no voltage peak, starting is more difficult and thermal switch starters are required. The lamp tends to deteriorate at one end due to the uniform direction of the current. This may be partially overcome by reversing the lamp position or the direction of current periodically.

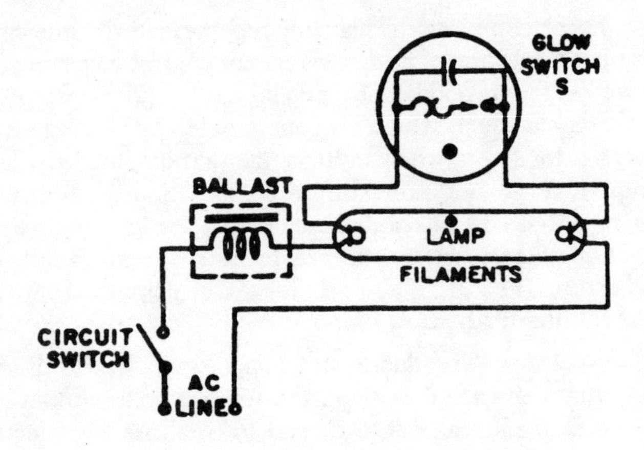

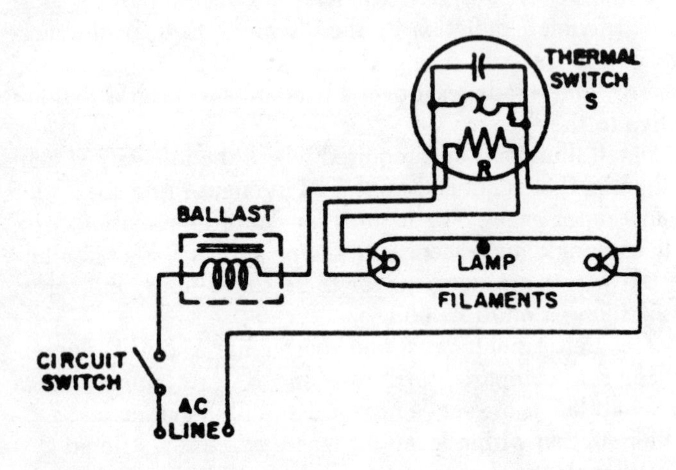

Fig. 22. Fluorescent schematic.

4. Because of the power lost in the resistance ballast box in the dc system, the overall lumens per watt efficiency of the dc system is about 60 percent of the ac system. Also, lamps operated on dc may provide as little as 80 percent of rated life.

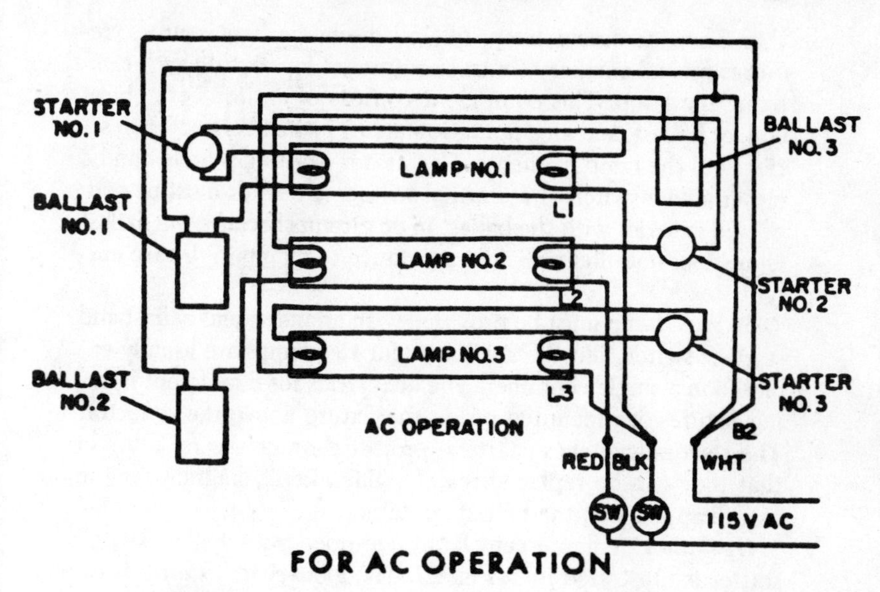

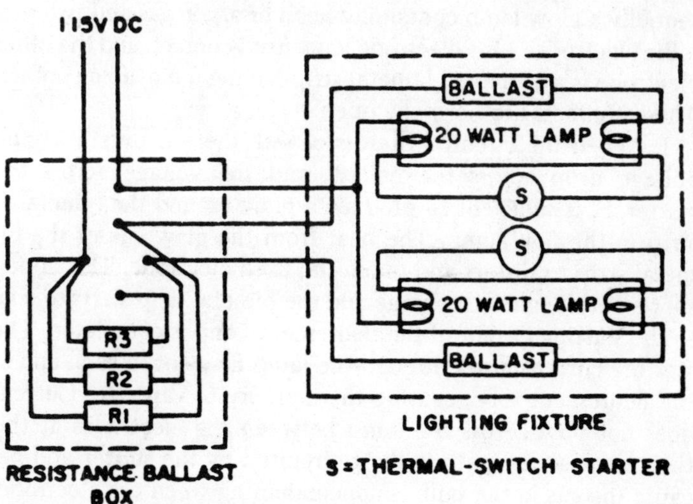

Fig. 22. (continued)

5. The fluorescent lamp, like all discharge light sources, requires special auxiliary control equipment for starting and stabilizing the lamp. This equipment consists of an iron-core choke coil, or ballast, and an automatic starting switch connected in series with the lamp filaments. The starter (starting switch) can be either a flow switch or a thermal switch. A resistor must be connected in series with the ballast in dc circuits because the ballast alone does not offer sufficient resistance to maintain the arc current steady.

6. Each lamp must be provided with an individual ballast and starting switch, but the auxiliaries for two lamps are usually enclosed in a single container. The auxiliaries for fluorescent lighting fixtures are mounted inside the fixture above the reflector. The starting switches (starters) project through the reflector so that they can be replaced readily. The circuit diagram for the fixture appears on the ballast container.

Operation. A fluorescent lamp equipped with a glow-switch starter is illustrated in 1, Fig. 22. The *glow-switch starter* is essentially a glow lamp containing neon or argon gas and two metallic electrodes. One electrode has a fixed contact, and the other electrode is a U-shaped bimetal strip having a movable contact. These contacts are normally open.

1. When the circuit switch is closed, there is practically no voltage drop across the ballast, and the voltage across the starter, S, is sufficient to produce a glow around the bimetallic strip in the flow lamp. The heat from the glow causes the bimetal strip to distort and touch the fixed electrode. This action shorts out the glow discharge and the bimetal strip starts to cool as the starting circuit of the fluorescent lamp is completed. The starting current flows through the lamp filament in each end of the fluorescent tube, causing the mercury to vaporize. Current does not flow across the lamp between the electrodes at this time because the path is short-circuited by the starter and because the gas in the bulb is nonconducting when the electrodes are cold. The preheating of the fluorescent tube continues until the bimetal strip in the starter cools sufficiently to open the starting circuit.

2. When the starting circuit opens, the decrease of current in the ballast produces an induced voltage across the lamp electrodes. The magnitude of this voltage is sufficient to ionize the mercury vapor and start the lamp. The resulting glow discharge (arc) through the fluorescent lamp produces a large amount of ultraviolet radiation that impinges on the phosphor, causing it to fluoresce and emit a relatively bright light. During normal operation the voltage across the fluorescent lamp is not sufficient to produce a glow in the starter. Hence, the contacts remain open and the starter consumes no energy.

3. A fluorescent lamp equipped with a thermal-switch starter is illustrated in 2, Fig. 22. The thermal-switch starter consists of two normally closed metallic contacts and a series resistance contained in a cylindrical enclosure. One contact is fixed, and the movable contact is mounted on a bimetal strip.

4. When the circuit switch is closed, the starting circuit of the fluorescent lamp is completed (through the series resistance, R) to allow the preheating current to flow through the electrodes. The current through the series resistance produces heat that causes the bimetal strip to bend and open the starting circuit. The accompanying induced voltage produced by the ballast starts the lamp. The normal operating current holds the thermal switch open.

5. The majority of thermal-switch starters use some energy during normal operation of the lamp. However, this switch insures more positive starting by providing an adequate preheating period and a higher induced starting voltage.

Characteristics. The failure of a hot-cathode fluorescent lamp usually results from loss of electron-emissive material from the electrodes. This loss proceeds gradually throughout the life of the lamp and is accelerated by frequent starting. The rated average life of the lamp is based on normal burning periods of three to four hours. Blackening of the ends of the bulb progresses gradually throughout the life of the lamp.

1. The efficiency of the energy conversion of a fluorescent lamp is very sensitive to changes in temperature of the bulb. The maximum efficiency occurs in the range of 100° F. to 120° F.,

which is the operating temperature that corresponds to an ambient room temperature range of 65° to 85° F. The efficiency decreases slowly as the temperature is increased above normal, but also decreases very rapidly as the temperature is decreased below normal. Hence, the fluorescent lamp is not satisfactory for locations in which it will be subjected to wide variations in temperature. The reduction in efficiency with low ambient room temperature can be minimized by operating the fluorescent lamp in a tubular glass enclosure so that the lamp will operate at more nearly the desired temperature.

2. Fluorescent lamps are relatively efficient compared with incandescent lamps. For example, a 40-watt fluorescent lamp produces approximately 2800 lumens, or 70 lumens per watt. A 40-watt fluorescent lamp produces six times as much light per watt as does the comparable incandescent lamp.

3. Fluorescent lamps should be operated at voltage within eight percent of their rated voltage. If the lamps are operated at lower voltages, uncertain starting may result, and if operated at higher voltages, the ballast may overheat. Operation of the lamps at either lower or higher voltage results in decreased lamp life. The characteristic curves for hot-cathode fluorescent lamps show the effect of variations from rated voltage on the condition of lamp operation. Also, the performance of fluorescent lamps depends to a great extend on the characteristics of the ballast, which determines the power delivered to the lamp for a given line voltage.

4. When lamps are operated on ac circuits, the light output executes cyclic pulsations as the current passes through zero. This reduction in light output produces a flicker that is more noticeable in fluorescent lamps than in incandescent lamps at frequencies of 50 to 60 cycles and may cause unpleasant stroboscopic effects when moving objects are viewed. The cyclic flicker can be minimized by combining two or three lamps in a fixture and operating the lamps on different phases of a three-phase system. Where only single-phase circuits are available, leading current may be supplied to one lamp and lagging current to another through a lead-lag ballast circuit so that the light pulsations compensate each other.

The fluorescent lamp is inherently a high power-factor device, but the ballast required to stabilize the arc is a low power-factor device. The voltage drop across the ballast is usually equal to the drop across the arc, and the resulting power factor for a single-lamp circuit with ballast is about 50 percent. The low power factor can be corrected in a single-lamp ballast circuit by a capacitor shunted across the line. This correction is accomplished in a two-lamp circuit by means of a "tulamp" auxiliary that connects a capacitor in series with one of the lamps to displace the lamp currents, and, at the same time, to remove the unpleasant stroboscopic effects when moving objects come into view.

Glow lamps. Glow lamps are electric discharge light sources, which are used as indicator or pilot lights for various instruments and on control panels. These lamps have relatively low light output, and thus are used to indicate when circuits are energized or to indicate the operation of electrical equipment installed in remote locations.

1. The glow lamp consists of two closely spaced metallic electrodes sealed in a glass bulb that contains an inert gas. The color of the light emitted by the lamp depends on the gas. Neon gas produces a blue light. The lamp must be operated in series with a current-limiting device to stabilize the discharge. This current-limiting device consists of a high resistance that is usually contained in the lamp base.

2. The glow lamp produces light only when the voltage exceeds a certain striking voltage. As the voltage is decreased somewhat below this value, the glow suddenly vanishes. When the lamp is operated on alternating current, light is produced only during a portion of each half cycle, and both electrodes are alternately surrounded with a glow. When the lamp is operated on direct current, light is produced continuously, and only the negative electrode is surrounded with a glow. This characteristic makes it possible to use the glow lamp as an indicator of alternating current and direct current. It has the advantages of small size, ruggedness, long life, and negligible current consumption, and can be operated on standard lighting circuits.

Transformers (*See* Chapter 20, Transformers.)

The *transformer* is a device for changing alternating current voltages into either high voltages for efficient powerline transmission or low voltages for consumption in lamps, electrical devices, and machines. Transformers vary in size according to their power-handling rating. Their selection is determined by input and output voltage and load current requirements. For example, the transformer used to furnish power for a doorbell reduces 115-volt alternating current to about six to ten volts. This is accomplished by two primary wire leads which are permanently connected to the 115-volt circuit and two secondary screw terminals from the low-voltage side of the transformer. Figure 23 shows a common type of signal system transformer. It is used to lower the building voltage of 120 volts or 240 volts ac to the 6, 12, 18, or 24 volts ac. The wires shown are input and output leads. In Fig. 23 the input leads are smaller than the output leads because the current in the output circuit is greater than in the input circuit.

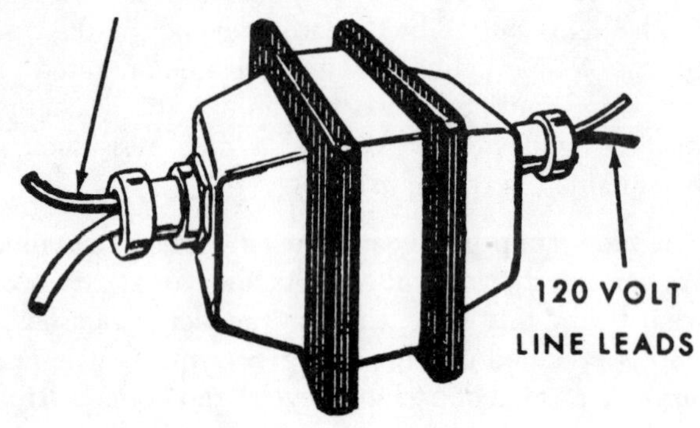

Fig. 23. Transformer.

Chapter 4

Electrical Conductors and Wiring Techniques

Since all electrical circuits utilize conductors of one type or another, it is essential that you know the basic physical features and electrical characteristics of the most common types of conductors.

Any substance that permits the free motion of a larger number of electrons is classed as a *conductor.* A conductor may be made from many different types of metals, but only the most commonly used types of materials will be discussed in this chapter.

To compare the resistance and size of one conductor with that of another, a standard or unit size of conductor must be established. A convenient unit of linear measurement, as far as the diameter of a piece of wire is concerned, is the mil (0.001 of an inch); and a convenient unit of wire length is the foot. The standard unit of size in most cases is the *mil-foot*; that is, a wire will have unit size if it has diameter of 1 mil and a length of 1 foot. The resistance in ohms of a unit conductor of a given substance is called the specific resistance, or specific resistivity, of the substance.

Gage numbers are a further convenience in comparing the diameter of wires. The gage commonly used is the American wire gage (AWG), formerly the Brown and Sharpe (B and S) gage.

MIL

Square Mil

The *square mil* is a convenient unit of cross-sectional area for square or rectangular conductors. A square mil is the area of a square, the sides of which are 1 mil, as shown in Fig. 1. To obtain the cross-sectional area in square mils of a square conductor, square one side measured in mils. To obtain the cross-sectional

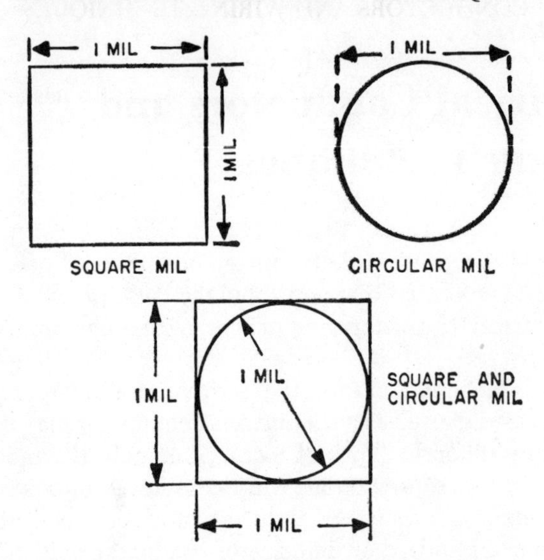

Fig. 1. Square mil; circular mil; and comparison of circular to square mil.

area in square mils of a rectangular conductor, multiply the length of one side by that of the other, each length being expressed in mils.

For example, find the cross-sectional area of a large rectangular conductor ⅜ inch thick and 4 inches wide. The thickness may be expressed in mils as 0.375 × 1,000 = 375 mils, and the width as 4 × 1,000, or 4,000 mils. The cross-sectional area is 375 × 4,000, or 1,500,000 square mils.

Circular Mil

The *circular mil* is the standard unit of wire cross-sectional area used in American and English wire tables. Because the diameters of round conductors, or wires, used to conduct electricity may be only a small fraction of an inch, it is convenient to express these diameters in mils, to avoid the use of decimals. For example, the diameter of a wire is expressed as 25 mils instead of 0.025 inch. A circular mil is the area of a circle having a diameter of 1 mil, as shown in Fig. 1. The area in circular mils of a round conductor is obtained by squaring the diameter measured in mils. Thus, a wire having a diameter of 25 mils has an area of 25^2 or 625 circular mils. By way of comparison, the basic formula for the area of a circle is $A = \pi R^2$ and in this example the area in square inches is

$$A = \pi R^2 = 3.14(0.0125)^2 = 0.00049 \text{ sq. in.}$$

If D is the diameter of a wire in mils, the area in square mils is

$$A = \pi \left(\frac{D}{2}\right)^2 = \frac{3.1416}{4} D^2 = 0.7854D^2 \text{ sq. mils}$$

Therefore, a wire 1 mil in diameter has an area of

$$A = 0.7854 \times 1^2 = 0.7854 \text{ sq. mils,}$$

which is equivalent to 1 circular mil. The cross-sectional area of a wire in circular mils is therefore determined as

$$A = \frac{0.7854D^2}{0.7854} = D^2 \text{ circular mils,}$$

where D is the diameter in mils. Thus, the constant $\frac{\pi}{4}$ is eliminated from the calculation.

In comparing square and round conductors it should be noted that the circular mil is a smaller unit of area than the square mil, and therefore there are more circular mils than square mils in any given area. The comparison is shown in Fig. 1. The area of a circular mil is equal to 0.7854 of a square mil. Therefore, to determine the circular-mil area when the square-mil area is given, divide the area in square mils by 0.7854. Conversely, to determine the square-mil area when the circular-mil area is given, multiply the area in circular mils by 0.7854.

For example, a No. 12 wire has a diameter of 80.81 mils. What is (1) its area in circular mils and (2) its area in square mils? Solution:

$$(1)\ A = D^2 = 80.81^2 = 6{,}530 \text{ circular mils}$$
$$(2)\ A = 0.7854 \times 6{,}530 = 5{,}128.7 \text{ square mils}$$

A rectangular conductor is 1.5 inches wide and 0.25 inch thick. (1) What is its area in square mils? (2) What size of round conductor in circular mils is necessary to carry the same current as the rectangular bar?
Solution:

$$\begin{aligned} (1)\quad 1.5'' &= 1.5 \times 1{,}000 = 1{,}500 \text{ mils} \\ 0.25'' &= 0.25 \times 1{,}000 = 250 \text{ mils} \\ A &= 1{,}500 \times 250 = 375{,}000 \text{ sq. mils} \end{aligned}$$

(2) To carry the same current, the cross-sectional area of the rectangular bar and the cross-sectional area of the round conductor must be equal. There are more circular mils than square mils in this area, and therefore

$$A = \frac{375{,}000}{0.7854} = 477{,}000 \text{ circular mils}$$

A wire in its usual form is a slender rod or filament of drawn metal. In large sizes, wire becomes difficult to handle, and its flexibility is increased by stranding. The strands are usually single wires twisted together in sufficient numbers to make up the necessary cross-sectional area of the cable. The total area in circular mils is determined by multiplying the area of one strand in circular mils by the number of strands in the cable.

Circular-Mil-Foot

A *circular-mil-foot,* as shown in Fig. 2, is actually a unit of volume. It is a unit conductor 1 foot in length and having a cross-sectional area of 1 circular mil. Because it is considered a unit conductor, the circular-mil-foot is useful in making comparisons between wires that are made of different metals. For example, a basis of comparison of the ***resistivity*** (to be treated later) of various substances may be made by determining the resistance of a circular-mil-foot of each of the substances.

In working with certain substances it is sometimes more convenient to employ a different unit volume. Accordingly, unit volume may also be taken as the centimeter cube, and specific

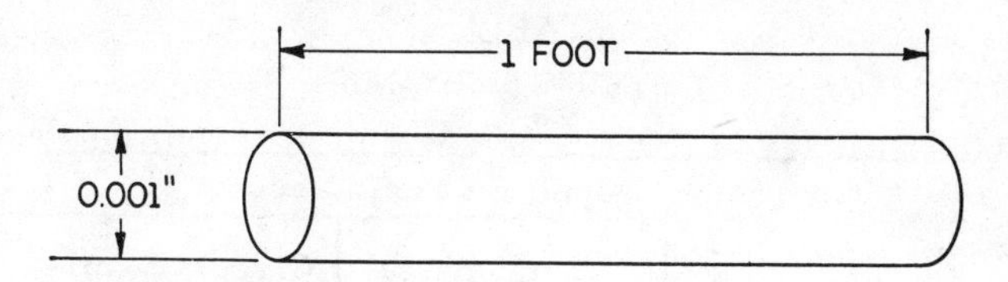

Fig. 2. Circular-mil foot.

resistance becomes the resistance offered by a cube-shaped conductor 1 cm. long and 1 sq. cm. in cross-sectional area. The inch cube may also be used. The unit of volume employed is given in tables of specific resistances.

SPECIFIC RESISTANCE (RESISTIVITY)

Specific resistance, or *resistivity,* is the resistance in ohms offered by unit volume (the circular-mil-foot) of a substance to the flow of electric current. *Resistivity* is the reciprocal of conductivity. A substance that has a high resistivity will have a low conductivity, and vice versa.

Thus, the specific resistance of a substance is the resistance of a unit volume of that substance. Many tables of specific resistance are based on the resistance in ohms of a volume of the substance 1 foot long and 1 circular mil in cross-sectional area. The temperature at which the resistance measurement is made is also specified. If the kind of metal of which a conductor is made is known, the specific resistance of the metal may be obtained from a table. The specific resistances of some common substances are given in Table 1.

The resistance of a conductor of uniform cross section varies directly as the product of the length and the specific resistance of the conductor and inversely as the cross-sectional area of the conductor. Therefore, the resistance of a conductor may be calculated if the length, cross-sectional area, and specific resistance of the substance are known. Expressed as an equation, the resistance, R in ohms, of a conductor is

$$R = \rho \frac{L}{A}$$

TABLE 1

SPECIFIC RESISTANCE

Substance	Specific resistance at 20° C.	
	Centimeter cube (microhms)	Circular-mil-foot (ohms)
Silver	1.629	9.8
Copper (drawn).	1.724	10.37
Gold	2.44	14.7
Aluminum . . .	2.828	17.02
Carbon (amorphous.)	3.8 to 4.1	. . . .
Tungsten	5.51	33.2
Brass	7.0	42.1
Steel (soft) . . .	15.9	95.8
Nichrome . . .	109.0	660.0

where ρ (Greek rho) is the specific resistance in ohms per circular-mil-foot, L the length in feet and A the cross-sectional area in circular mils.

For example, what is the resistance of 1,000 feet of copper wire having a cross-sectional area of 10,400 circular mils (No. 10 wire), the wire temperature being 20°C?

Solution:

The specific resistance, from Table 1, is 10.37. Substituting the known values in the preceding equation, the resistance, R, is determined as

$$R = \rho \frac{L}{A} = 10.37 \times \frac{1{,}000}{10{,}400} = 1 \text{ ohm, approximately}$$

If R, ρ, and A are known, the length may be determined by a simple mathematical transposition. This is of value in many applications. For example, when it is desired to locate a ground in a telephone line, special test equipment is used that operates on the principle that the resistance of a line varies directly with its length. Thus, the distance between the test point and a fault can be computed accurately.

Conductance (G) is the reciprocal of resistance. When R is in ohms, the conductance is expressed in mhos. Whereas resistance is opposition to flow, conductance is the ease with which the current flows. Conductance in mhos is equivalent to the number of amperes flowing in a conductor per volt of applied emf. Expressed in terms of the specific resistance, length, and cross section of a conductor,

$$G = \rho \frac{A}{L}$$

The conductance, G, varies directly as the cross-sectional area, A, and inversely as the specific resistance, ρ, and the length, L. When A is in circular mils, ρ is in ohms per circular-mil-foot, L is in feet, and G is in mhos.

The relative conductance of several substances is given in Table 2.

TABLE 2

RELATIVE CONDUCTANCE

Substance	Relative conductance (Silver = 100%)
Silver	100
Copper	98
Gold	78
Aluminum	61
Tungsten	32
Zinc	30
Platinum	17
Iron	16
Lead	15
Tin	9
Nickel	7
Mercury	1
Carbon	0.05

WIRE MEASURE

Relation Between Wire Sizes

Wires are manufactured in sizes numbered according to a table known as the American wire gage (AWG). As may be seen in Table 3, the wire diameters become smaller as the gage numbers become larger. The largest wire size shown in Table 3 is 0000 (read "4 naught"), and the smallest is number 40. Larger and smaller sizes are manufactured but are not commonly used. The ratio of the diameter of one gage number to the diameter of the next higher gage number is a constant, 1.123. The cross-sectional area varies as the square of the diameter. Therefore, the ratio of the cross section of one gage number to that of the next higher gage number is the square of 1.123, or 1.261. Because the cube of 1.261 is very nearly 2, the cross-sectional area is approximately halved, or doubled, every three gage numbers. Also because 1.261 raised to the 10th power is very nearly equal to 10, the cross-sectional area is increased or decreased ten times every 10 gage numbers.

A No. 10 wire has a diameter of approximately 102 mils, a cross-sectional area of approximately 10,400 circular mils, and a resistance of approximately 1 ohm per 1,000 feet. From these facts it is possible to estimate quickly the cross-sectional area and the resistance of any size copper wire without referring directly to a wire table.

For example, to estimate the cross-sectional area and the resistance of 1,000 feet of No. 17 wire, the following reasoning might be employed. A No. 17 wire is three sizes removed from a No. 20 wire and therefore has twice the cross-sectional area of a No. 20 wire. A No. 20 wire is ten sizes removed from a No. 10 wire and therefore has one-tenth the cross section of a No. 10 wire. Therefore, the cross-sectional area of a No. 17 wire is $2 \times 0.1 \times 10{,}000 = 2{,}000$ circular mils. Since resistance varies inversely with the cross-sectional area, the resistance of a No. 17 wire is $10 \times 1 \times 0.5 = 5$ ohms per 1,000 feet.

A *wire gage* is shown in Fig. 3. It will measure wires ranging in size from number 0 to number 36. The wire whose size is to be

TABLE 3

STANDARD ANNEALED SOLID COPPER WIRE (AMERICAN WIRE GAGE—BROWN & SHARPE)

Gage number	Diameter (mils)	Cross section		Ohms per 1,000 ft.		Ohms per mile	Pounds per 1,000 ft.
		Circular mils	Square inches	25°C. (=77°F.)	65°C. (=149°F.)	25°C. (=77°F.)	
0000	460.0	212,000.0	0.166	0.0500	0.0577	0.264	641.0
000	410.0	168,000.0	.132	.0630	.0727	.333	508.0
00	365.0	133,000.0	.105	.0795	.0917	.420	403.0
0	325.0	106,000.0	.0829	.100	.116	.528	319.0
1	289.0	83,700.0	.0657	.126	.146	.665	253.0
2	258.0	66,400.0	.0521	.159	.184	.839	201.0
3	229.0	52,600.0	.0413	.201	.232	1.061	159.0
4	204.0	41,700.0	.0328	.253	.292	1.335	126.0
5	182.0	33,100.0	.0260	.319	.369	1.685	100.0
6	162.0	26,300.0	.0206	.403	.465	2.13	79.5
7	144.0	20,800.0	.0164	.508	.586	2.68	63.0
8	128.0	16,500.0	.0130	.641	.739	3.38	50.0
9	114.0	13,100.0	.0103	.808	.932	4.27	39.6
10	102.0	10,400.0	.00815	1.02	1.18	5.38	31.4
11	91.0	8,230.0	.00647	1.28	1.48	6.75	24.9
12	81.0	6,530.0	.00513	1.62	1.87	8.55	19.8
13	72.0	5,180.0	.00407	2.04	2.36	10.77	15.7
14	64.0	4,110.0	.00323	2.58	2.97	13.62	12.4
15	57.0	3,260.0	.00256	3.25	3.75	17.16	9.86
16	51.0	2,580.0	.00203	4.09	4.73	21.6	7.82
17	45.0	2,050.0	.00161	5.16	5.96	27.2	6.20
18	40.0	1,620.0	.00128	6.51	7.51	34.4	4.92
19	36.0	1,290.0	.00101	8.21	9.48	43.3	3.90
20	32.0	1,020.0	.000802	10.4	11.9	54.9	3.09
21	28.5	810.0	.000636	13.1	15.1	69.1	2.45
22	25.3	642.0	.000505	16.5	19.0	87.1	1.94
23	22.6	509.0	.000400	20.8	24.0	109.8	1.54
24	20.1	404.0	.000317	26.2	30.2	138.3	1.22
25	17.9	320.0	.000252	33.0	38.1	174.1	0.970
26	15.9	254.0	.000200	41.6	48.0	220.0	0.769
27	14.2	202.0	.000158	52.5	60.6	277.0	0.610
28	12.6	160.0	.000126	66.2	76.4	350.0	0.484
29	11.3	127.0	.0000995	83.4	96.3	440.0	0.384
30	10.0	101.0	.0000789	105.0	121.0	554.0	0.304
31	8.9	79.7	.0000626	133.0	153.0	702.0	0.241
32	8.0	63.2	.0000496	167.0	193.0	882.0	0.191
33	7.1	50.1	.0000394	211.0	243.0	1,114.0	0.152
34	6.3	39.8	.0000312	266.0	307.0	1,404.0	0.120
35	5.6	31.5	.0000248	335.0	387.0	1,769.0	0.0954
36	5.0	25.0	.0000196	423.0	488.0	2,230.0	0.0757
37	4.5	19.8	.0000156	533.0	616.0	2,810.0	0.0600
38	4.0	15.7	.0000123	673.0	776.0	3,550.0	0.0476
39	3.5	12.5	.0000098	848.0	979.0	4,480.0	0.0377
40	3.1	9.9	.0000078	1,070.0	1,230.0	5,650.0	0.0299

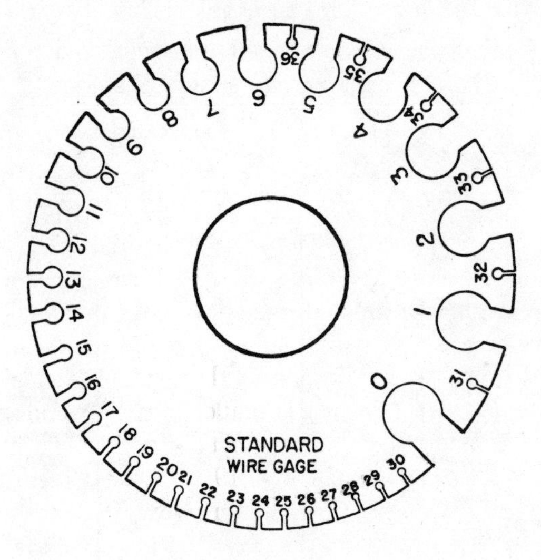

Fig. 3. Wire gage.

measured is inserted in the smallest slot that will just accommodate the bare wire. The gage number corresponding to that slot indicates the wire size. The slot has parallel sides and should not be confused with the semicircular opening at the end of the slot. The opening simply permits the free movement of the wire all the way through the slot.

Stranded Wires and Cables

A *wire* is a slender rod or filament of drawn metal. This definition restricts the term to what would ordinarily be understood as "solid wire." The word "slender" is used because the length of a wire is usually large in comparison with the diameter. If a wire is covered with insulation, it is properly called an insulated wire. Although the term "wire" properly refers to the metal, it is generally understood to include the insulation.

A *conductor* is a wire or combination of wires not insulated from one another, suitable for carrying an electric current.

A *stranded conductor* is a conductor composed of a group of wires or of any combination of groups of wires. The wires in a stranded conductor are usually twisted together.

A *cable* is either a stranded conductor (single-conductor cable) or a combination of conductors insulated from one another (multiple-conductor cable). The term cable is a general one, and in practice it is usually applied only to the larger sizes of conductors. A small cable is more often called a stranded wire or cord. Cables may be bare or insulated. The insulated cables may be sheathed (covered) with lead or protective armor.

Figure 4 shows some of the different types of cables used.

Conductors are stranded mainly to increase their flexibility. The arrangement of the wire strands in concentric-layer cables is as follows:

The first layer of strands around the center is made up of six conductors; the second layer is made up of 12 conductors; the third layer is made up of 18 conductors; and so on. Thus, standard cables are composed of 1, 7, 19, 37, and so forth, strands.

The overall flexibility may be increased by further stranding of the individual strands.

Figure 5 shows a typical 37-strand cable. It also shows how the total circular-mil cross-sectional area of a stranded cable is determined.

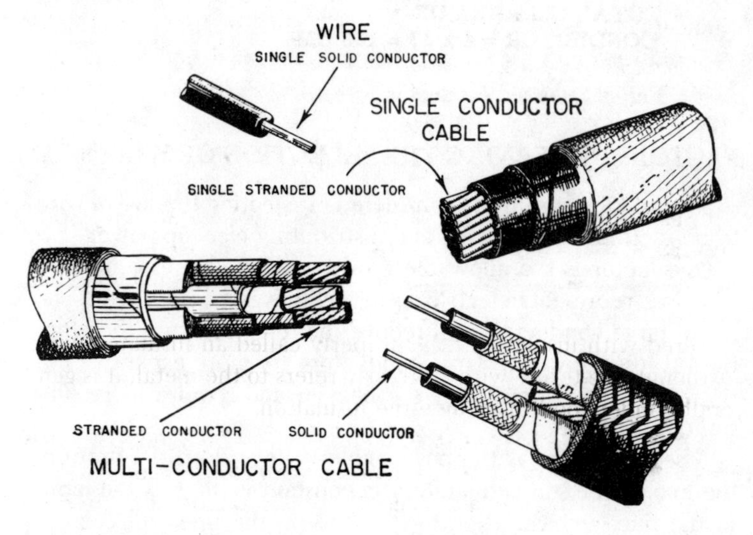

Fig. 4. Conductors.

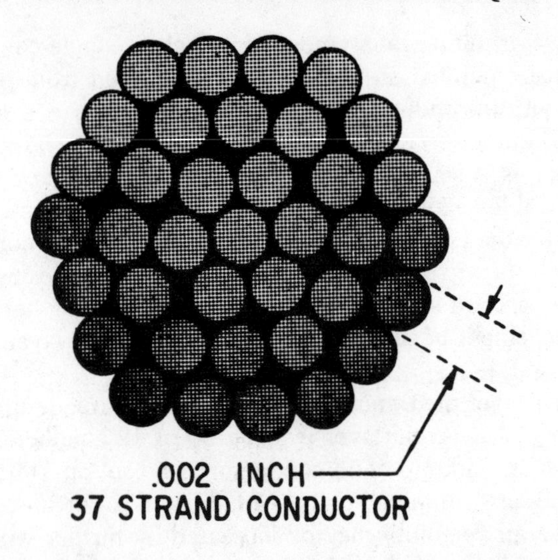

DIAMETER OF EACH STRAND = .002 INCH
DIAMETER OF EACH STRAND, MILS = 2 MILS
CIRCULAR MIL AREA OF
EACH STRAND = D^2 = 4 CM
TOTAL CM AREA OF
CONDUCTOR = 4 x 37 = 148 CM

Fig. 5. Stranded conductor.

FACTORS GOVERNING THE SELECTION OF WIRE SIZE

Several factors must be considered in selecting the size of wire to be used for transmitting and distributing electric power.

One factor is the allowable power loss (I^2R loss) in the line. This loss represents electrical energy converted into heat. The use of large conductors will reduce the resistance and therefore the I^2R loss. However, large conductors are more expensive initially than small ones; they are heavier and require more substantial supports.

A second factor is the permissible voltage drop (IR drop) in the line. If the source maintains a constant voltage at the input to the line, any variation in the load on the line will cause a variation in line current, and a consequent variation in the IR drop in the line. A wide variation in the IR drop in the line causes poor voltage regulation at the load. The obvious remedy is to reduce either I or R. A reduction in load current lowers the amount of power being transmitted, whereas a reduction in line resistance increases the size and weight of conductors required. A compromise is generally reached whereby the voltage variation at the load is within tolerable limits and the weight of line conductors is not excessive.

A third factor is the *current-carrying ability* of the line. When current is drawn through the line, heat is generated. The temperature of the line will rise until the heat radiated, or otherwise dissipated, is equal to the heat generated by the passage of current through the line. If the conductor is insulated, the heat generated in the conductor is not so readily removed as it would be if the conductor were not insulated. Thus, to protect the insulation from too much heat, the current through the conductor must be maintained below a certain value. Rubber insulation will begin to deteriorate at relatively low temperatures. Varnished cloth insulation retains its insulating properties at higher temperatures, and insulation such as asbestos or silicon is effective at still higher temperatures.

Electrical conductors may be installed in locations where the ambient (surrounding) temperature is relatively high. In this case the heat generated by external sources constitutes an appreciable part of the total conductor heating. Due allowance must be made for the influence of external heating on the allowable conductor current, and each case has its own specific limitations. The maximum allowable operating temperature of insulated conductors is specified in tables and varies with the type of conductor insulation being used.

Tables have been prepared by the National Board of Fire Underwriters giving the safe current ratings for various sizes and types of conductors covered with various types of insulation. For example, the allowable current-carrying capacities of copper conductors at not over 30°C (86°F) room temperature for single conductors in free air are given in Table 4.

TABLE 4

CURRENT-CARRYING CAPACITIES (IN AMPERES) OF SINGLE COPPER CONDUCTORS AT AMBIENT TEMPERATURE OF BELOW 30° C

Size	Rubber or thermo-plastic	Thermoplastic asbestos, var-cam, or asbestos var-cam	Impregnated asbestos	Asbestos	Slow-burning or weather-proof
0000	300	385	475	510	370
000	260	330	410	430	320
00	225	285	355	370	275
0	195	245	305	325	235
1	165	210	265	280	205
2	140	180	225	240	175
3	120	155	195	210	150
4	105	135	170	180	130
6	80	100	125	135	100
8	55	70	90	100	70
10	40	55	70	75	55
12	25	40	50	55	40
14	20	30	40	45	30

COPPER VS. ALUMINUM CONDUCTORS

Although silver is the best conductor, its cost limits its use to special circuits where a substance with high conductivity is needed.

The two most generally used conductors are copper and aluminum. Each has characteristics that make its use advantageous under certain circumstances. Likewise, each has certain disadvantages.

Copper has a higher conductivity, is more ductile (can be drawn out), has relatively high tensile strength, and can be easily soldered. It is more expensive and heavier than aluminum.

Although aluminum has only about 60 percent of the conductivity of copper, its lightness makes possible long spans, and its relatively large diameter for a given conductivity reduces corona—that is, the discharge of electricity from the wire when it has a high potential. The discharge is greater when smaller diameter wire is used than when larger diameter wire is used. However, aluminum conductors are not easily soldered, and alu-

TABLE 5

CHARACTERISTICS OF COPPER AND ALUMINUM

Characteristics	Copper	Aluminum
Tensile strength (lb/in^2).	55,000	25,000
Tensile strength for same conductivity (lb).	55,000	40,000
Weight for same conductivity (lb).	100	48
Cross section for same conductivity (C.M.).	100	160
Specific resistance (Ω/mil ft).	10.6	17

minum's relatively large size for a given conductance does not permit the economical use of an insulation covering.

A comparison of some of the characteristics of copper and aluminum is given in Table 5.

TEMPERATURE COEFFICIENT

The resistance of pure metals—such as silver, copper, and aluminum—increases as the temperature increases. However, the resistance of some alloys—such as constantan and manganin—changes very little as the temperature changes. Measuring instruments use these alloys because the resistance of the circuits must remain constant if accurate measurements are to be achieved.

In Table 1 the resistance of a circular-mil-foot of wire (the specific resistance) is given at a specific temperature, 20°C in this case. It is necessary to establish a standard temperature because, as has been stated, the resistance of pure metals increases with an increase in temperature, and a true basis of comparison cannot be made unless the resistances of all the substances being compared are measured at the same temperature. The amount of increase in the resistance of a 1-ohm sample of the conductor per degree rise in temperature above 0°C is called the *temperature coefficient of resistance*. For copper, the value is approximately 0.00427 ohm. For pure metals, the temperature coefficient of resistance ranges between 0.003 and 0.006 ohm.

Thus, a copper wire having a resistance of 50 ohms at an initial temperature of 0°C will have an increase in resistance of 50 × 0.00427, or 0.215 ohms for the entire length of wire for each degree of temperature rise above 0°C. At 20°C the increase in resistance is approximately 20 × 0.214, or 4.28 ohms. The total resistance at 20°C is 50 + 4.28, or 54.28 ohms.

CONDUCTOR INSULATION

To be useful and safe, electric current must be forced to flow only where it is needed. It must be "channeled" from the power source to a useful load. In general, current-carrying conductors must not be allowed to come into contact with one another, their supporting hardware, or personnel working near them. To prevent such contact, conductors are coated or wrapped with various materials. These materials have such a high resistance that they are, for all practical purposes, nonconductors. They are generally referred to as insulators or insulating material.

Because of the expense of insulation and its stiffening effect, together with the great variety of physical and electrical conditions under which the conductors are operated, only the necessary minimum of insulation is applied for any particular type of cable designed to do a specific job. Therefore, there is a wide variety of insulated conductors available to meet the requirements of any job.

Two fundamental properties of insulation materials (such as rubber, glass, asbestos, and plastic) are insulation resistance and dielectric strength. These are different and distinct properties.

Insulation resistance is the resistance to current leakage through and over the surface of insulation materials. Insulation resistance can be measured by means of a megger without damaging the insulation, and information so obtained serves as a useful guide in appraising the general condition. However, the data obtained in this manner may not give a true picture of the condition of the insulation. Clean-dry insulation having cracks or other faults may show a high value of insulation resistance but would not be suitable for use.

Dielectric strength is the ability of the insulator to withstand potential difference and is usually expressed in terms of the voltage at which the insulation fails because of the electrostatic stress. Maximum dielectric strength values can be measured by raising the voltage of a *test sample* until the insulation breaks down.

Rubber

One of the most common types of insulation is ***rubber***. The voltage that may be applied to a rubber-covered pair of conductors (twisted pair) is dependent on the thickness and the quality of the rubber covering. Other factors being equal, the thicker the insulation the higher may be the applied voltage. Figure 6 shows two types of rubber-covered wire. One is a single solid conductor, and the other is a 2-conductor cable in which each stranded conductor is covered with rubber insulation. In each case the rubber serves the same purpose—to confine the current to its conductor.

It may be seen from the enlarged cross-sectional view that a thin coating of tin separates the copper conductor from the rubber insulation. If the thin coating of tin were not used, chemical action would take place and the rubber would become soft and gummy where it makes contact with the copper. When small, solid, or stranded conductors are used, a winding of cotton threads is applied between the conductors and the rubber insulation.

Plastics

Plastic has become one of the more common types of material used as insulation for electrical conductors. It has good insulating, flexibility, and moisture-resistant qualities under various conditions. There are various types of plastics used as insulating material, thermoplastic being one of the most common. With the use of the thermoplastic, the conductor temperature can be higher than with some other types of insulating materials without damage to the insulating quality of the material.

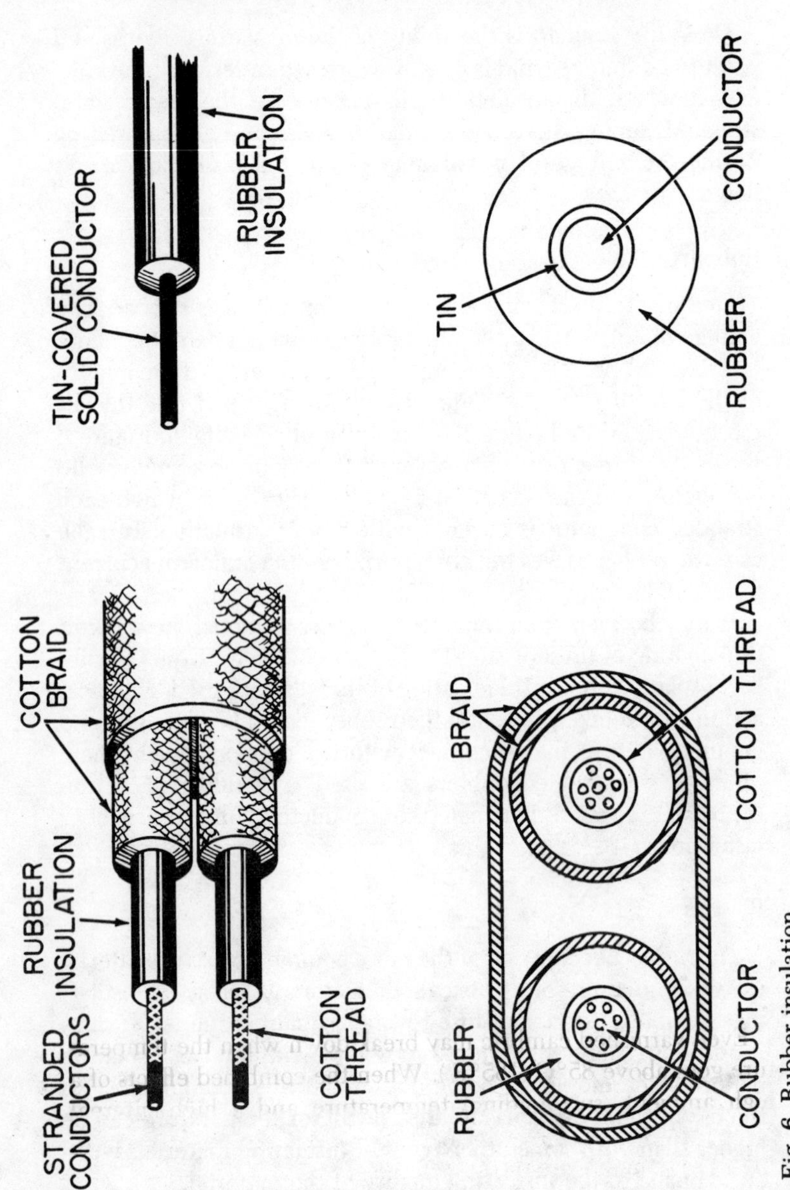

Fig. 6. Rubber insulation.

Varnished Cambric

Heat is developed when current flows through a wire, and when a large amount of current flows, considerable heat may be developed. The heat can be dissipated if air is circulated freely around the wire. If a cover of insulation is used, the heat is not removed so readily and the temperature may reach a high value.

Rubber is a good insulator at relatively low voltage as long as the temperature remains low. Too much heat will cause even the best grade of rubber insulation to become brittle and crack. *Varnished cambric insulation* will stand much higher temperatures than rubber insulation. Varnished cambric is cotton cloth that has been coated with an insulating varnish. Figure 7 shows some of the detail of a cable covered with varnished cambric insulation. The varnished cambric is in tape form and is wound around the conductor in layers. An oily compound is applied between each layer of the tape. This compound prevents water from seeping through the insulation. It also acts as a lubricant between the layers of tape, so that they will slide over each other when the cable is bent.

This type of insulation is used on high-voltage conductors associated with switch gear in substations and power houses and other locations subjected to high temperatures. It is also used on high-voltage generator coils and leads, and also on transformer leads because it is unaffected by oils or grease and because it has a high dielectric strength. Varnished cambric and paper insulation for cables are the two types of insulating materials most widely used at voltages above 15,000 volts, but such cables are always lead-covered to keep the moisture out.

Asbestos

Even varnished cambric may break down when the temperature goes above 85°C (185°F.). When the combined effects of a high ambient (surrounding) temperature and a high internal temperature due to large current flow through the wire makes the total temperature of the wire go above 85°C, *asbestos insulation* is used.

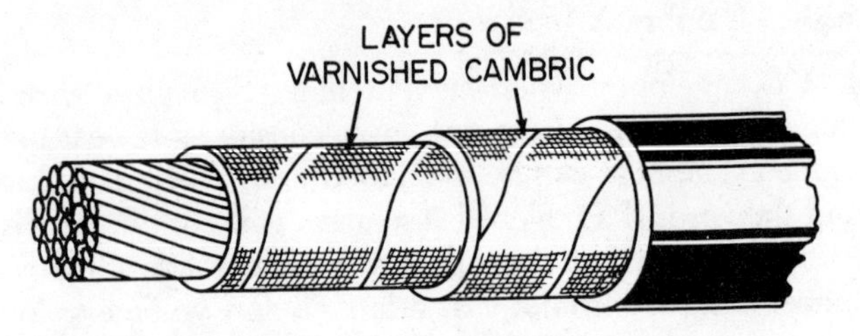

Fig. 7. Varnished cambric insulation.

Asbestos is a good insulation for wires and cables used under very high temperature conditions. It is fire resistant and does not change with age. One type of asbestos-covered wire is shown in Fig. 8. It consists of a stranded copper conductor covered with felted asbestos, which is, in turn, covered with asbestos braid. This type of wire is used in motion-picture projectors, arc lamps, spotlights, heating element leads, and so forth.

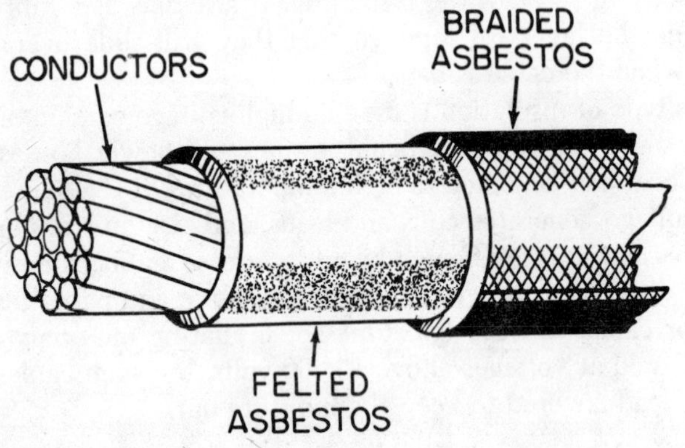

Fig. 8. Asbestos insulation.

Another type of asbestos-covered cable is shown in Fig. 9. It serves as leads for motors and transformers that sometimes must operate in hot, wet locations. The varnished cambric prevents moisture from reaching the innermost layer of asbestos. Asbestos loses its insulating properties when it becomes wet, and will in fact become a conductor. The varnished cambric prevents this from happening because it resists moisture. Although this insu-

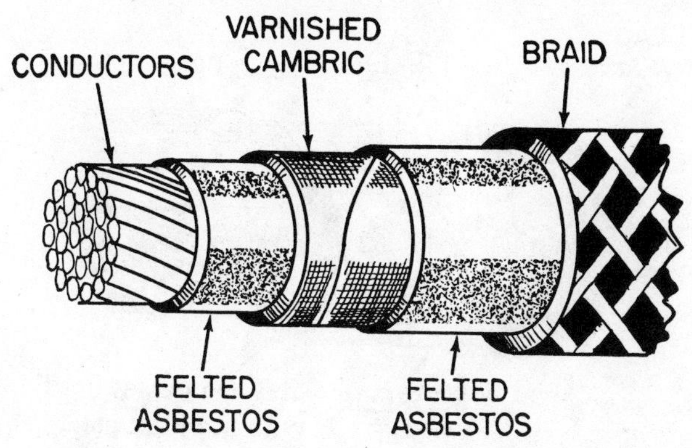

Fig. 9. Asbestos and varnished cambric insulation.

lation will withstand some moisture, it should not be used on conductors that may at times be partly immersed in water, unless the insulation is protected with an outer lead sheath.

Paper

Paper has little insulation value alone, but when impregnated with a high grade of mineral oil, it serves as a satisfactory insulation for high-voltage cables. The oil has a high dielectric strength, and tends to prevent breakdown of paper insulation that is thoroughly saturated with it. The thin paper tape is wrapped in many layers around the conductors, and it is then soaked with oil.

The *3-conductor cable* shown in Fig. 10 consists of paper insulation on each conductor with a spirally wrapped nonmagnetic metallic tape over the insulation. The space between conductors is filled with a suitable spacer to round out the cable and another nonmagnetic metal tape is used to secure the entire cable, and then a lead sheath is applied over all. This type of cable is used on voltages from 10,000 volts to 35,000 volts.

Silk and Cotton

In certain types of circuits—for example, communications circuits—a large number of conductors are needed, perhaps as

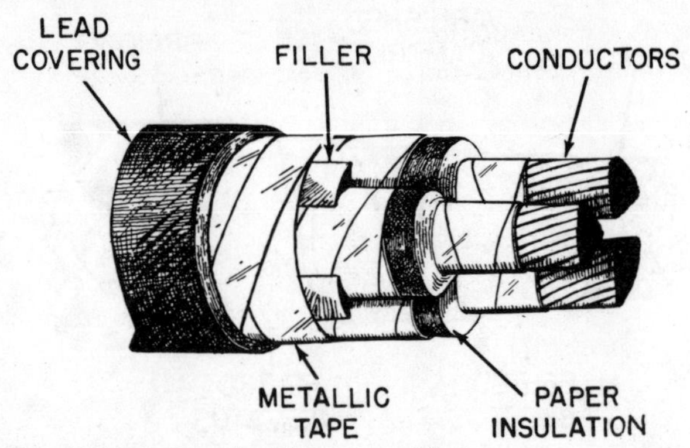

Fig. 10. Paper-insulated power cables.

many as several hundred. Figure 11 shows a cable containing many conductors, each insulated from the others by silk and cotton threads.

The use of silk and cotton as insulation keeps the size of the cable small enough to be handled easily. The silk and cotton threads are wrapped around the individual conductors in reverse directions, and the covering is then impregnated with a special wax compound.

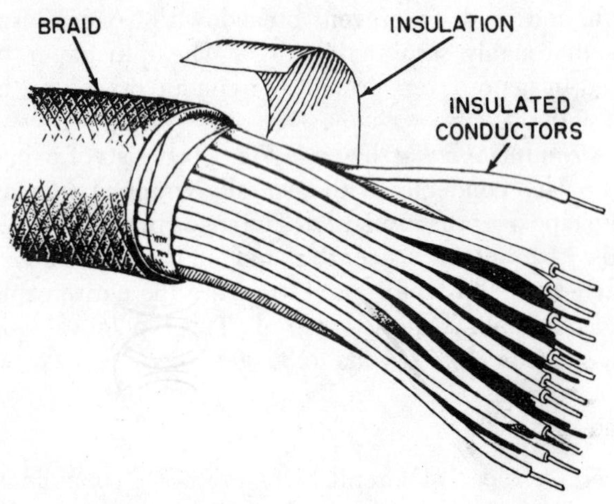

Fig. 11. Silk and cotton insulation.

Because the insulation in this type of cable is not subjected to high voltage, thin layers of silk and cotton are used.

Enamel

The wire used on the coils of meters, relays, small transformers, and so forth, is called *magnet wire.* This wire is insulated with an enamel coating. The enamel is a synthetic compound of cellulose acetate (wood pulp and magnesium). In the manufacturing process, the bare wire is passed through a solution of the hot enamel and then cooled. This is repeated until the wire acquires from 6 to 10 coatings. Enamel has a higher dielectric strength than rubber for equal thickness. It is not practical for large wires because of the expense and because the insulation is easily fractured when large wires are bent.

Figure 12 shows an enamel-coated wire. Enamel is the thinnest insulating coating that can be applied to wires. Hence,

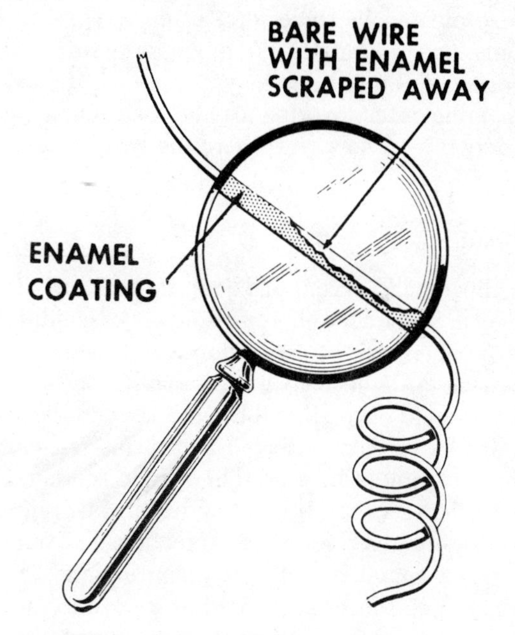

Fig. 12. Enamel insulation.

enamel-insulated magnet wire makes smaller coils. Enameled wire is sometimes covered with one or more layers of cotton covering to protect the enamel from nicks, cuts, or abrasions.

CONDUCTOR PROTECTION

Wires and cables are generally subject to abuse. The type and amount of abuse depends on how and where they are installed and the manner in which they are used. Cables buried directly in the ground must resist moisture, chemical action, and abrasion. Wires installed in buildings must be protected against mechanical injury and overloading. Wires strung on crossarms on poles are kept far enough apart so that they do not touch; but snow, ice, and strong winds necessitate the use of conductors having high tensile strength and substantial supporting frame structures.

Generally, except for overload transmission lines, wires or cables are protected by some form of covering. The covering may be some type of insulator like rubber or plastic. Over this an outer covering of fibrous braid may be applied. If conditions require, a metallic outer covering may be used. The type of outer covering depends on how and where the wire or cable is to be used.

Fibrous Braid

Cotton, linen, silk, rayon, and jute are materials for *fibrous braids*. They are used for outer covering under conditions where the wires or cables are not exposed to heavy mechanical injury. Interior wiring for lights or power is usually done with impregnated cotton-braid-covered, rubber-insulated wire. Generally, the wire will be further protected by a flame-resistant nonmetallic outer covering or by a flexible or rigid conduit.

Figure 13 shows a typical building wire. In this instance two braid coverings are used for extra protection. The outer braid is soaked with a compound that resists moisture and flame.

Impregnated cotton braid is used as a covering for outdoor overhead conductors to afford protection against abrasion. For

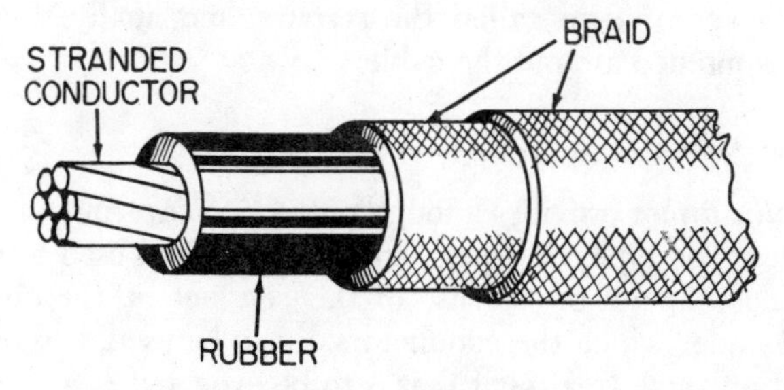

Fig. 13. Fibrous braid covering.

example, the service wires from the transformer secondary mains to the service entrance and also the high-voltage primary mains to the transformer are protected in this manner.

Lead Sheath

Subway-type cables or wires that are continually subjected to water must be protected by a watertight cover. This cover is made of either a continuous lead jacket or a rubber sheath molded around the cable.

Figure 14 is an example of a lead-sheathed cable used in power work. The cable shown is a stranded 3-conductor type. Each conductor is insulated and then wrapped with a layer of rubberized tape. The conductors are twisted together and fillers or rope are added to form a rounded core. Over this is wrapped a

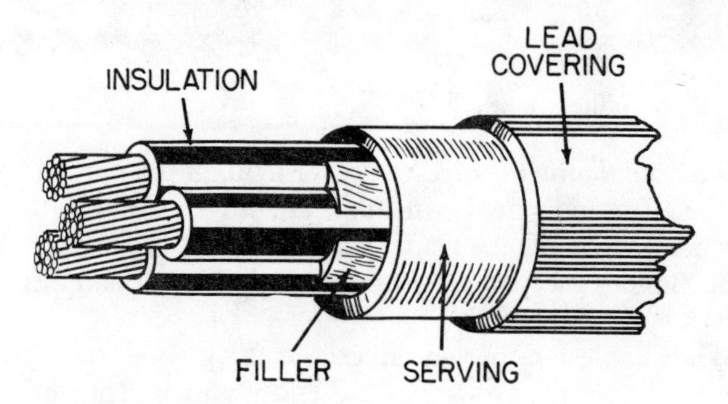

Fig. 14. Lead sheathed cable.

second layer of tape called the *serving*, and finally the lead sheath is molded around the cable.

Metallic Armor

Metallic armor provides a tough protective covering for wires or cables. The type, thickness, and kind of metal used to make the armor depend on the use of the conductors, the circumstances under which the conductors are to be used, and on the amount of rough treatment that is to be expected.

Four types of metallic armor cables are shown in Fig. 15.

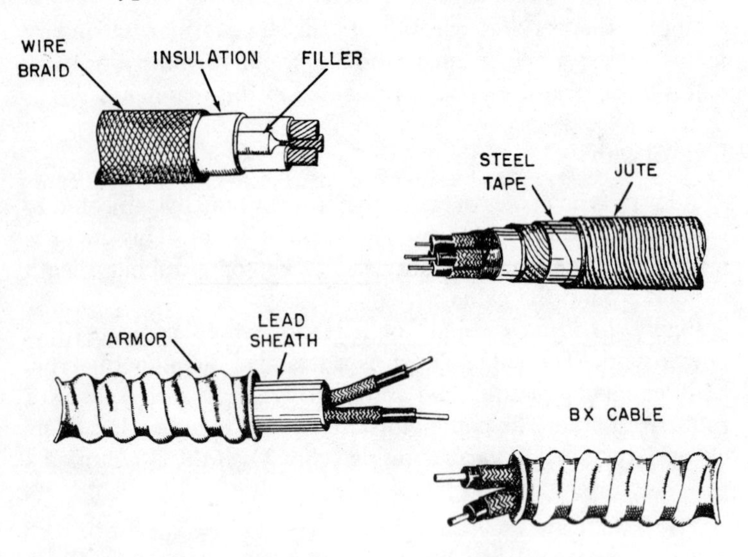

Fig. 15. Metallic armor.

Wire braid armor is used wherever light, flexible protection is needed. The individual wires that are woven together to form the metal braid may be made of steel, copper, bronze, or aluminum. Besides mechanical protection, the wire braid also presents a static shield.

When cables are buried directly in the ground, they risk injury from two sources—moisture and abrasion. They are protected from moisture by a *lead sheath*, and from abrasion by *steel tape* or interlocking armor covers. The steel tape covering, as shown in Fig. 15, is wrapped around the cable and then covered with a serving of jute. It is known as Parkway cable. The interlocking-armor covering can withstand impacts better than steel tape can. Interlocking armor-covered wire (*BX cable*) without the lead sheath is frequently used.

Armor wire is the best type of covering to withstand severe wear and tear. Underwater leaded cable usually has an outer armor wire cover.

All wires and cables do not have the same type of protective covering. Some coverings are designed to withstand moisture, others to withstand mechanical strain, and so forth. A cable may have a combination of each type, each doing its own job.

CONDUCTOR SPLICES AND TERMINAL CONNECTIONS

Conductor splices and *connections* are essential parts of any electric circuit. When conductors join each other, or connect to a load, splices or terminals must be used. It is important that they be properly made, since any electric circuit is only as good as its weakest link. The basic requirement of any splice or connection is that it be both mechanically and electrically as strong as the conductor or device with which it is used. High-quality workmanship and materials must be employed to insure lasting electrical contact, physical strength, and insulation (if required). The most common methods of making splices and connections in electric cables will now be discussed.

The first step in making a splice is preparing the wires or conductor. Insulation must be removed from the end of the conductor and the exposed metal cleaned. In removing the insulation from the wire, a sharp knife is used in much the same manner as in sharpening a pencil. That is, the knife blade is moved at a small angle with the wire to avoid "nicking" the wire. This produces a taper on the cut insulation, as shown in Fig. 16. The insulation may also be removed by using a plierlike hand-operated wire stripper. After the insulation is removed, the bare wire ends should then be scraped bright with the back of a knife blade or rubbed clean with fine sandpaper.

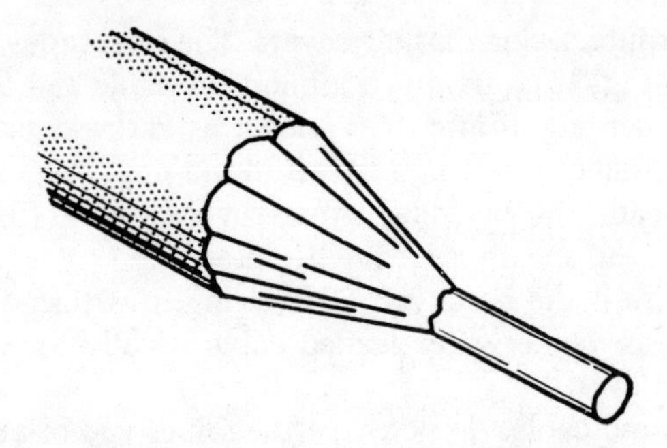

Fig. 16. Removing insulation from a wire.

Western Union Splice

Small, solid conductors may be joined together by a simple connection known as the *Western Union splice.* In most instances the wires may be twisted together with the fingers and the ends clamped into position with a pair of pliers.

Figure 17 shows the steps in making a Western Union splice. First, the wires are prepared for splicing by removing sufficient insulation and cleaning the conductor. Next, the wires are brought to a crossed position and a long twist or bend is made in each wire. Then one of the wire ends is wrapped four or five times around the straight portion of the wire. The other end wire is wrapped in a similar manner. Finally, the ends of the wires should be pressed down as close as possible to the straight

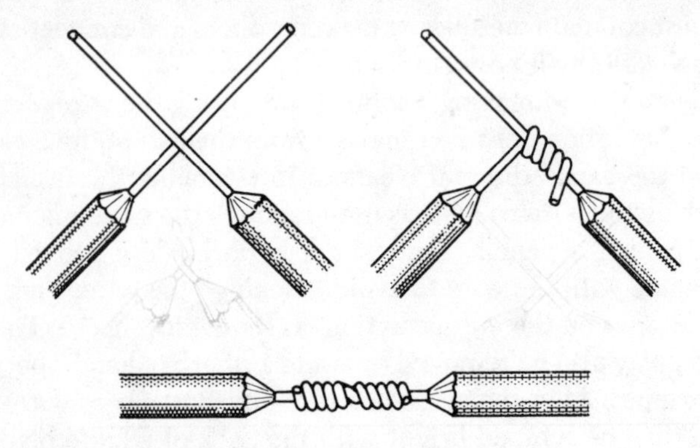

Fig. 17. Western Union splice.

portion of the wire; this prevents the sharp ends from puncturing the tape covering that is wrapped over the splice.

Staggered Splice

Joining small, multiconductor cables together presents something of a problem. Each conductor must be spliced and taped; if the splices are directly opposite each other, the overall size of the joint becomes large and bulky. A smoother, less bulky joint may be made by staggering the splices.

Figure 18 shows how a 2-conductor cable is joined to a similar cable by means of the *staggered splice.* Take care to ensure that a short wire is connected to a long wire, and that the sharp ends are clamped firmly down on the conductor.

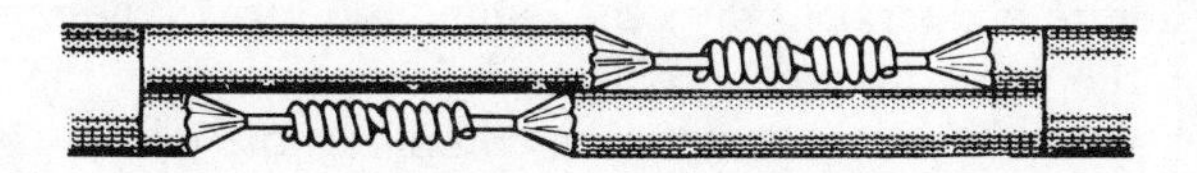

Fig. 18. Staggered splice.

Rattail Joint

Wiring that is installed in buildings is usually placed inside long lengths of steel pipe (conduit). Whenever branch circuits are required, junction or pull boxes are inserted in the conduit. One type of splice that is used for branch circuits is the rattail joint shown in Fig. 19.

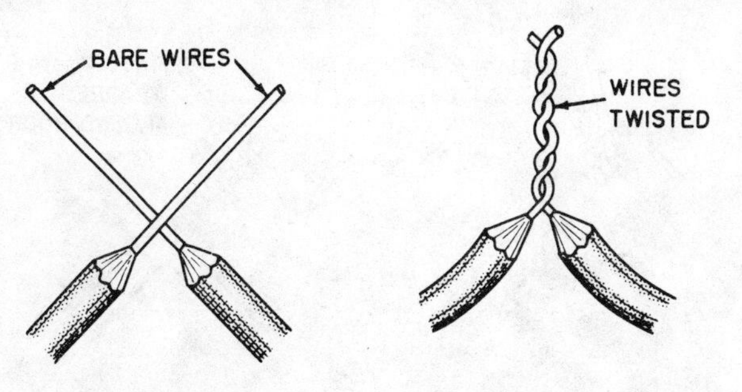

Fig. 19. Rattail joint.

The ends of the conductors to be joined are stripped of insulation. The wires are then twisted to form the rattail effect.

Fixture Joint

A *fixture joint* is used to connect a light fixture to the branch circuit of an electrical system where the fixture wire is smaller in diameter than the branch wire. Like the rattail joint, it will not stand much mechanical strain.

The first step is to remove the insulation from the wires to be joined. Figure 20 shows the steps in making a fixture joint.

After the wires are prepared, the fixture wire is wrapped a few times around the branch wire, as shown in Fig. 20. The wires are not twisted as in the rattail joint. The end of the branch wire is then bent over the completed turns. The remainder of the bare fixture wire is then wrapped over the bent branch wire. Soldering and taping completes the job.

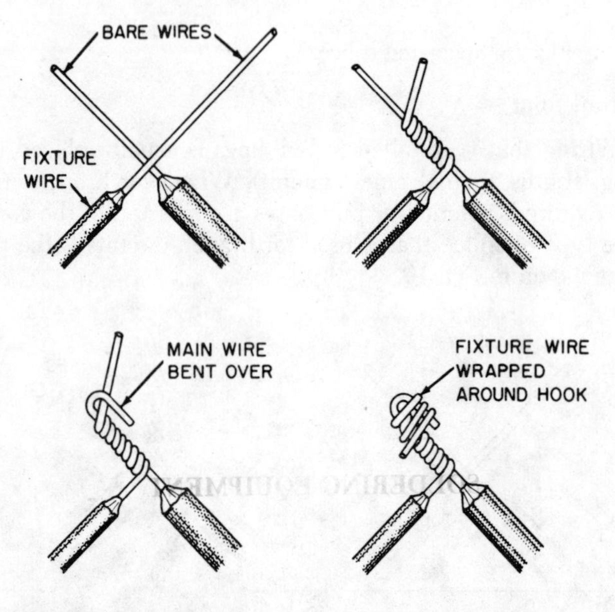

Fig. 20. Fixture joint.

Knotted Tap Joint

All of the splices described up to this point are known as *butted splices*. Each was made by joining the free ends of the conductors together. Sometimes, however, it is necessary to join a conductor to a continuous wire, and such a junction is called a *tap joint*.

The main wire, to which the branch wire is to be tapped, has about one inch of insulation removed. The branch wire is stripped of about three inches of insulation. The steps in making the tap are shown in Fig. 21.

The branch wire is crossed over the main wire, as shown in Fig. 21, with about three-fourths of the bare portion of the branch wire extending above the main wire. The end of the branch wire is bent over the main wire, brought under the main wire, around the branch wire, and then over the main wire to form a knot. It is then wrapped around the main conductor in short, tight turns and the end is trimmed off.

The knotted tap is used where the splice is subject to strain or slip. When there is no mechanical strain, the knot may be eliminated.

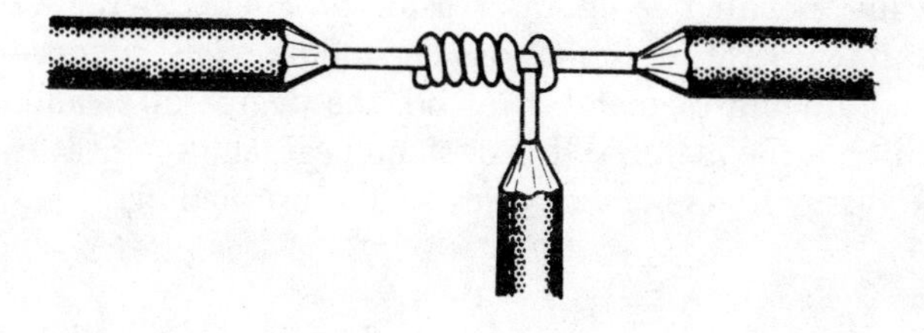

Fig. 21. Knotted tap joint.

SOLDERING EQUIPMENT

Soldering operations are a vital part of electrical/electronics maintenance procedures. It is a manual skill which can and must be learned by all those who work in the field of electricity or by men and women who can do their own work. Practice is required to develop proficiency in the techniques of soldering. However, practice serves no useful purpose unless it is founded on a thorough understanding of basic principles. This discussion is devoted to providing information regarding some important aspects of soldering operations.

Both the solder and the material to be soldered must be heated to a temperature which allows the solder to flow. If either is heated inadequately, "cold" solder joints result. Such joints do not provide either the physical strength or the electrical conductivity required. Appreciably exceeding the flow point temperature, however, is likely to cause damage to the parts being soldered. Various types of solder flow at different temperatures. In soldering operations it is necessary to select a solder that will flow at a temperature low enough to avoid damage to the part being soldered or to any other part or material in the immediate vicinity.

The duration of heat is almost as important as the temperature. Insulation and many other materials in electrical equipment are susceptible to heat. They are damaged if exposed to excessively high temperatures, or deteriorate if exposed to less drastically elevated temperatures for prolonged periods. The time and temperature limitations depend on many factors—the kind and amount of metal involved, the degree of cleanliness, the ability of the material to withstand heat, and the heat transfer and dissipation characteristics of the surroundings.

Solder

The three *grades of solder* generally used for electrical work are 40-60, 50-50, and 60-40 solder. The first figure is the percentage of tin, while the other is the percentage of lead. The higher the percentage of tin content, the lower the temperature required for melting. Also, the higher the tin content, the easier the flow, the less the time required to harden, and generally the easier it is to do a good soldering job.

In addition to the solder, there must be flux to remove any oxide film on the metal being joined; otherwise they cannot fuse.

The flux enables the molten solder to wet the metals so the solder can stick. The two types of flux are acid flux and rosin flux. Acid flux is more active in cleaning metals, but is corrosive. Rosin is always used for the light soldering work in making wire connections. Generally, the rosin is in the hollow core of solder intended for electrical work, so that a separate flux is unnecessary. Such rosin-core solder is the type generally used. It should be noted, though, that the flux is not a substitute for cleaning the metals to be soldered. The metal must be shiny clean for the solder to stick.

Soldering Process

Cleanliness is a prerequisite for efficient, effective soldering. Solder will not adhere to dirty, greasy, or oxidized surfaces. Heated metals tend to oxidize rapidly, and the oxide must be removed prior to soldering. Oxides, scale, and dirt can be removed by mechanical means (such as scraping or cutting with an abrasive) or by chemical means. Grease or oil films can be removed by a suitable solvent. Cleaning should be accomplished immediately prior to the actual soldering operation.

Items to be soldered should normally be tinned before making mechanical connection. Tinning is the coating of the material to be soldered with a light coat of solder. When the surface has been properly cleaned, a thin, even coating of flux may be placed over the surface to be tinned to prevent oxidation while the part is being heated to soldering temperature. Rosin core solder is usually preferred in electrical work, but a separate rosin flux may be used instead. Separate rosin flux is frequently used when tinning wires in cable fabrication.

The tinning on a wire should extend only far enough to take advantage of the depth of the terminal or receptacle. Tinning or solder on wires subject to flexing causes stiffness, and may result in breakage.

The tinned surfaces to be joined should be shaped and fitted, then mechanically joined to make good mechanical and electrical contact. They must be held still with no relative movement of the parts. Any motion between parts will likely result in a poor solder connection.

SOLDERING TOOLS

Soldering Irons

All high-quality irons operate in the temperature range of 500° to 600°F. Even the little 25-watt midget irons produce this temperature. The important difference in iron sizes is not temperature, but thermal inertia (the capacity of the iron to generate and maintain a satisfactory soldering temperature while giving up heat to the joint to be soldered). Although it is not practical to try to solder a heavy metal box with the 25-watt iron, that iron is quite suitable for replacing a half-watt resistor in a printed circuit. An iron with a rating as large as 150 watts would be satisfactory for use on a printed circuit, provided that suitable soldering techniques are used. One advantage of using a small iron for small work is that it is light and easy to handle and has a small tip which is easily inserted into close places. Also, even though its temperature is high, it does not have the capacity to transfer large quantities of heat.

Some irons have built-in thermostats. Others are provided with thermostatically controlled stands. These devices control the temperature of the soldering iron, but are a source of trouble. A well-designed iron is self-regulating by virtue of the fact that the resistance of its element increases with rising temperature, thus limiting the flow of current. For critical work, it is convenient to have a variable transformer for fine adjustment of heat; but for general-purpose work, no temperature regulation is needed.

Soldering Gun

The *soldering gun* has gained great popularity in recent years because it heats and cools rapidly. It is especially well adapted to maintenance and troubleshooting work where only a small part of the technician's time is spent actually soldering. A soldering iron, if kept hot constantly, oxidizes rapidly and is therefore difficult to keep clean.

A transformer in the soldering gun supplies approximately 1 volt at high current to a loop of copper which acts as the tip. It heats to soldering temperature in three to five seconds, but may overheat to the point of incandescence if left on over 30 seconds. The gun is operated with a finger switch so that the gun heats only while the switch is depressed.

Since the gun normally operates for only short periods at a time, it is comparatively easy to keep clean and well tinned; thus, little oxidation is allowed to form. However, the tip is made of pure copper, and is susceptible to pitting which results from the dissolving action of the solder.

Tinning of the tip is desirable unless it has already been done. The gun or iron should always be kept tinned in order to permit proper heat transfer to the work to be soldered. Tinning also provides adequate control of the heat to prevent thermal spillover to nearby materials. Tinning of the tip of a gun may be somewhat more difficult than tinning the tip of an iron. Maintaining the proper tinning on either one, however, may be made easier by tinning with silver solder. The temperature at which the bond is formed between the copper tip and the silver solder is considerably higher than with lead-tin solder. This tends to decrease the pitting action of the solder on the copper tip.

Pitting of the tip indicates the need for retinning, after first filing away a portion of the tip. Retinning too often results in using up the tip too fast.

Overheating can easily occur when the gun is used to solder delicate wiring. It takes practice, but the heat can be accurately controlled by pulsing the gun on and off with its trigger. For most jobs, even the LOW position of the trigger overheats the soldering gun after 10 seconds; the HIGH position is used only for fast heating and for soldering heavy connections.

Heating and cooling cycles tend to loosen the nuts or screws which retain the replaceable tips on soldering irons or guns. When the nut on a gun is loosened, the resistance of the tip connection increases, and the temperature of the connection is increased. Continued loosening may eventually cause an open circuit. Therefore, the nut should be tightened periodically.

Resistance Soldering

A time-controlled resistance soldering set is now available. The set consists of a transformer that supplies three or six volts at high current to stainless steel or carbon tips. The transformer is turned ON by a foot switch and OFF by an electronic timer. The timer can be adjusted for as long as three seconds soldering time. This set is especially useful for soldering cables to plugs and similar connectors—even the smallest types available.

In use, the double-tip probes of the soldering unit are adjusted to straddle the connector cup to be soldered. One pulse of current heats it for tinning and, after the wire is inserted, a second pulse of current completes the job. Since the soldering tips are hot only during the brief period of actual soldering, burning of wire insulation and melting of connector inserts are greatly minimized.

The greatest difficulty with this device is keeping the probe tips free of rosin and corrosion. A cleaning block is mounted on the transformer case for this purpose. Some technicians prefer fine sandpaper for cleaning the double tips. *Caution*: Do not use steel wool. It is dangerous when used around electrical equipment.

Pencil Iron and Special Tips

An almost indispensable item is the pencil type soldering iron with an assortment of tips (Fig. 22). Miniature soldering irons, with wattage ratings of less than 40 watts, are easy to use and are recommended. In an emergency, larger irons can be converted and used on subminiature equipment as described later in this section.

One type of iron is equipped with several different tips that range from one-fourth to one-half inch in diameter and are of various shapes. This feature makes it adaptable to a variety of jobs. Unlike most tips, which are held in place by setscrews,

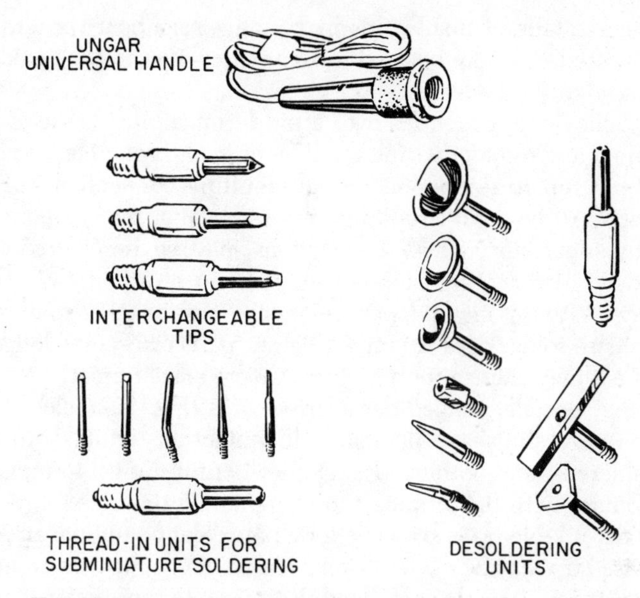

Fig. 22. Pencil iron kit with special tips.

these tips have threads and screw into the barrel. This feature provides excellent contact with the heating element, thus improving heat transfer efficiency. A pad of "antiseize" compound is supplied with each iron. This compound is applied to the threads each time a tip is removed and another is to be inserted.

A special feature of this iron is the soldering pot that screws in like a tip and holds about a thimbleful of solder. It is useful for tinning the ends of large numbers of wires.

The interchangeable tips are of various sizes and shapes for specific applications. Extra tips may be obtained and shaped to serve special purposes. The thread-in units are useful in soldering subminiature items. The desoldering units are specifically designed for performing special and individual functions.

Another advantage of the pencil soldering iron is its possible use as an improvised light source for inspections. Simply remove the soldering tip and insert a 120-volt, 6-watt, type 6S6, candelabra screwbase lamp bulb into the socket.

If leads, tabs, or small wires are bent against a board or terminal, slotted tips may be used to simultaneously melt the solder and straighten the leads.

A hollow tip, which fits over a pin terminal, may be used to desolder and resolder wiring at cables or feed-through terminals.

Many miniature components have multiple connections, all of which must be desoldered to permit removal of the component in one operation. These connections may be desoldered individually by heating each connection and brushing away the solder. With this method, particular care must be taken to insure that loose solder does not stick to other parts or become lodged where it may cause a short circuit. A more efficient method is to use the specially shaped desoldering units (Fig. 22). Select the proper size and shape tip that will contact all terminals to be desoldered—and nothing else. Do not permit the tip to remain in contact with the terminals too long at one time.

If no suitable tip is available for a particular operation, an improvised tip may be made. Wrap a length of copper wire around one of the regular tips and bend the wire into the proper shape for the purpose. This method also serves to reduce tip temperature when a larger iron must be used on minature components. (*See* Fig. 23.)

In connection with the discussion of soldering tools and devices, the selection of solder and flux is also critical. A small-diameter rosin-core solder with a high tin-lead (60–40) ratio is normally preferred in miniature circuits where heat is critical.

Soldering Aids

Several devices other than the soldering iron and its tips are required in soldering miniature circuits. Several of these (brushes, probes, scrapers, knives, and the like) have been mentioned previously. (*See* Chapter 2, Electrician's Tools and Equipment.)

Some type of thermal shunt is essential in all soldering operations which involve heat-sensitive components. Pliers, tweezers, or hemostats may be used for some applications, but their effectiveness is limited. A superior *heat shunt*, as shown in Fig. 24,

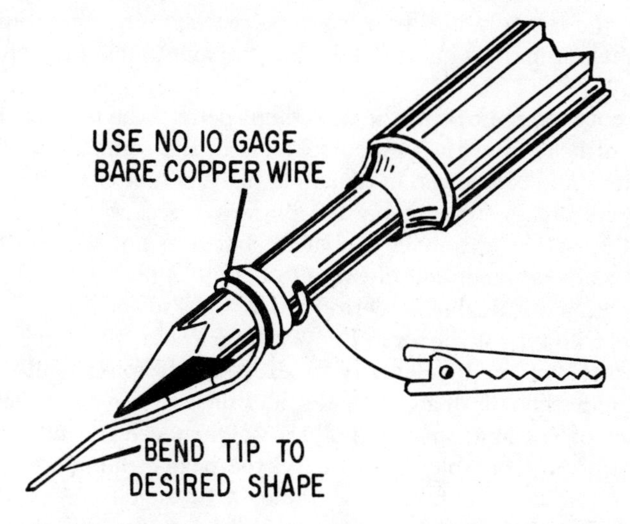

Fig. 23. Improvised tip to reduce tip temperature.

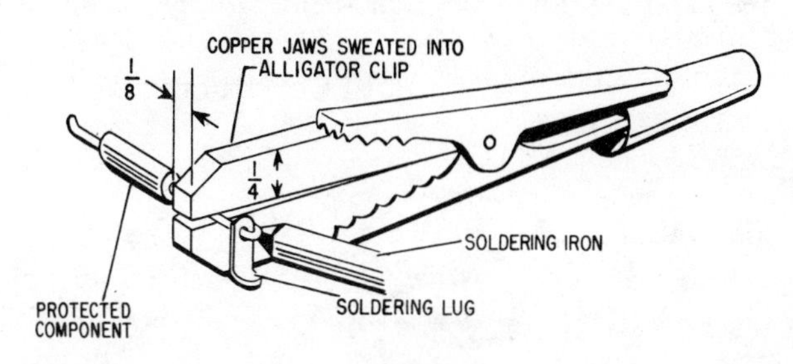

Fig. 24. Heat shunt.

permits soldering of the leads of component parts without overheating the part itself.

For maximum effectiveness, any protective coating should be removed before applying the heat shunt. The shunt should be attached carefully to prevent damage to the leads, terminals, or component parts. The shunt should be clipped to the lead, between the joint and the part being protected. As the joint is heated, the shunt absorbs the excess heat before it can reach the part and cause damage.

A small piece of beeswax may be placed between the protected unit and the heat shunt. When the beeswax begins to melt, the temperature limit has been reached. The heat source should be removed immediately, but the shunt should be left in place.

Premature removal of the heat shunt permits the unrestricted flow of heat from the melted solder into the component. The shunt should be allowed to remain in place until it cools to room temperature. A clip-on shunt is preferred because it requires positive action to remove the shunt, but does not require that the technician maintain pressure to hold it in place.

Another invaluable soldering aid is the "solder sucker" syringe. One type is shown in Fig. 25. Its purpose is to "suck up" excess solder (and incidentally the excess heat) from a joint. The only requirements of an efficient solder sucker are a controllable source of vacuum (squeeze bulb), a solder receiver, and a tip. The tip must be able to withstand the heat of molten solder. Teflon is ideal, but may be difficult to acquire. A silicon rubber-covered fiberglass sleeving with an inner diameter of 0.162 inch and the bulb from a medicine dropper makes a suit-

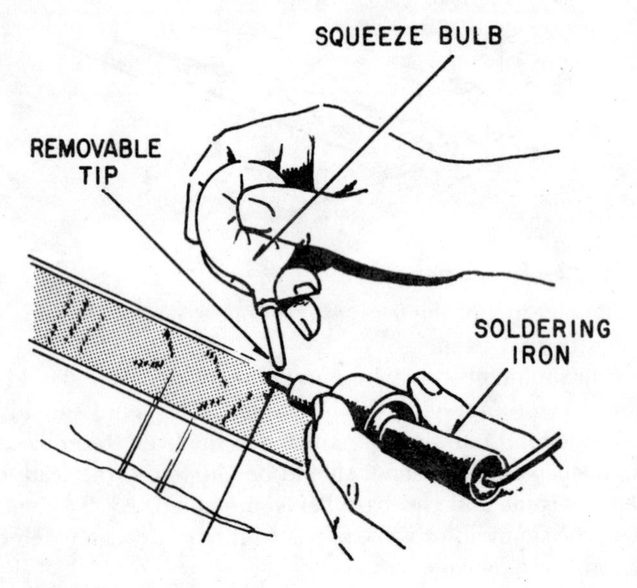

Fig. 25. Solder sucker.

able syringe. (The glass or plastic tip of the medicine dropper cannot withstand the heat.)

SOLDER CONNECTIONS

Frequent arguments occur in electrical shops, and among do-it-yourself individuals, concerning the proper method of making soldered connections to terminals and binding posts. For many years it was considered necessary to wrap the lead tightly around the terminal, so as to provide maximum mechanical support and strength. General specification for soldering process states that the parts to be joined shall be held together in such a manner that the parts shall not move in relation to one another during the soldering process. The joint must not be disturbed until the solder has completely solidified.

Electronics laboratories tested many standard capacitors and resistors solder to terminals of various types. The joints were then subjected to vibrations far in excess of those normally encountered in electrical and electronic equipment. The connections were made with various degrees of wrapping around the terminals, with main reliance for physical strength being placed on the solder. As a result of these tests and others conducted by other organizations, the joints illustrated in Fig. 26 are recommended. Wrappings of three-eighths to three-fourths turn are usually recommended so that the joint need not be held during the application and cooling of the solder.

Excessive wrappings of leads results in increased heat requirements, more strain on parts, greater difficulty of inspection, greater difficulty of assembly and disassembly of the joints, and increased danger of breaking the parts or terminals during desoldering operations. Insufficient wrapping may result in poor solder joints due to movement of the lead during the soldering operation.

The areas to be joined must be heated to or slightly above the flow temperature of the solder. The application of heat must be carefully controlled to prevent damage to components of the assembly, insulation, or nearby materials. Solder is then applied to

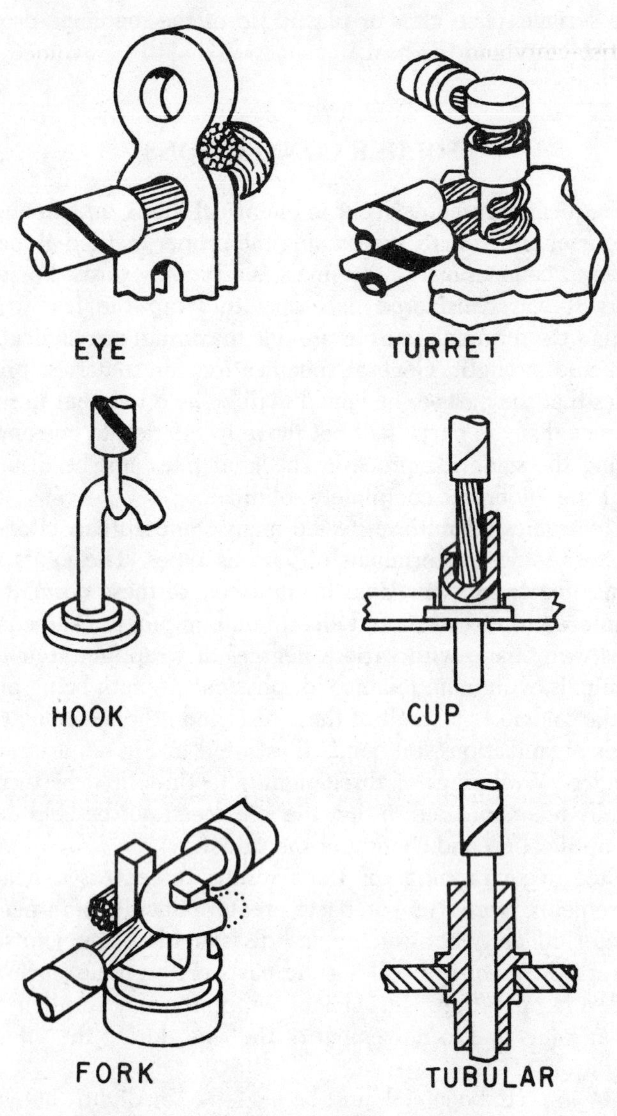

Fig. 26. Wrapping of terminals for soldering.

the heated area. Only enough solder should be used to make a satisfactory joint. Heavy fillets or beads must be avoided.

Solder should not be melted with the soldering tip and allowed to flow onto the joint. The joint should be heated and the solder applied to the joint. When the joint is adequately heated, the solder will flow evenly. Excessive temperature tends to carbonize flux, thus hindering the soldering operation.

No liquid should be used to cool a solder joint. By using the proper tools and soldering technique, a joint should not become so hot that rapid cooling is needed.

If, for any reason, a satisfactory joint is not initially obtained, the joint must be taken apart, the surfaces cleaned, excess solder removed, and the entire soldering operation (except tinning) repeated.

After the joint has cooled, all flux residues should be removed. Any flux residue remaining on the surface of electrical contacts may collect dirt and promote arcing at a later time. This cleaning is necessary even when rosin-core solder is used.

Connections should never be soldered or desoldered while equipment power is on or while the circuit is under test. Always discharge any capacitors in the circuit prior to any soldering operation.

Solder Splicers

The *solder-type splicer* is essentially a short piece of metal tube. Its inside diameter is just large enough to allow the tip of a stranded conductor to be inserted in either end, after the conductor tip has been stripped of insulation. This type of splicer is shown in Fig. 27.

The splicer is first heated and filled with solder. While still molten, the solder is then poured out, leaving the inner surfaces tinned. When the conductor tips are stripped, the length of exposed strands should be long enough so that the insulation butts against the splicer when the conductors are tinned and fully inserted. (*See* Fig. 27.) When heat is applied to the connection and the solder melts, excess solder will be squeezed out through the

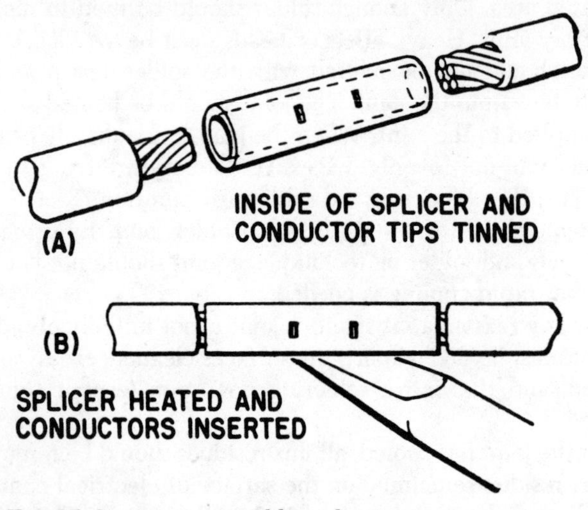

Fig. 27. Steps in using solder splicer.

vents. This must be cleaned away. After the splice has cooled, insulating material must be wrapped or tied over the joint.

Solder Terminal Lugs

In addition to being joined or spliced to one another, conductors are often connected to other objects, such as motors and switches. Since this is where a length of conductor ends (terminates), such connections are referred to as terminal points. In some cases, it is allowable to bend the end of the conductor into a small "eye" and put it around a terminal binding post. Where a mounting screw is used, the screw is passed through the eye. The conductor tip which forms the eye should be bent as shown in Fig. 28. Note that when the screw or binding nut is tightened, it also tends to tighten the conductor eye.

This method of connection is sometimes not desirable. When design requirements are more rigid, terminal connections are made by using special hardware devices called terminal lugs. There are terminal lugs of many different sizes and shapes, but all are essentially the same as the type shown in Fig. 29.

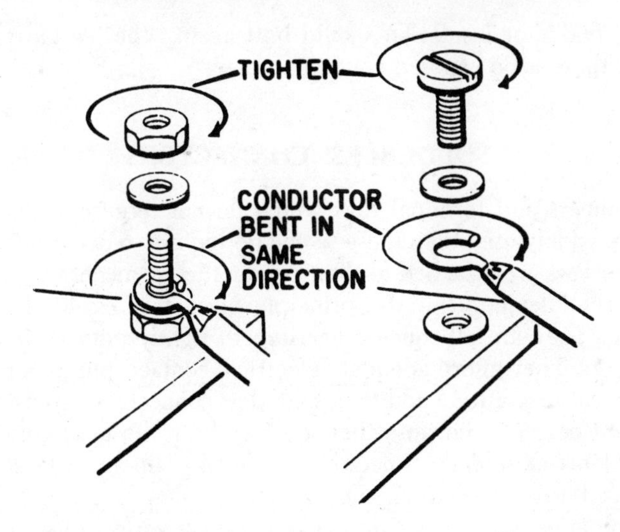

Fig. 28. Conductor terminal connection.

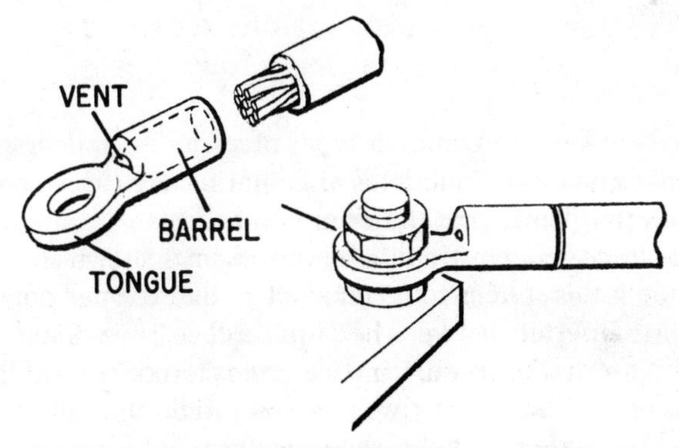

Fig. 29. Solder-type terminal lug.

Each type of lug has a barrel (sleeve) which is wedged, crimped, or soldered to its conductor. There is also a tongue with a hole or slot in it to receive the terminal post or screw. When mounting a solder-type terminal lug to a conductor, first tin the inside of the barrel. The conductor tip is stripped and also tinned, then inserted in the preheated lug. When mounted, the conductor insulation should butt against the lug barrel, so that there is no exposed conductor.

SOLDERLESS CONNECTORS

Splicers and terminal lugs which do not require solder are more widely used than those which do require solder. *Solderless connectors* are attached to their conductors by means of several different devices, but the principle of each is essentially the same. They are all squeezed (crimped) tightly onto their conductors. They afford adequate electrical contact, plus great mechanical strength. In addition, solderless connectors are easier to mount correctly because they are free from the most common problems of solder connector mounting—namely, cold solder joints, burned insulation, and so forth.

Solderless connectors are made in a great variety of sizes and shapes, and for many different purposes. Only a few are discussed here.

Solderless Splicers

Three of the most common types of solderless splicers, classified according to their methods of mounting, are the split-sleeve, split-tapered-sleeve, and the crimp-on splicers.

Split-sleeve-splicer. A split-sleeve-splicer is shown in Fig. 30. To connect this splicer to its conductor, the stripped conductor tip is first inserted between the split-sleeve jaws. Using a tool designed for that purpose, the slide ring is forced toward the end of the sleeve. The sleeve jaws are closed tightly on the conductor, and the slide ring holds them securely.

Split-tapered-sleeve splicer. A cross-sectional view of a split-tapered-sleeve splicer is shown in Fig. 31. To mount this type of splicer, the conductor is stripped and inserted in the split-tapered sleeve. The threaded sleeve is turned or screwed into the tapered bore of the body. As the sleeve is turned in, the split segments are squeezed tightly around the conductor by the narrowing bore. The finished splice (Fig. 31) must be covered with insulation.

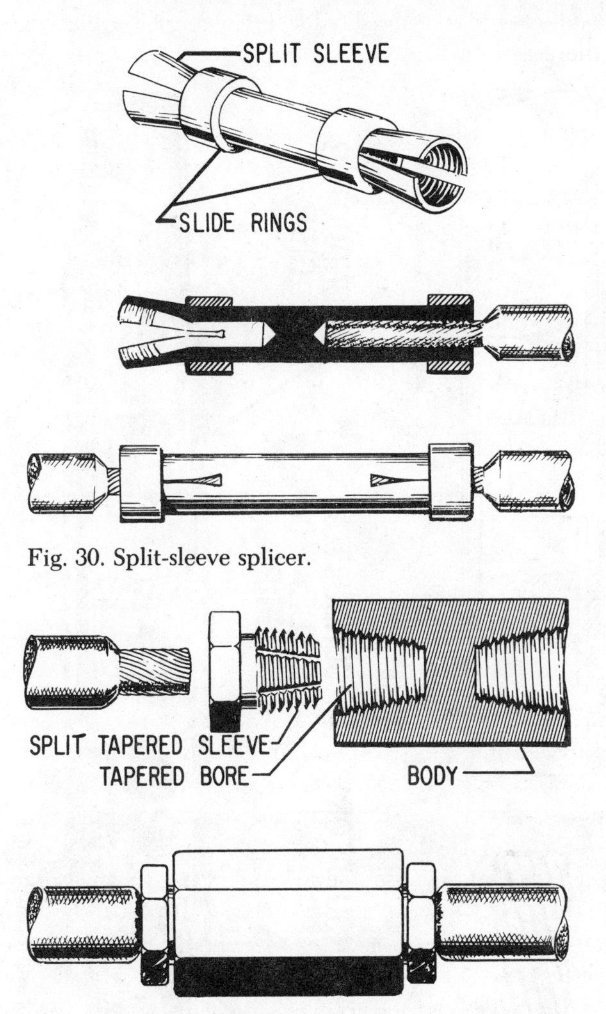

Fig. 30. Split-sleeve splicer.

Fig. 31. Split-tapered-sleeve, splice-cross-sectional view, and finished splice.

Crimp-on-splicer. The crimp-on-splicer (Fig. 32) is the simplest of the splicers discussed. The type shown is preinsulated, though uninsulated types are manufactured. These splicers are mounted with a special plierlike hand-crimping tool designed

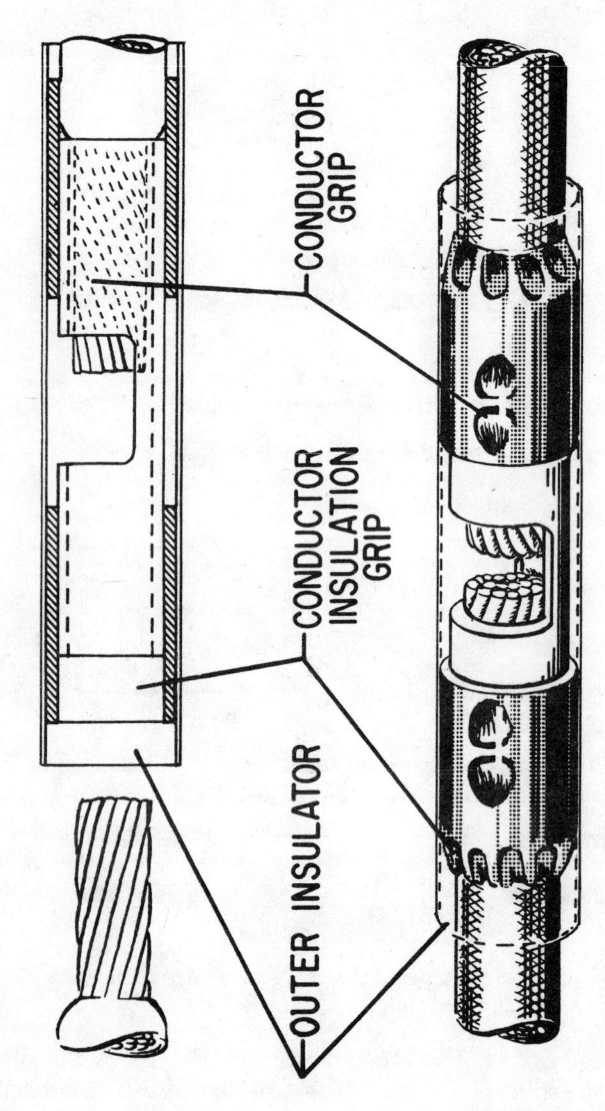

Fig. 32. Crimp-on splicer.

for that purpose. The stripped conductor tips are inserted in the splicer, which is then squeezed tightly closed. The insulating sleeve grips the outer insulated conductor, and the metallic internal splicer grips the bare conductor strands.

Solderless Terminal Lugs

Solderless terminal lugs are used more widely than solder terminal lugs. They afford adequate electrical contact, plus great mechanical strength. In addition, solderless lugs are easier to attach correctly, because they are free from the most common problems of solder terminal lugs; namely, cold solder joints, burned insulation, and so forth. There are many sizes and shapes of these lugs, each intended for a different type of service or conductor size. Only a few are discussed here.

These are classified according to their method of mounting. They are the split-tapered-sleeve (wedge-type), split-tapered-sleeve (threaded type), and crimp-on.

Split-tapered-sleeve terminal lug (wedge). This type lug is shown in Fig. 33. It is commonly referred to as a "wedge-on," because of the manner in which it is secured to a conductor. The stripped conductor is inserted through the hole in the split sleeve. When the sleeve is forced or "wedged" down into the barrel, its tapered segments are squeezed tightly around the conductor.

Split-tapered-sleeve terminal lug (threaded). This lug (Fig. 34) is attached to a conductor in exactly the same manner as a split-sleeve splicer. The segments of the threaded split sleeve squeezes tightly around the conductor as it is turned into the tapered bore of the barrel. For this reason, the lug is commonly referred to as a "screw-wedge."

Crimp-on terminal lug. The crimp-on lug is shown in Fig. 35. This lug is simply squeezed or "crimped" tightly onto a conductor. This is done by using the same tool used with the crimp-on splicer. The lug shown is preinsulated, but uninsulated types are manufactured. When mounted, both the conductor and its insulation are gripped by the lug.

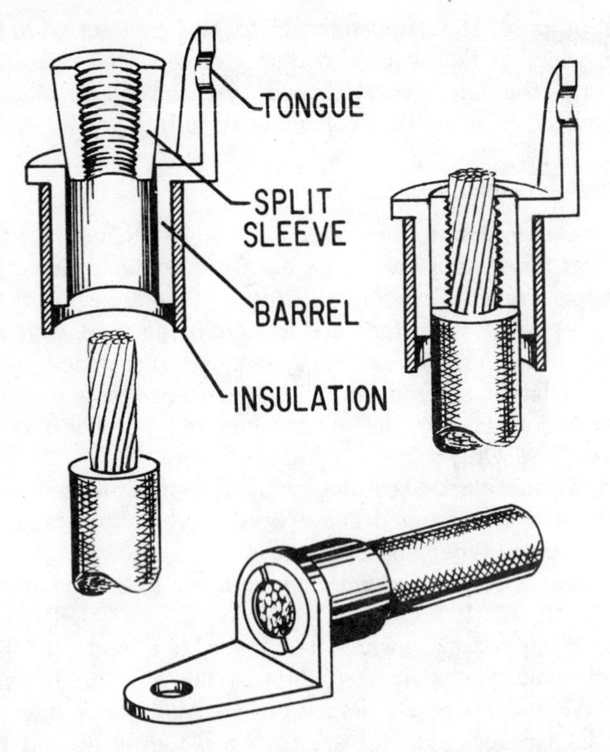

Fig. 33. Split-tapered-sleeve terminal lug (wedge type).

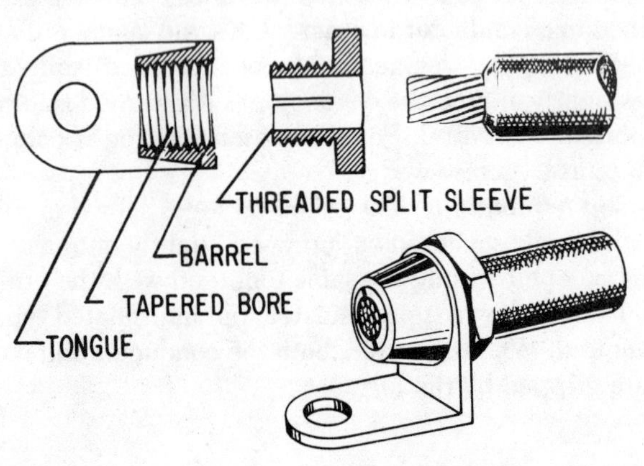

Fig. 34. Split-tapered-sleeve terminal lug (threaded).

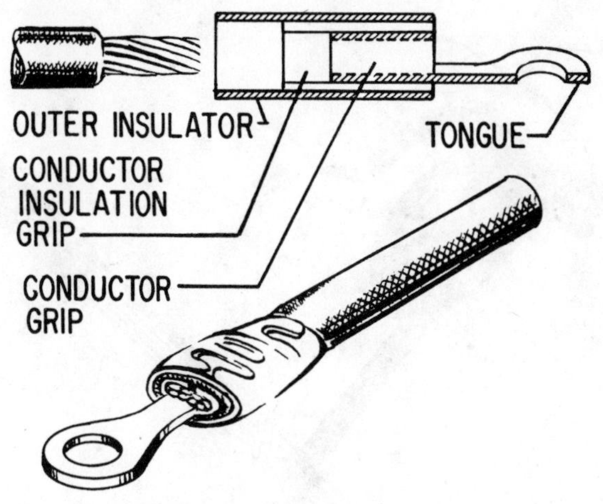

Fig. 35. Crimp-on terminal lug.

TAPING A SPLICE

The final step in completing a splice or joint is the placing of insulation over the bare wire. The insulation should be of the same basic substance as the original insulation. Usually a rubber splicing compound is used.

Rubber Tape

Latex (rubber) tape is a splicing compound. It is used where the original insulation was rubber. The tape is applied to the splice with a light tension so that each layer presses tightly against the one underneath it. This pressure causes the rubber tape to blend into a solid mass. When the application is completed, an insulation similar to the original has been restored.

Between each layer of latex tape, when it is in roll form, there is a layer of paper or treated cloth. This layer prevents the latex from fusing while still on the roll. The paper or cloth is peeled off and discarded before the tape is applied to the splice.

Figure 36 shows the correct way to cover a splice with rubber insulation. The rubber splicing tape should be applied smoothly and under tension so that there will be no air spaces between the

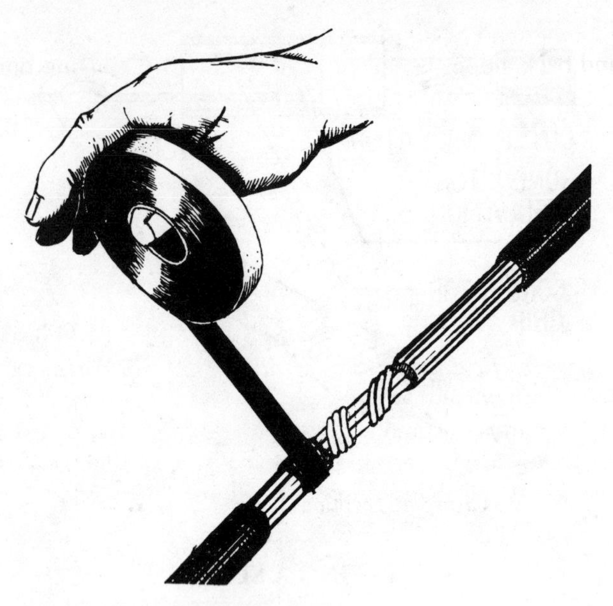

Fig. 36. Applying rubber tape.

layers. In putting on the first layer, start near the middle of the joint instead of the end. The diameter of the completed insulated joint should be somewhat greater than the overall diameter of the original cable, including the insulation.

Friction Tape

Putting rubber tape over the splice means that the insulation has been restored to a great degree. It is also necessary to restore the protective covering. *Friction tape* is used for this purpose; it also affords a minor degree of electrical insulation.

Friction tape is a cotton cloth that has been treated with a sticky rubber compound. It comes in rolls similar to rubber tape except that no paper or cloth separator is used. Friction tape is applied like rubber tape; however, it does not stretch.

The friction tape should be started slightly back on the original braid covering. Wind the tape so that each turn overlaps the one before it, and extend the tape over onto the braid covering at the other end of the splice. From this point a second layer is wound back along the splice until the original starting point is reached. Cutting the tape and firmly pressing down the end complete the job. When proper care is taken, the splice can take as much abuse as the rest of the wire.

Weatherproof wire has no rubber insulation, just a braid covering. In that case, no rubber tape is necessary; only friction tape need be used.

Plastic Electrical Tape

Plastic electrical tape has come into wide use in recent years. It has certain advantages over rubber and friction tape. For example, it will withstand higher voltages for a given thickness. Single thin layers of certain commercially available plastic tape will stand several thousand volts without breaking down. However, to provide an extra margin of safety, several layers are usually wound over the splice. Because the tape is very thin, the extra layers add only a very small amount of bulk; but at the same time the added protection, normally furnished by friction tape, is provided by the additional layers of plastic tape. In the choice of plastic tape, the factor of expense must be balanced against the other factors involved.

Plastic electrical tape normally has a certain amount of stretch so that it easily conforms to the contour of the splice without adding unnecessary bulk. The lack of bulkiness is especially important in some junction boxes where space is at a premium.

For high temperatures—for example, above 175°F.—a special type of tape backed with glass cloth is used.

at the other end of the [illegible] [illegible] [illegible] [illegible] [illegible] [illegible] [illegible] the [illegible] [illegible] [illegible] [illegible] [illegible] [illegible] [illegible] and [illegible] [illegible] down the [illegible] [illegible]

[illegible]

[illegible] [illegible] [illegible] [illegible] [illegible] [illegible] [illegible] [illegible] [illegible] [illegible] [illegible]

Plastic Insulation

An electrical [illegible] [illegible] [illegible] [illegible] [illegible] [illegible] [illegible] rubber and [illegible] [illegible] [illegible] [illegible] [illegible] [illegible] [illegible] [illegible] [illegible] [illegible] [illegible] [illegible] [illegible] [illegible] available plastic [illegible] normally [illegible] by friction tape [illegible] [illegible] [illegible] [illegible] [illegible] of plastic tape. In the choice of plastic tape, the factor of [illegible] must be balanced against the other factors involved.

Plastic [illegible] [illegible] [illegible] [illegible] [illegible] a certain amount [illegible]

[illegible] [illegible] [illegible] [illegible] [illegible] [illegible] [illegible] [illegible] [illegible] [illegible] [illegible] [illegible] [illegible]

Section II

Installation Methods and Procedures

This section provides practical information in the design, layout, installation, and maintenance of electrical wiring systems.

Chapter 5

Design and Layout of Interior Wiring

The different wiring systems in common use are often called open-wire, cable, and conduit systems. Chapters 6, 7, 8, and 9 cover the installation details for each of these systems. Many installation methods and procedures used in the wiring processes are common to all systems, and these are described in this chapter.

TYPES OF DISTRIBUTION

The electrical power load in any building cannot be properly circuited until the type and voltage of the central power-distribution system is known. The voltage and the number of wires from the powerlines to the buildings are normally shown or specified on the blueprints. However, you should check the voltage and type of distribution at the power-service entrance to every building in which wiring is to be done. This is especially necessary when you are altering or adding circuits. The voltage checks are usually made with an indicating voltmeter at the service-entrance switches or at the distribution load centers. The type of distribution is determined by visual check of the number of wires entering the building. (*See* Appendix 2, Table 7, Characteristics of Electrical Systems.)

If only two wires enter the building, the service is either direct current or single-phase alternating current. The voltage is determined by an indicating voltmeter.

When three wires enter a building from an alternating current distribution system, the service can be either single-phase, three-phase or two ungrounded conductors and a neutral of a three-phase system (V-phase).

1. If the service is single-phase, two of the conductors are hot and the third is ground. A voltmeter reading between the two hot conductors will be twice as great as the reading between either hot conductor and the neutral or ground conductor.

2. If the service is three-phase, the voltage between any two of the conductors is the same. Normally one of the conductors is grounded to establish a ground reference voltage for the system.

3. Two ungrounded conductors and a neutral or a V-phase system is the most common service for theater of operations construction. The distribution system is described in the following paragraph. The voltage between the two hot conductors will be the $\sqrt{3}$ or 1.732 times greater than the voltage between either hot conductor and the neutral or ground.

Four-wire distribution denotes three-phase and neutral service. When tested, voltages between the neutral conductor and each of the three hot conductors should be all the same. The voltage readings between any two of these three wires are similar and should equal the neutral to hot wire voltage multiplied by 1.732. Common operating voltages for this type of service are 120 and 208 volts or 277 and 480 volts.

LOAD PER OUTLET

The first step in planning the circuit for any wiring installation is to determine connected load per outlet. It is best to use volt-amperes as the method of determining electrical needs. This eliminates power factor considerations. The power needed for each outlet or load per outlet is used to find the number of circuits. It is also used to find the power needed for the whole building. The load per outlet can be obtained in several different ways.

1. The most accurate method of determining load per outlet is to obtain the stated value from the blueprints or specifications.

Commonly, the lighting outlets shown on the blueprints are listed in the specifications along with their wattage ratings. If the lights used are of the incandescent type, this figure represents the total wattage of the lamp.

When fluorescent lights are specified, the wattage drain (also called load per outlet) should be increased approximately 20 percent to provide for the ballast load. For example, when the fixture is rated as a two-lamp, 40-watt unit, the actual wattage drain is 80 watts, plus approximately 16 watts for both lamp ballasts, or a total load of 96 watts.

2. If the specifications are not available, the blueprints in many cases designate the type of equipment to be connected to specific outlets. Though the equipment ultimately used in the outlet may come from a different manufacturer, equipment standards provide the electrician with assurance that the outlets will use approximately the same wattage. If the equipment is available, the nameplate will list the wattage used or ampere drain. If not, Table 20, in Appendix 2, may be used to obtain the average wattage consumption of electrical appliances.

To provide adequate wiring for systems where the blueprints or specifications do not list any special or appliance loads, the following general rules will apply:

1. For heavy-duty outlets or mogul-size lampholders, the load per outlet should be figured at 5 amperes each.

2. For all other outlets, both ceiling and wall, the wattage drain (load per outlet) should be computed at 1.5 amperes per outlet.

The total outlet load may also be determined on a watts-per-square-foot basis. In this load-determination method the floor area of the building to be wired is computed from the outside dimensions of the building. This square footage area is then multiplied by the standard watts-per-square-foot requirement based on the type of building to be wired.

CIRCUITING THE LOAD

If all the power load in a building were connected to a single pair of wires and protected by a single fuse, the entire establishment would be without power in case of a breakdown, a short circuit, or a fuse blowout. In addition, the wires would have to be large enough to handle the entire load, and, therefore, too large in some cases to make connections to individual devices. Consequently, the outlets in a building are divided into small groups known as branch circuits. These circuits normally are rated in amperes, as shown in Table 21, Appendix 2. This table contains a comparison of the various ampere requirements of the branch circuits with the standard circuit components. Normally the total load per circuit should not exceed 80 percent of the circuit rating.

The method of circuiting the building load varies with the size of the building and the power load.

In a *small* building with little load, the circuit breakers or fuses are installed at the power-service entrance and the individual circuits are run from this location.

For buildings of *medium* size with numerous wiring circuits, the fuse box should be located at the center of the building load so that all the branch runs are short, minimizing the voltage drop in the lines.

When buildings are *large* or have the loads concentrated at several remote locations, ideal circuiting would place fuse boxes at each individual load center. It is assumed that the branch circuits would be radially installed at each of these centers to minimize the voltage drops in the runs.

The number of circuits required for adequate wiring can be determined by adding the connected load in watts and dividing the total by the wattage permitted on the size of branch circuit selected. This method should not include special heavy loads, such as air conditioners, requiring separate circuits. The total wattage is obtained from the sum of the loads of each individual outlet determined by one of the three methods outlined previously in the section on Load Per Outlet. For example, if 20-ampere, 120-volt circuits are to be used, 80 percent of this rating, or 16 amperes per circuit, are allowed. The maximum wattage permitted on each circuit equals 16 times 120, or 1920 watts. If the total connected load is assumed to be 18,000 watts, 18,000 divided by 1920 shows that 9.375 circuits are required. Since we can have only whole circuits, ten 20-ampere circuits should be used to carry the connected load. The number of circuits determined by this method should be the basic minimum. For long-range planning in permanent installations, the best practice requires the addition of several circuits to the minimum required, or the installation of the next larger modular-size fusing panel to allow for future wiring additions. If additional circuits over the minimum required are used, the number of outlets per circuit can be reduced, making the electrical installation more efficient because the voltage drop in the system is reduced.

Motors on portable appliances are normally disconnected from the power source either by removal of the appliance plug from its receptacle or by the operation of a built-in switch. Some large-horsepower motors, however, require a permanent power installation with special controls. Motor switches, two of which are shown in Fig. 1, are rated in horsepower capacity. In a single motor installation a separate circuit must be run from the fuse or circuit breaker panel to the motor, and individual fuses or circuit breakers installed. For multiple motor installations the National Electrical Code requires that "Two or more motors may be connected to the same branch circuit, protected at not more than 20 amperes at 125 volts or less or 15 amperes at 600 volts or less, if each does not exceed one horsepower in rating and each

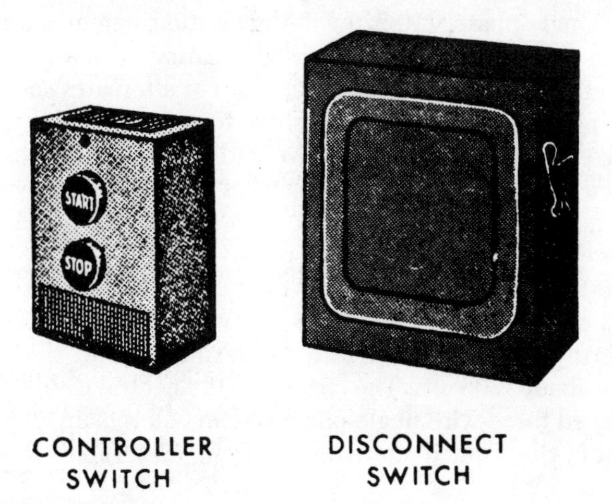

Fig. 1. Motor switches.

does not have a full load rating in excess of six amperes. Two or more motors of any rating, each with individual overcurrent protection (provided integral with the motor start switches or an individual unit), may be connected on one branch circuit provided each motor controller and motor-running overcurrent device be approved for group installation and the branch circuit fusing rating be equal to the rating required for the largest motor plus an amount equal to the sum of the full load ratings of the other motors."

BALANCING THE POWER LOAD ON A CIRCUIT

The *ideal wiring system* is planned so that each wiring circuit will have the same ampere drain at all times. Since this can never be achieved, the circuiting is planned to divide the connected load as evenly as possible. Thus, each individual circuit uses approximately the average power consumption for the total system. This will allow minimum service interruption. Figure 2 demonstrates the advantage of a balanced circuit when a three-wire, single-phase, 110-220 volt distribution system is used.

The current used is known as alternating current, or ac, because the current in each wire changes or alternates continually from positive to negative to positive to negative and so on. The change from positive to negative and back again to positive is known as a "cycle." Usually this takes place 60 times every second, and such current is then known as 60-cycle current. Sixty times every second each wire is positive, and 60 times every second it is negative, and 120 times every second there is no voltage at all on the wire. The voltage is never constant but is always gradually changing from zero to maximum, with the average being about 120 volts. The current in the neutral conductor of a balanced three-wire single-phase system will remain at zero as a result of the applied alternating current. (*See* Tables 18 and 19, Appendix 2.)

LOAD PER BUILDING

Maximum Demand

In some building installations, the total possible power load may be connected at the same time. In this case the demand on the power supply, which must be kept available for these buildings, is equal to the connected load. In the majority of building installations, the maximum load which the system is required to service is much less than the connected load. This power load, which is set at some arbitrary figure below the possible total connected load, is called the "maximum demand" of the building.

Demand Factor

The ratio of maximum demand to total connected load in a building expressed as a percentage is termed "demand factor." The determination of building loads can be obtained by the use of standard demand factors.

BALANCING THE POWER LOAD OF A BUILDING

The standard voltage distribution system from a generating station to individual building installation is the three- or four-wire, three-phase type. Distribution transformers on the power line poles change the voltage to 120 or 240, and are designed to deliver three-wire single-phase service. These transformers are then connected across the distribution phase leads in a balanced arrangement. Consequently, for maximum transformer efficiency, the building loads assumed for power distribution should also be balanced, as illustrated in Fig. 2.

WIRE SIZE

Wire sizes No. 14 and larger are classified according to their maximum allowable current-carrying capacity based on their physical behavior when subjected to the stress and temperatures

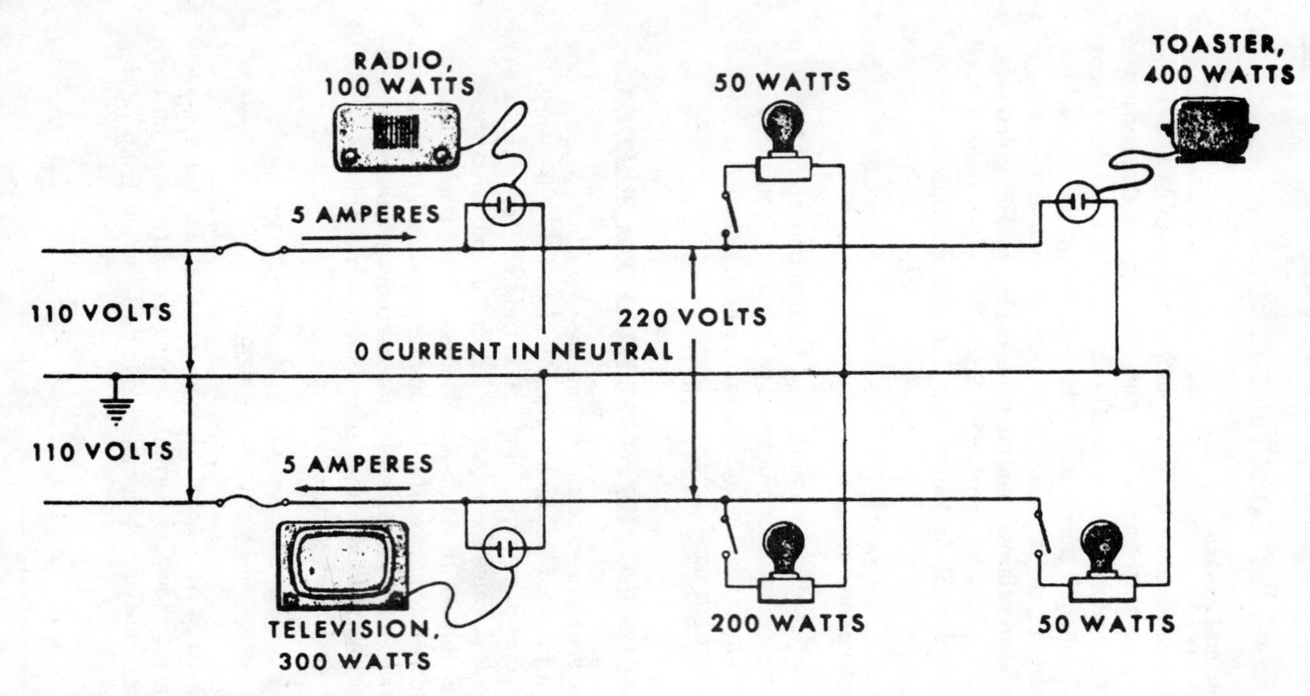

Fig. 2. Diagram showing circuit balancing.

of operating conditions. Fourteen-gage wire is the smallest wire size permitted in interior systems.

The determination of the conductor size to be used in feeder and branch circuits is dependent on the maximum allowable current-carrying capacity and the voltage drop coincident with each wire size. The size of the conductors for branch circuits (that portion of the wiring system extending beyond the last overcurrent device protecting the circuit) should be such that the voltage drop will not exceed 3 percent to the farthest outlet for power, heating, or lighting loads. The maximum voltage drop for feeders is also 3 percent, provided that the total voltage drop for both feeder and branch circuit does not exceed 5 percent. Table 22, in Appendix 2, which is based on an allowable 3 percent voltage drop, lists the wire sizes required for various distances between supply and load at different amperages. Table 22 may be used for branch circuits originating at the service entrance. This is the common house or small building circuit. Appendix 3, Design Procedures for Electrical Wiring, gives a more detailed method of determining wire size. (*See* Table 28, Values of Resistance, Appendix 3.)

The minimum gage for service-wire installation is No. 8. The service-wire sizes are increased because they must not only meet the voltage-drop requirement but also be inherently strong enough to support their own weight, plus any additional loading caused by climatic conditions (ice, branches, and so on).

GROUNDING AND BONDING

Requirements

1. All electrical systems must have the neutral conductor grounded if the voltage between the hot lead and the ground is less than 150 volts.
2. It is recommended that all systems have a grounded neutral where the voltage to ground does not exceed 300 volts.
3. *Interior circuits* operating at less than 50 volts need not be grounded, provided the transformer supplying the circuit is connected to a grounded system.

Types of Grounding

A *system ground* is the ground applied to a neutral conductor. It reduces the possibility of fire and shock by reducing the voltage of one of the wires of a system to 0 volts potential above ground.

An *equipment ground* is an additional ground that should be attached to all appliances and machinery. An equipment ground is advantageous because the appliances and machinery can be maintained at zero voltage, and if a short circuit does occur in a hot load, the fuse protection opens the circuit and prevents serious injury to operating personnel.

Methods of Grounding

A system ground is provided by placing a No. 6 copper (or No. 4 aluminum) wire between the neutral wire, service box, bonding wire, and a grounding electrode at the building service entrance. The grounding electrode may be a buried water pipe, the metal frame of a building, a local underground system, or a fabricated device. If more than one electrode is used, they must be placed a minimum of six feet apart.

Water pipe. An underground water piping system will always be used as the grounding electrode if such a piping system is available. If the piping system is less than 10 feet deep, supplemental electrodes will be used. Interior metallic cold water piping systems will always be bonded to the grounding electrode or electrodes.

Plate electrodes. Each plate electrode will have at least two square feet of surface exposed to the soil. Iron or steel electrodes shall be at least ¼-inch thick and nonferrous metal shall be at least $^{3}/_{5}$-inch thick. Plates should be buried below the permanent moisture level if possible.

Pipe electrodes. Clean metallic pipe or conduit of at least ¾-inch trade size may be used. Each pipe must be driven to a depth of at least eight feet. If this cannot be done, the electrodes may be buried in a horizontal trench. In this case the electrode must be at least eight feet long.

Rod electrodes. Rod electrodes of steel or iron must be at least ⅝-inch in diameter. Rods of nonferrous material must be at least ½-inch in diameter. Installation of rod electrodes is the same as stated previously for pipe electrodes. Figure 3 shows typical grounding fixtures.

Ground Resistance

Electrodes should have a resistance to ground of 25 ohms or less. Underground piping systems and metal frames of buildings normally have resistances to ground of less than 25 ohms. Resistances to ground of fabricated electrodes will vary greatly depending on the soil and method of installation. Burying an electrode below the permanent moisture level of the soil normally reduces the resistance to ground to acceptable values. Grounding systems should be tested by using a "Megger" ground tester. This device may be found in certain types of organizations.

The voltage between the neutral or grounded conductor of a grounded system and ground (water pipes, metal frame of building, ground rod), should be zero at all times. The detection of any voltage indicates a faulty wiring system. Short circuits which are commonly called grounds will be discussed in Chapter 11, Preventive Maintenance and Miscellaneous Equipment Maintenance.

Bonding

Bonding is a method of providing a continuous and separate electrical circuit between all metallic circuit elements (conduit, boxes, and so on) and the service entrance ground. The ground or neutral conductor is attached to the bonding circuit *only* at the service entrance. At this point the neutral wire and a bonding wire from the service box are attached to the grounding electrode. All other metallic circuit elements are connected to this grounded point through physical connections of the metallic conduit, BX, or flexible metal conduit. A nonmetallic conduit or cable must have a conductor in it to provide this bonding cir-

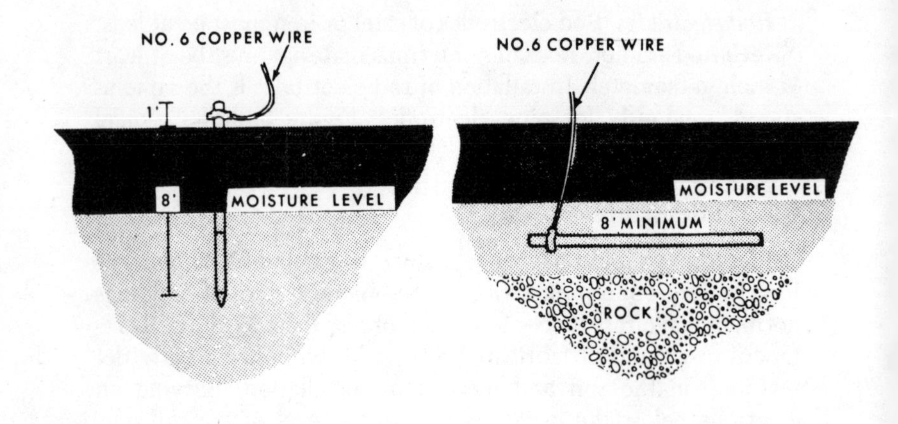

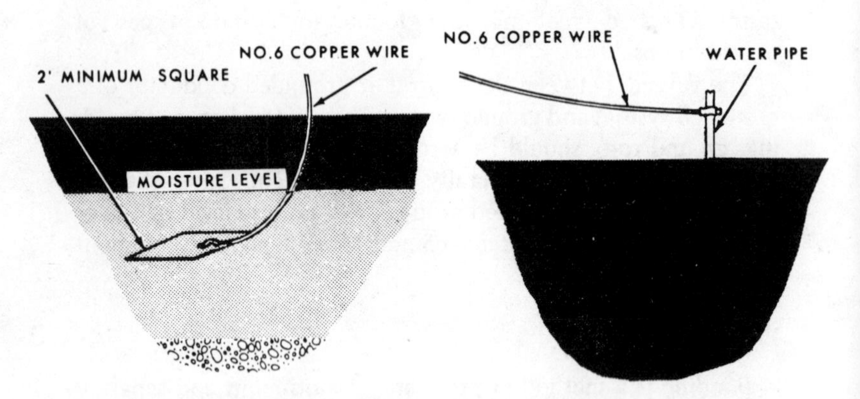

Fig. 3. Typical grounding fixtures.

cuit. This conductor would be attached to the metal boxes or fixtures where the nonmetallic cable or conduit terminates.

To insure a continuous bonding circuit, all metal-to-metal connections must be made up tight. When the electrical continuity of a metal-to-metal connection is doubtful, a wire jumper must be used between the two metal pieces. As an example, the equipment ground wire of a receptacle (the green wire) is connected to the bonding circuit through the two screws fastening the receptacle to the box. If the box is recessed rather than flush, the connecting screws are not tightened but are used as adjusting screws to properly align the receptacle. In this case a wire must be placed between the box and the equipment ground terminal of the receptacle. Proper bonding prevents shocks from metal surfaces of an electrical system.

WIRING FOR HAZARDOUS LOCATIONS

Special materials and procedures must be used when installing electrical systems in areas where a spark could cause a fire or explosion. Typical areas of this type are acetylene storage or production facilities. Hazardous locations are divided into three classes. Each class is further subdivided into two divisions. In division one of each class, the hazardous material is present in free air so that the atmosphere is dangerous. In division two of each class, the hazardous material is in containers and dangerous mixtures in air occur only through accidents. The following is a description of the classes.

Class I. Locations in which flammable gases or vapors are or may be present in the air in quantities sufficient to produce explosive or ignitable mixtures.

Class II. Locations where combustible dust is present.

Class III. Locations where easily ignitable fibers or flyings are present but are not suspended in air in sufficient quantities to produce ignitable mixtures.

INSTALLATION IN HAZARDOUS LOCATIONS

Detailed information on installation material and procedures must be obtained from the National Electrical Code. Figure 4 shows typical explosion-proof fittings. The information below gives general guidance as to the type of installation required for each class and division of hazardous locations.

Class I. In division-one locations, all wiring must be in threaded rigid metal conduit with explosion-proof fittings or mineral-insulated metal-sheathed (type MI) cable. Division-two locations may have flexible metal fittings, flexible metal conduit,

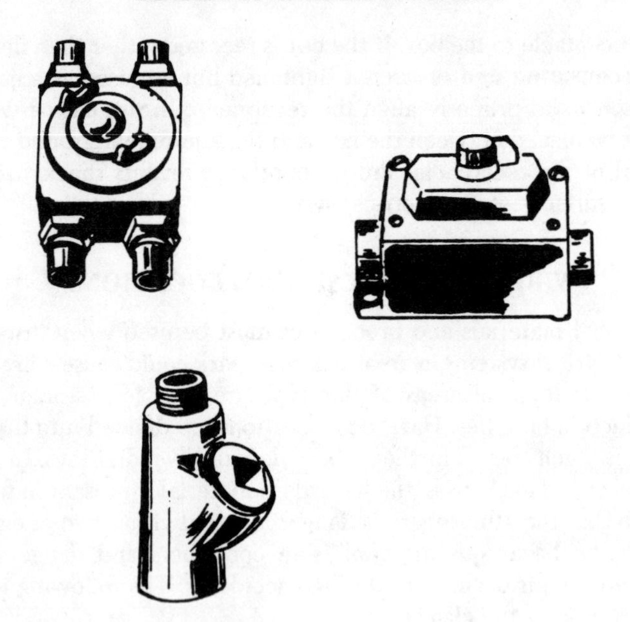

Fig. 4. Explosion-proof fittings.

or flexible cord approved for hard usage. All equipment such as generators, controllers, motors, fuses, and circuit breakers must be enclosed in explosion-proof housings.

Class II. In division one, wiring must be in threaded rigid conduit or type MI cable, with flexible metal conduit and threaded fittings where necessary. Equipment must be in dust-proof cabinets with motors and generators in totally enclosed fan-cooled housings. In division-two locations, electrical metallic tubing may also be used.

Class III. In division one the same requirements exist as for class I, division one. In division two, open wiring may be permitted. Motors and generators must be totally enclosed.

INSTALLATION OF SIGNAL EQUIPMENT

Signal equipment may occasionally be supplied for 120-volt operation, in which case it must be installed in the same manner as outlets and sockets operating on this voltage. Most bells and buzzers are rated to operate on 6, 12, 18, or 24 volts ac or dc. They can be installed with minimum consideration for circuit insulation, since there is no danger of shock to personnel or fire due to short circuits. The wire commonly used is insulated with several layers of paraffin-impregnated cotton or with a thermoplastic covering. Upon installation, these wires are attached to building members with small insulating staples and are threaded through building construction members without insulators.

BATTERY OPERATION

Early installations of low-voltage signal systems were powered by 6-volt dry cells. For example, two of these batteries were installed in series to service a 12-volt system. If the systems involved a number of signals over a large area, one or more batteries were added in series to offset the voltage drop. Though this type of alarm or announcing system is still being used and installed in some areas, it is a poor system because the batteries used as a power source require periodic replacement. (*See* Chapter 14, Batteries.)

TRANSFORMER OPERATION

The majority of our present-day buzzer and bell signal systems operate from a transformer power source. The transformers are equipped to be mounted on outlet boxes and are constructed so that the 120-volt primary-winding leads normally extend from the side of the transformer adjacent to box mounting. These leads are permanently attached to the 120-volt power circuits, and the low-voltage secondary-winding leads of the transformer are connected to the bell circuit in a manner similar to a switch-and-light combination (Fig. 5). If more than one buzzer and pushbutton are to be installed, they are paralleled with the first signal installation. A typical wiring schematic diagram for this type of installation is shown in Fig. 6. (*See* Chapter 15, Transformers.)

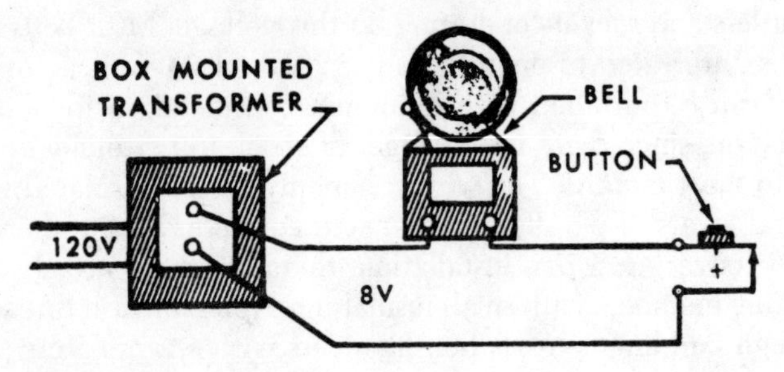

Fig. 5. Bell and buzzer wiring.

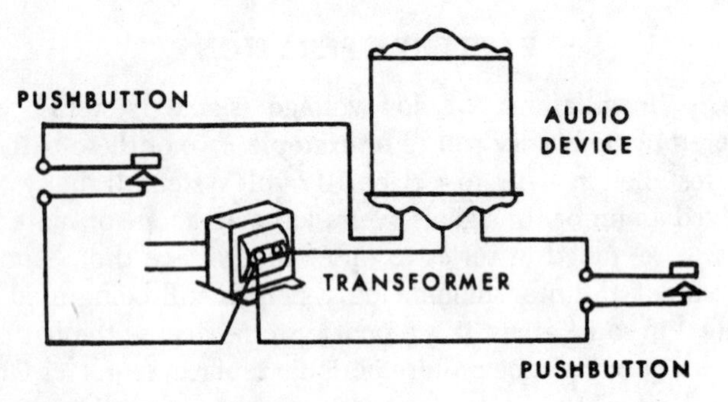

Fig. 6. Two pushbutton system.

SPECIAL SWITCHES

Three-Way Switching

A single-throw switch controls a light or a receptacle from only one location. When lights have to be controlled from more than one location, a double-throw commonly called a *three-way switch* is used. Three-way switches can be identified by a common terminal, normally color-coded darker than the other terminals and located alone at the end of the switch housing. A schematic wiring diagram of a three-way switch with a three-wire cable is shown in Fig. 7. In the diagram, terminals A and A′ are the common terminals, and switch operation connects them either to B or C or B′ or C′, respectively. Either switch will operate to close or open the circuit, turning the lights on or off.

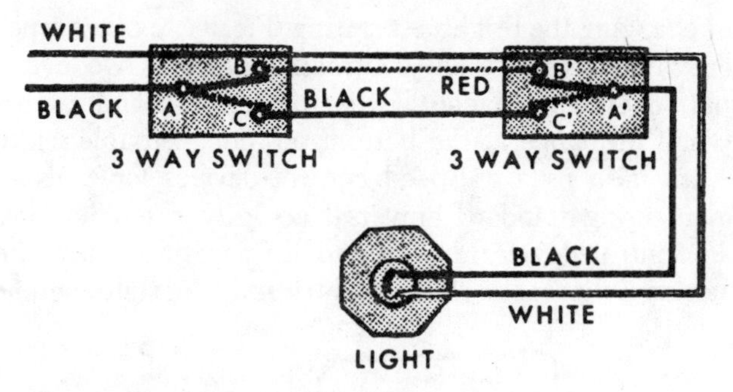

Fig. 7. Three-way switch wiring.

Four-Way Switching

Occasionally it is necessary to control an outlet or light from more than two locations. Two three-way switches plus a four-way switch will provide control at three locations, as shown in Fig. 8. The switches must be installed with the four-way units connected between the two three-way units and the three-wire cable installed between the switches.

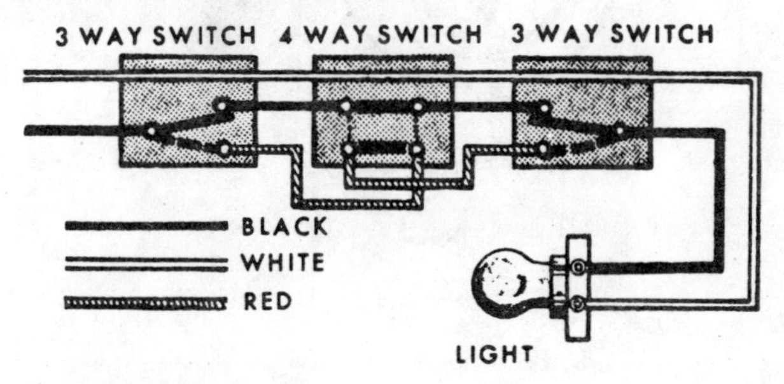

Fig. 8. Four-way switch wiring.

Variable Control Devices

Electrical dimmers provide a full range of light from bright to dim for incandescent lighting. Special electronic dimmers are used with fluorescent lighting. A turn of the dial adjusts the brightness level, and a push-knob switch turns the light on or off without changing the brightness setting (Fig. 9). Another type of dimmer switch has a high-low control that provides two levels of illumination—full brilliance at the top position and approximately half brilliance at the bottom position. Variable control devices are also used as speed control devices for tools and equipment using standard universal ac-dc type motors. Some variable control devices are electronic in nature, since their construction utilizes solid state circuitry and solid state switches.

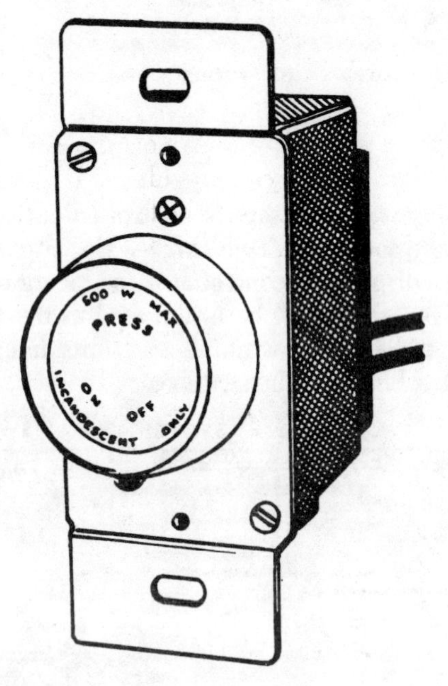

Fig. 9. Push-pull rotary switch.

ADDITIONS TO EXISTING WIRING

Circuit Capacity

In the installation of additions to existing wiring in a building, the electrician first determines the available extra capacity of the present circuits. This can readily be obtained by ascertaining the fused capacity of the building and subtracting the present connected load. If all the outlets do not have connected loads, their average load should be used to obtain the connected load figure. When the existing circuits have available capacity for new outlets and are located near the additional outlet required, they should be extended and connected to the new outlets. Consideration must be given to the additional voltage drop created by extending the circuit. The proper wire size may then be determined.

New Circuits

When the existing outlets cannot handle an additional load, and a spare circuit has been provided in the local fuse or circuit breaker panel, a new circuit is installed. This is also done if the new outlet or outlets are so located that a new circuit can be installed more economically than can an existing circuit extension. Moreover, the installation of a new circuit will generally decrease the voltage drop on all circuits, resulting in an increase in appliance operating efficiency. Figure 10 illustrates the addition of a new circuit from the spare circuit No. 4 in the circuit breaker panel.

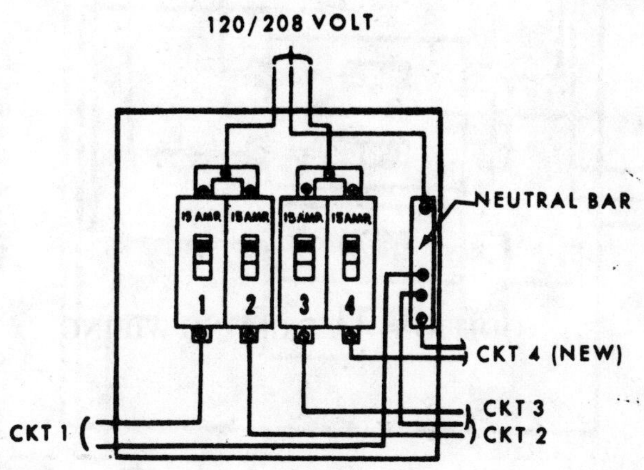

Fig. 10. Addition of a new circuit.

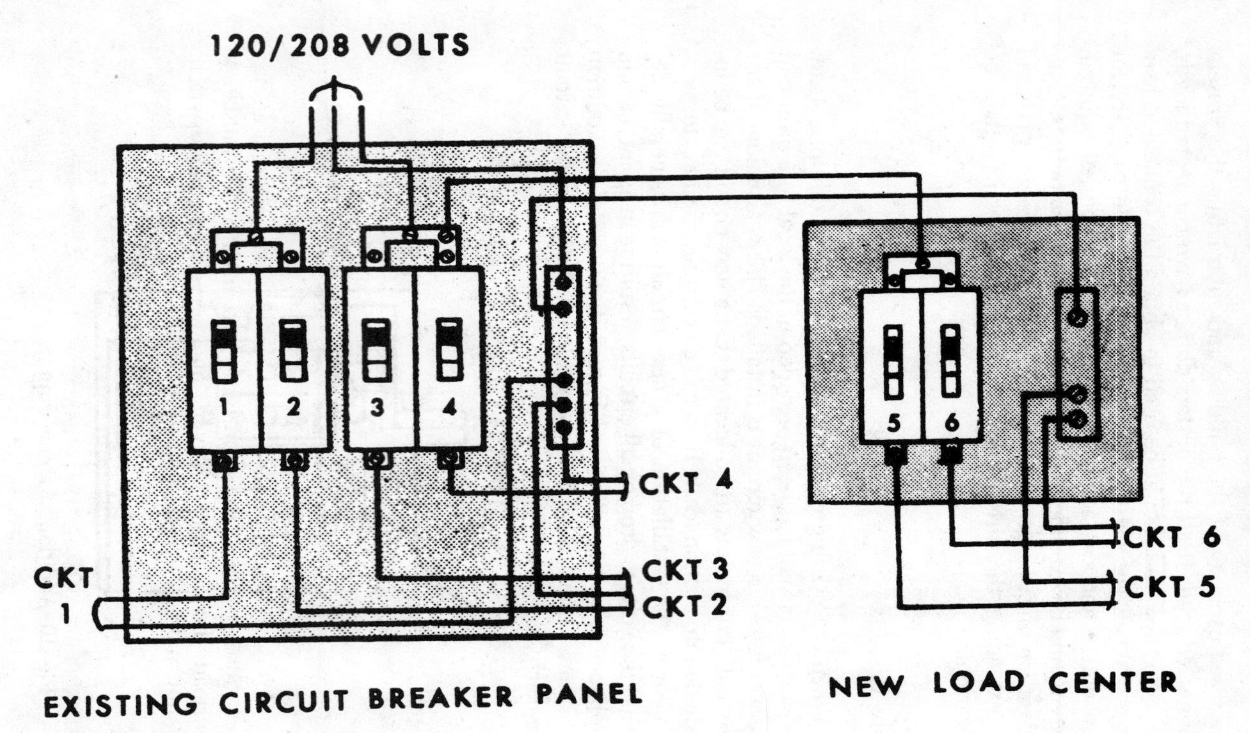

Fig. 11. Addition of new load center.

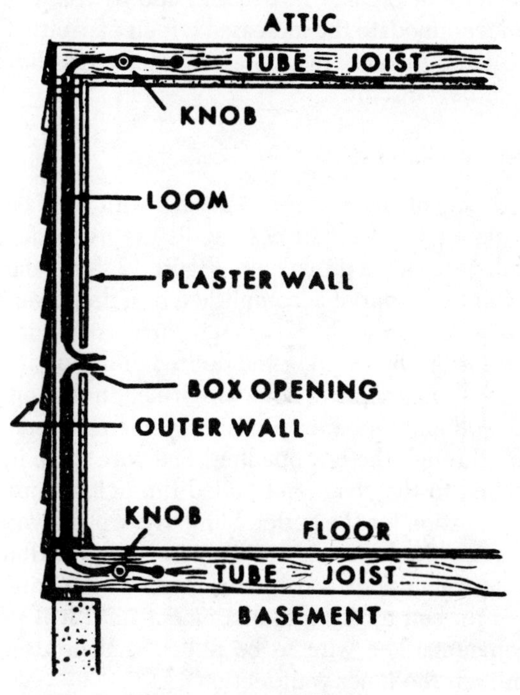

Fig. 12. Concealed wire addition.

New Load Center

In many wiring installations, no provisions are made for spare circuits in the fuse panel. Moreover, the location of the new circuit required is often remote from the existing fusing or circuit breaker panel. In this case the most favorable method of providing service to the circuit is to install a new load center at a location close to the circuit outlets. This installation must not overload the incoming service and service-entrance switch. Should such an overload be indicated, the service equipment should also be changed to suit the new requirements. This sometimes can be accomplished in two-wire systems by pulling in an additional wire from the powerline. This changes the service from two-wire to three-wire at 120 to 240 volts. In these cases the fuse or circuit breaker box should also be changed and enlarged to accommodate the increased circuit capacity. Figure 11 schematically illustrates the installation of an additional load center for a new circuit.

Concealed Installations

The addition of outlets in a building with finished interior walls having enclosed air spaces entails the use of a stiff wire, called a fish wire, and a drop chain. Figure 12 shows the addition of a wire run for an outlet accomplished by a drop from the attic space or as a riser from the basement. First an opening is made in the interior finished wall at the desired outlet position. If the attic circuit is to be tapped, holes are drilled in the top plates of the wood studding and a drop chain is lowered inside the wall and pulled through the box opening. The wire to be installed is then attached to the chain and pulled through, completing the rough-in operation for the outlet. Similarly, when a wire is to be pulled in from the basement, a fish wire is used. After drilling through the rough floor and bottom plate of the studding, the fish wire is pushed up from the basement until it is grasped at the box opening. The wire to be pulled is then attached and pulled through the inner wall section.

Chapter 6

Open Wiring, Knobs, and Tubes

Open wiring is permitted by the National Electrical Code for interior use. A cost comparison of the four basic types of wiring indicates that open wiring is the most economical. This is true only because the costs of the materials used in installation are low when compared to the other systems. If the labor costs were computed, this system could be equal to or higher than the other methods of installation in cost, especially when a great amount of damage-protection installation is needed. Installations of open wiring, however, are very common during times of material shortages.

INSTALLATION

During open wiring installation planning, strong consideration should be given to nonmetallic sheathed cable to expedite installation and eliminate looms and porcelain insulators normally required by open wiring.

Materials

Conductors. Conductors for open wiring should be selected from Table 8 through Table 13, Appendix 2, based on the operating temperature and application.

Insulators. Insulators should be free of projections or sharp edges that might cut into and injure the insulation. They are commonly made of porcelain. Loom, which is a flexible nonmetallic tubing, is also used to protect the electrical conductors.

Boxes and devices. Boxes and devices used with open wiring are described in Chapter 3, Wiring Materials, section on electrical boxes.

Wire Spacing

In an *exposed installation* of knob-and-tube wiring, the wires must be separated from each other by at least 2½ inches. They must be spaced at least ½ inch from the building surface in a dry location, and at least 1 inch when in a wet or damp location. In a *concealed installation* the wires must be separated at least 3 inches and must be supported at least 1 inch from the mounting surface. The minimum spacing of wires in straight runs and at right-angle turns is illustrated in Figs. 1 and 2.

Support Spacing

Run spacing. When wiring is run over exposed flat surfaces, the knobs and cleats should be spaced no further than 4½ feet apart, as shown in Fig. 3.

Tap spacing. A support should be installed within 6 inches of a tap or takeoff. The wire of the tap circuit should always be secured to this support to insure a strain-free tap.

Support spacing from boxes. Supports should be installed within 12 inches of an outlet box. The wires to the box should be installed loosely so that there is no strain on the terminal connections.

INSTALLATION OF KNOBS, CLEATS, AND WIRE PROTECTORS

Knobs and Cleats

Split knobs are used to support wire sizes 10 through 14 and can support one or two wires. They are used as two-wire supports at splices and taps. Figure 4 illustrates the use of split knobs.

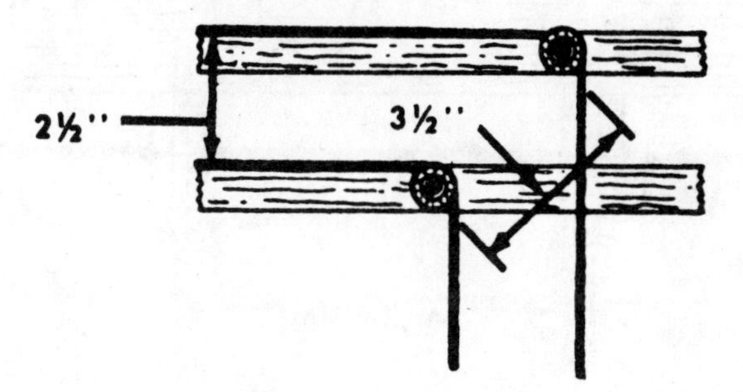

Fig. 1. Wire spacing for exposed work.

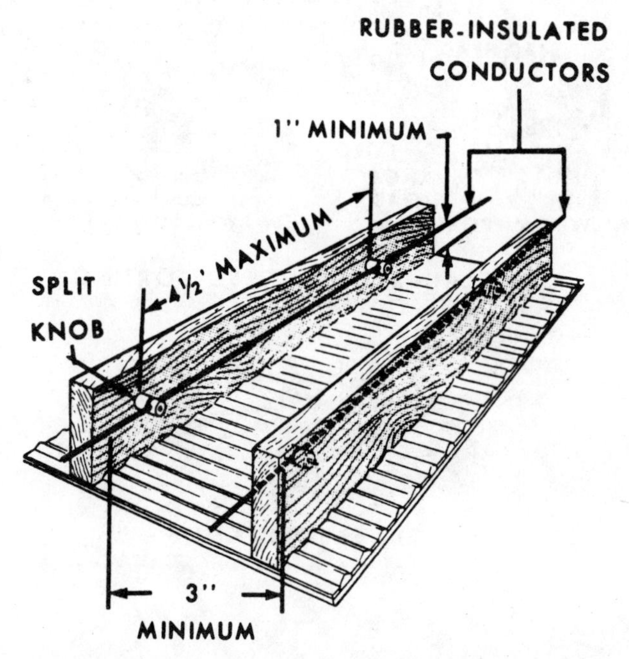

Fig. 2. Minimum wire spacing for concealed installation.

Solid knobs are employed to support wire size No. 8 or larger. The wires must be supported on the solid knobs by tying. A porcelain solid knob is shown in Fig. 4.

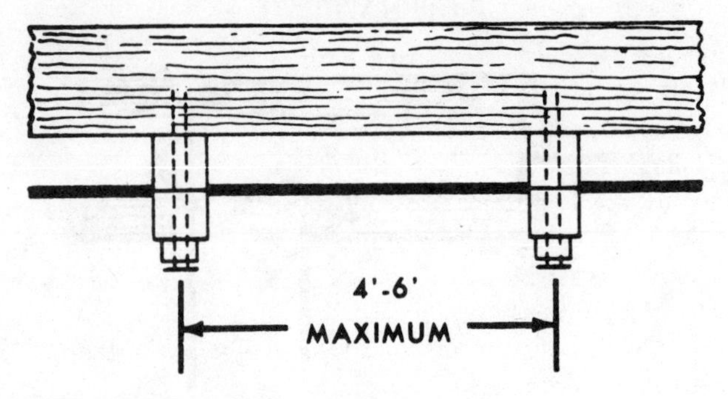

Fig. 3. Knob and cleat spacing.

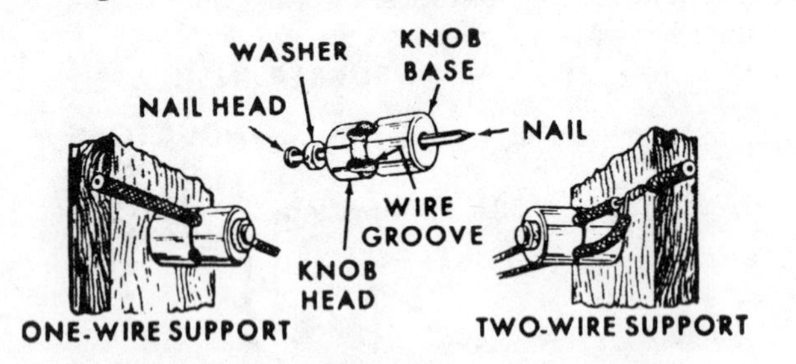

PORCELAIN SPLIT KNOB SUPPORTING ONE OR TWO WIRES

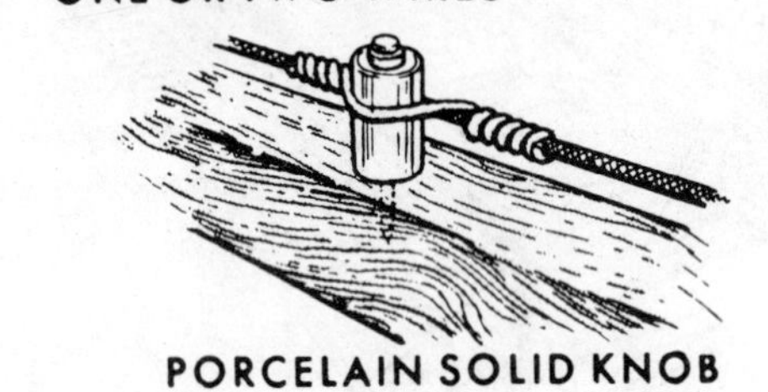

PORCELAIN SOLID KNOB

PORCELAIN CLEATS SUPPORTING TWO PARALLEL WIRES

Fig. 4. Knob and cleat installation.

Two-or-three-wire cleats are also used in supporting wire sizes No. 10 to 14. Single cleats must be used for wire size No. 8 or larger. Cleats are available that support the wire at distances of ½ to 1 inch from the surface on which the cleats are mounted.

The installation steps used in mounting the split knobs or cleats for supporting wires are shown in Fig. 4. In the first operation, leather washers are threaded on the nails of a two-wire cleat to cushion the porcelain. In the second step two wires are placed in the grooves of the cleat base section and the cleat head and nails are positioned above the wires. The third step shows the cleat in supporting position after the wires have been pulled tight and the nails driven firmly into the wood.

Wire Protectors

Tubes. When conductors pass through studs, joists, floors, walls, or partitions, they must be protected by porcelain tubes installed in the hole through the supporting members. These tubes are available in standard sizes ranging from 1 to 24 inches long and ⅜- to 1½-inch inner diameter. The tubes must be long enough to extend through the entire wall. If the wall is too thick for porcelain bushings, standard iron pipe or conduit may be used, provided insulated bushings are installed at each end of the pipe. All conductors of the circuit should be contained in the same piece of pipe if magnetic material is used. (*See* section on connection to devices later in this chapter.) The holes in which the tubes are to be installed should be drilled at an angle so that the tube head can be placed on the high side of the hole to prevent it from being dislodged by gravity. The tubes may also be used to protect wires at points of crossover. As the tube is installed on the wire closest to the supporting surface, it is always installed on the inner wire, thus preventing the outer wires from making contact with the mounting surface. Figure 5 illustrates the proper and improper use of tubes at points of wire crossover, and also the use of a tube installation for protecting an electrical conductor passing over a pipe. Conductors passing through timber cross braces in plastered partitions must be protected by an additional tube extending at least three inches above the timber.

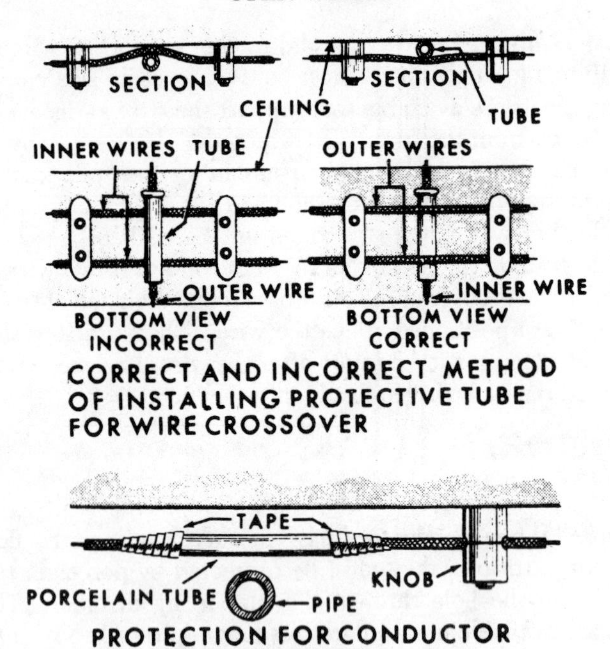

Fig. 5. Protective tubes for conductors.

The extra tubes (Fig. 6) protect the conductors from plaster accumulation, which collects on the horizontal cross members when plastering.

Loom. In some installations where it is difficult to support wires on knobs and cleats, the wires may be encased in a continuous flexible tubing, commonly called loom. This tubing, which is fabricated of woven varnished cambric, should be supported on the building by means of knobs spaced approximately 18 inches apart. Any such run should not exceed a distance of 15 feet. Figure 7 illustrates typical uses of loom. Loom is used to insulate wires at crossovers when they are installed closer than ½ inch to supporting timbers, when two or more wires are spaced less than 2½ inches apart, or upon entry to an outlet box. Outlet boxes used in open wiring are designed for the secure clamping of the loom wire to the box.

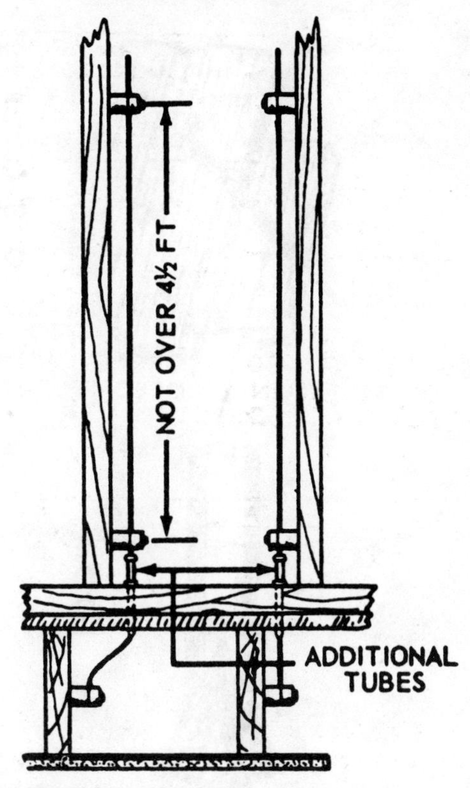

Fig. 6. Additional tubes to protect against plaster accumulation.

Damage Protection

Running boards. When conductors are installed where they may be subject to mechanical damage, protective shields called running boards must be used. Exposed open wiring located within 7 feet of the floor is considered to be subject to mechanical injury. The required installation of a running board on the rafters and below joists for preventing such injury is shown in Fig. 8. Running boards must be at least ½ inch thick and must extend at least 1 inch but not more than 2 inches outside the conductors. This method of installation is used when the wires are threaded through the joists and rafters. In some installations

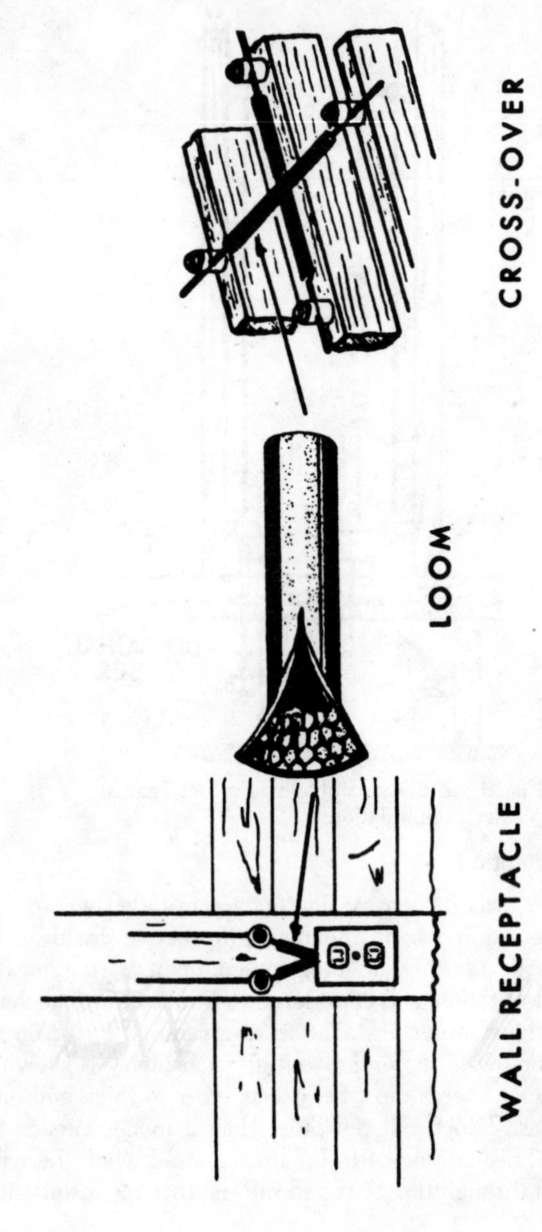

Fig. 7. Typical insulation of wires with loom.

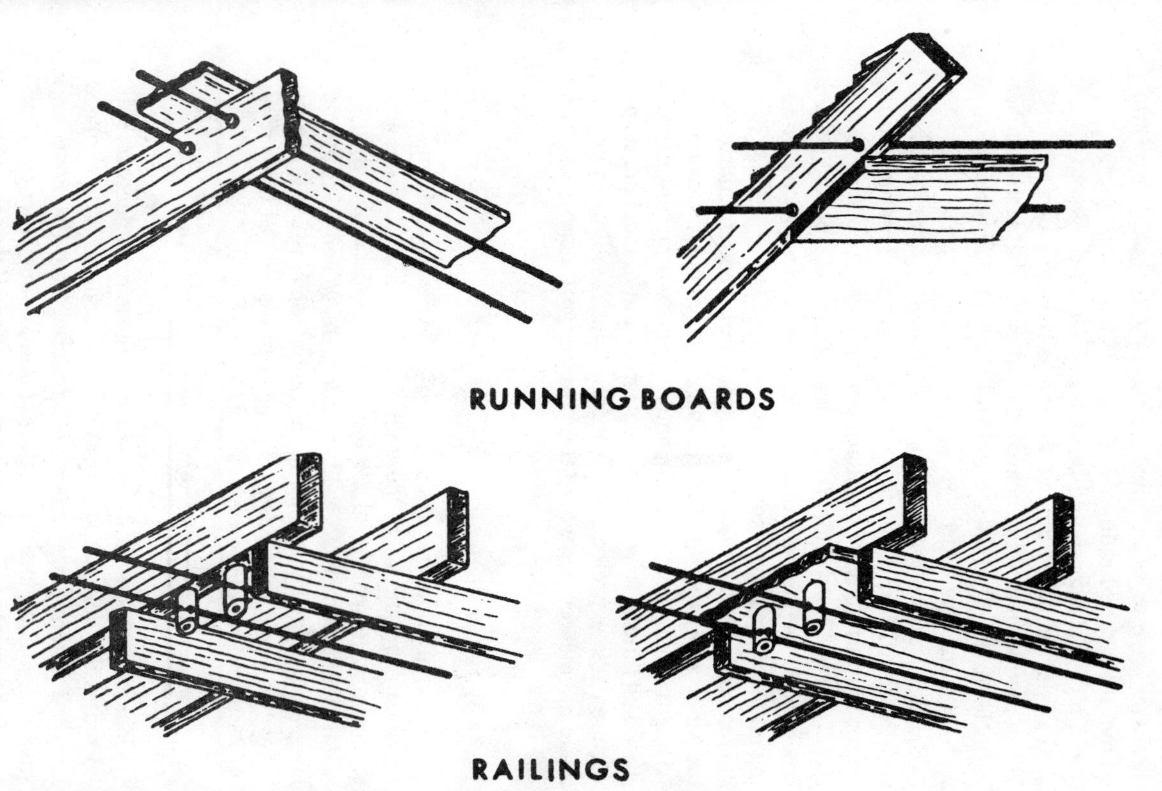

Fig. 8. Protection for wiring subject to damage.

the wires have to be installed on the running boards with protective sides called railings.

Railings. Railings should be at least ⅞ inch thick and, when used alone, should be at least as high as the insulating supports. When used with running boards, they should be at least 2 inches high. Figure 8 illustrates the installation of railings with and without a running board.

Boxing. The preferred method of protecting open wiring on walls within 7 feet of the floor is called *boxing*. This method requires the installation of railings with a cover spaced at least 1 inch from the conductor. In this installation the boxing should be closed at the top and bushings installed to protect the entering and leaving wire leads.

Protection limitations. As previously outlined and illustrated, the labor and expense of installing damage protection in open wiring is extensive. Consequently, open wiring installations should be limited to wiring layouts whose outlet locations do not require damage protection. Nonconforming installations may be made in emergencies where the possibility of mechanical damage is not present.

Three-Wire Installations

The installation of wires in groups of three on joists and running boards requires that those surfaces be at least 7 inches wide to insure wire spacing of 2½ inches and a space of 1 inch for wood clearance beyond each outside wire. When joists are not large enough, one wire may be run on the top of the joist and the other two wires on the sides. Typical installations of three wires on joists and running boards are shown in Fig. 9.

Concealed Installation

Concealed knob-and-tube wiring consists of conductors supported in the hollow spaces of walls and ceilings. The wiring is installed in buildings under construction after the floors and studdings are in place, but before lathing or any other construction is completed. The wires are attached to devices in boxes that must have their front edges mounted flush with the finished

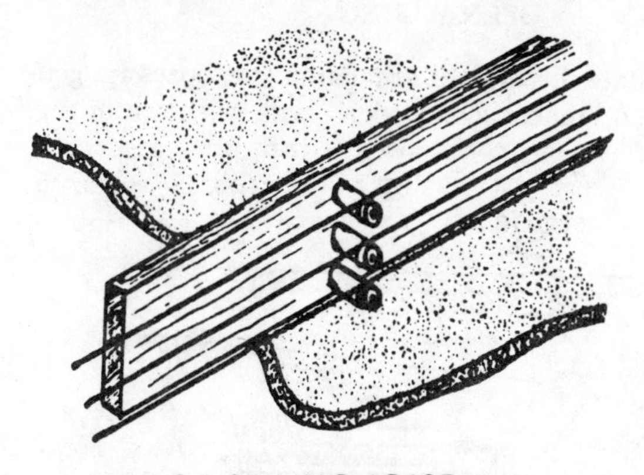

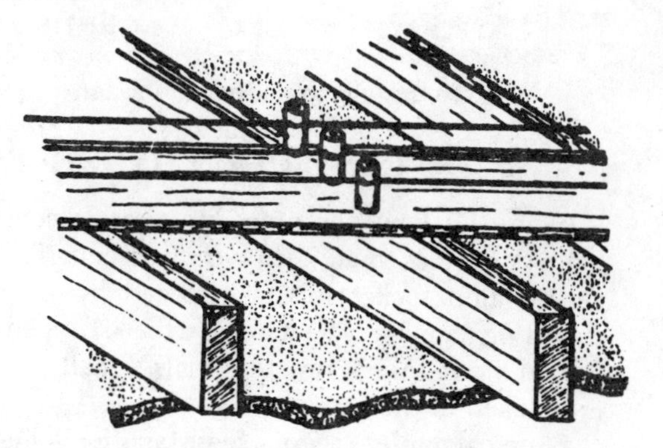

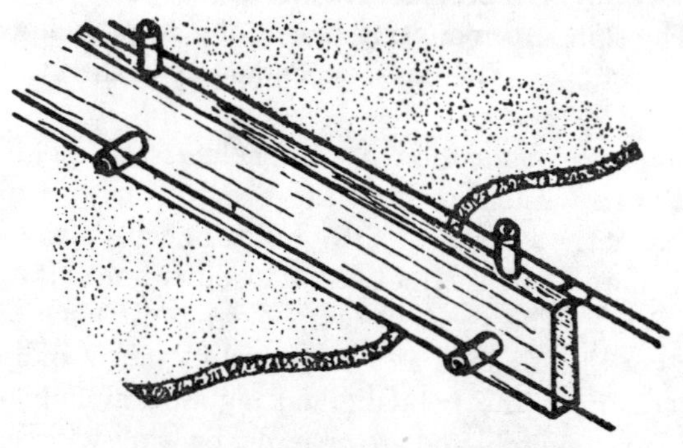

Fig. 9. Knob mounting for three-wire circuits.

surface. To facilitate this type of installation, the boxes are generally mounted on brackets or wooden cleats, as shown in Fig. 10.

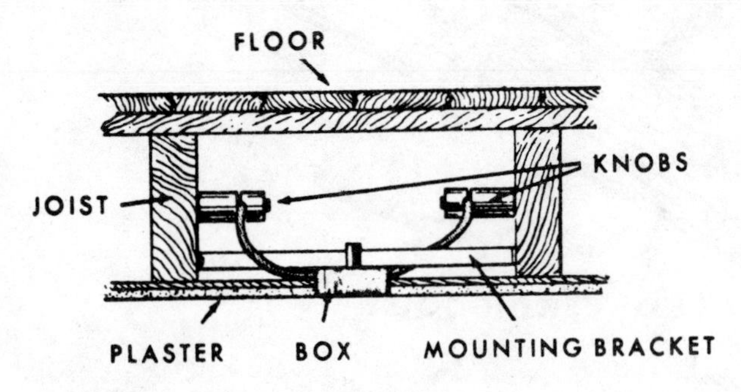

Fig. 10. Installation of box in plaster.

CONNECTION TO DEVICES

Figure 11 shows the procedure used in connecting electrical lighting devices to an open wiring circuit. The base of the porcelain lamp socket is first fastened by wood screws to the mounting member. The wires are then stripped of insulation and looped around the screw terminals. Finally, the porcelain head is attached to the base.

A typical duplex receptacle installation for an open wiring installation (Fig. 12) illustrates the required knob mounting 12 inches from the box, and the placement of loom over the wire at the box entry. The standard mounting height of a receptacle is either 1 foot or 4 feet above the floor, depending upon the location of the outlet.

The installation and connection of lampholders commonly used in exposed open wiring are shown in Fig. 13. The pigtail socket has permanently attached leads of No. 14 wire size or larger. These are paired, but are not twisted together unless they are longer than 3 feet. The pendant lampholder is a device to which the lamp cord is attached and supported by means of an Underwriters knot. Both the pendant and pigtail lampholder sockets are keyless (no switch) and are operated by wall switches to prevent additional strain on the lead wires supporting the sockets.

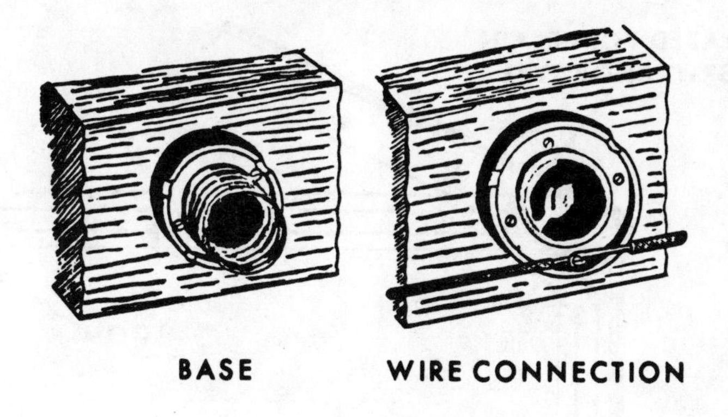

Fig. 11. Porcelain fillings used with knob and tube wiring.

Figure 14 illustrates a typical service-entrance installation, and Fig. 15 shows the procedures in circuit breaker wiring. If a service-entrance switch were used instead of a main circuit breaker, a separate fuse cabinet would be required to provide individual circuit protection. The wires from the powerline should be secured to the building at least 10 feet from the ground for normal installations. When the service entrance is located above a roadway, this height should be increased to 18 feet. If the building is not high enough to meet these requirements, the entrance height may be less, provided all conductors within eight feet of the ground are rubber-insulated. The line wires at the service entrance to a building should be spaced at least 6 inches

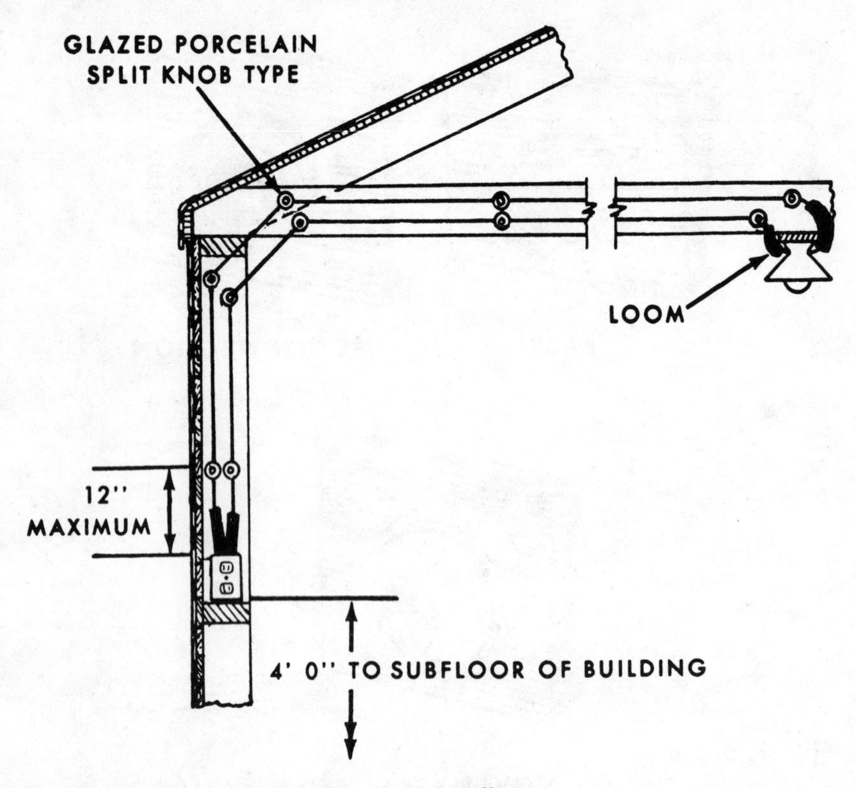

Fig. 12. Typical duplex receptacle installation.

apart and should be supported at least 2 inches from the building by service-entrance insulators or brackets. Upon entering the building, the line wires should be threaded upward through slanting noncombustible tubes so that moisture will not follow the conductor into the service-entrance switch.

Motors are often located with permanent power leads of exposed open wiring, requiring extensive damage protection. To minimize both time and expense, the tap from the open-wiring ceiling circuits should be made with armored cable or conduit. Figure 16 shows a diagrammatic installation of the power connections and operating switch for a three-phase motor connected to exposed knob-and-tube wiring.

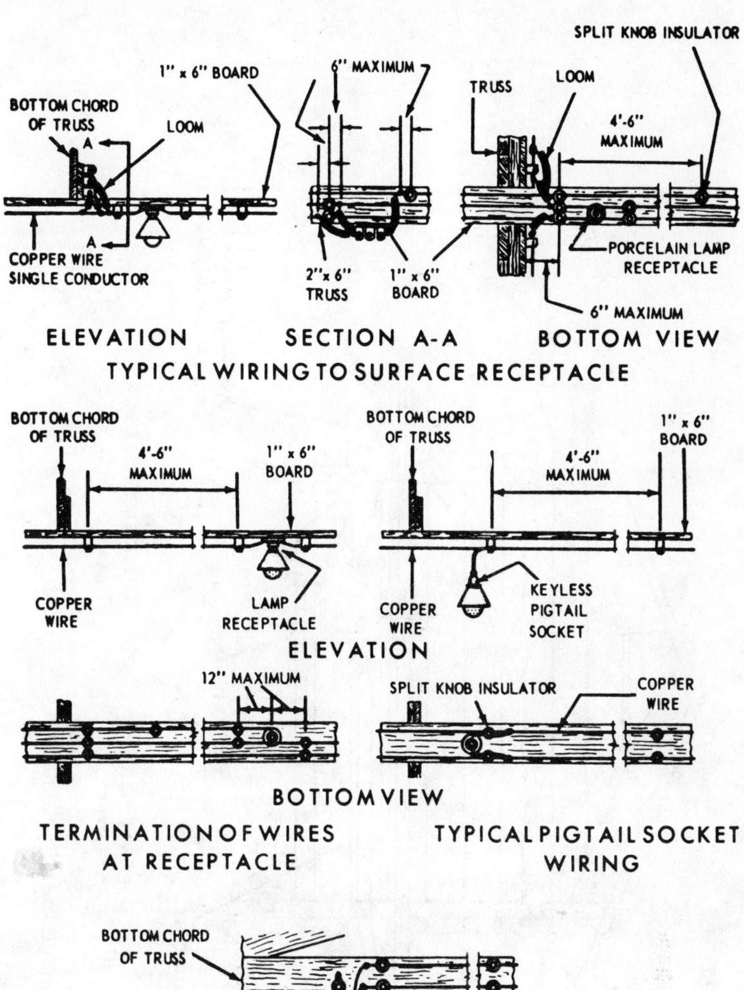

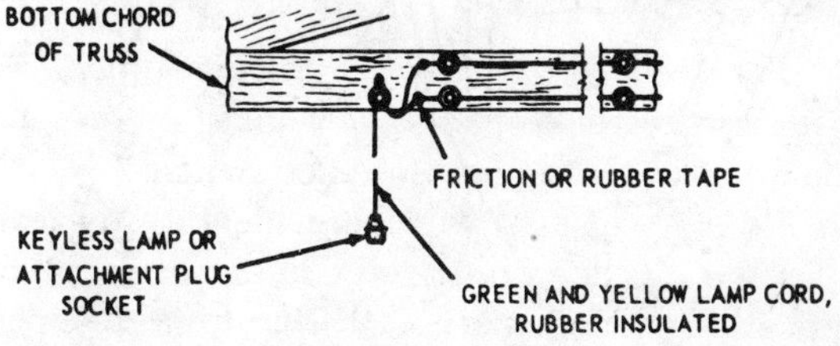

ELEVATION

PENDANT-CONNECTOR INSTALLATION DETAIL

Fig. 13. Lampholder installations.

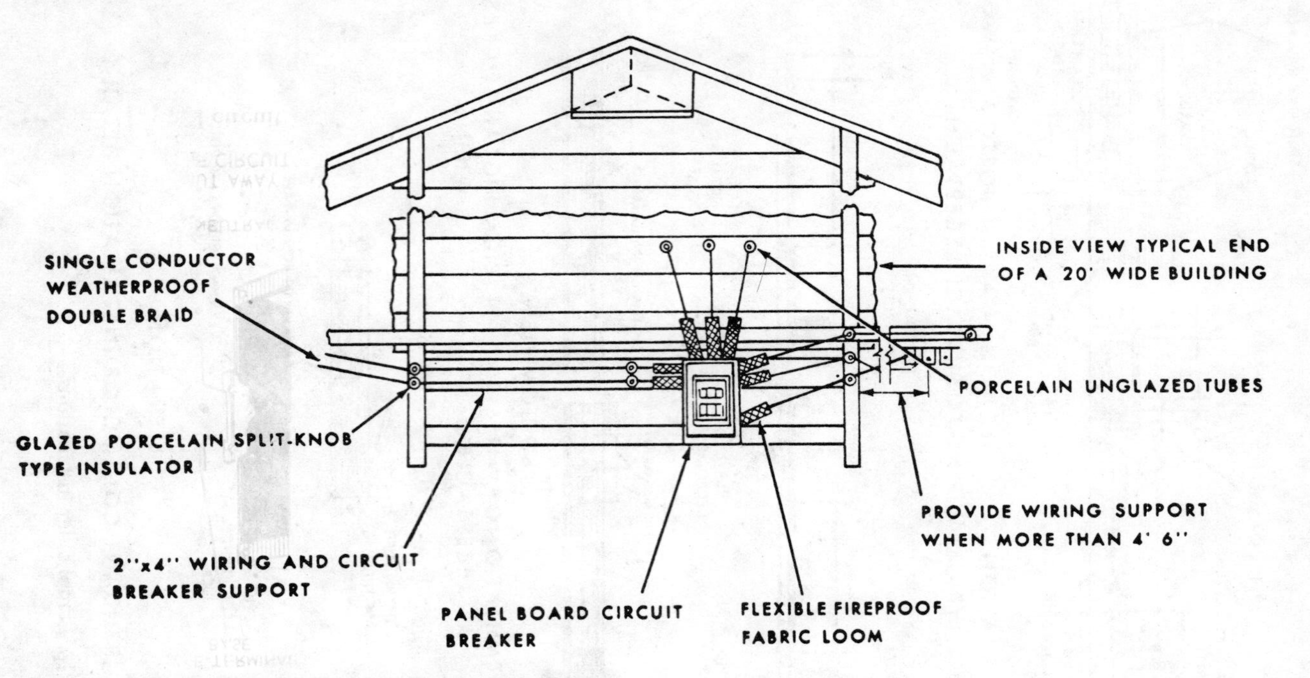

Fig. 14. Typical main circuit breaker installation.

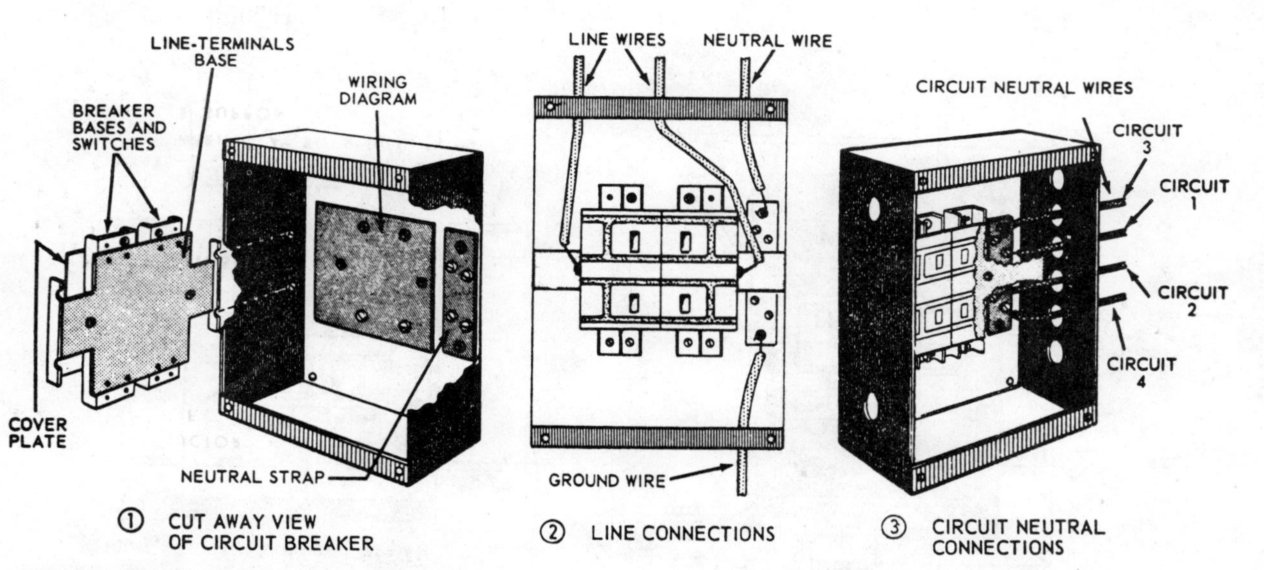

Fig. 15. Typical circuit breaker wiring.

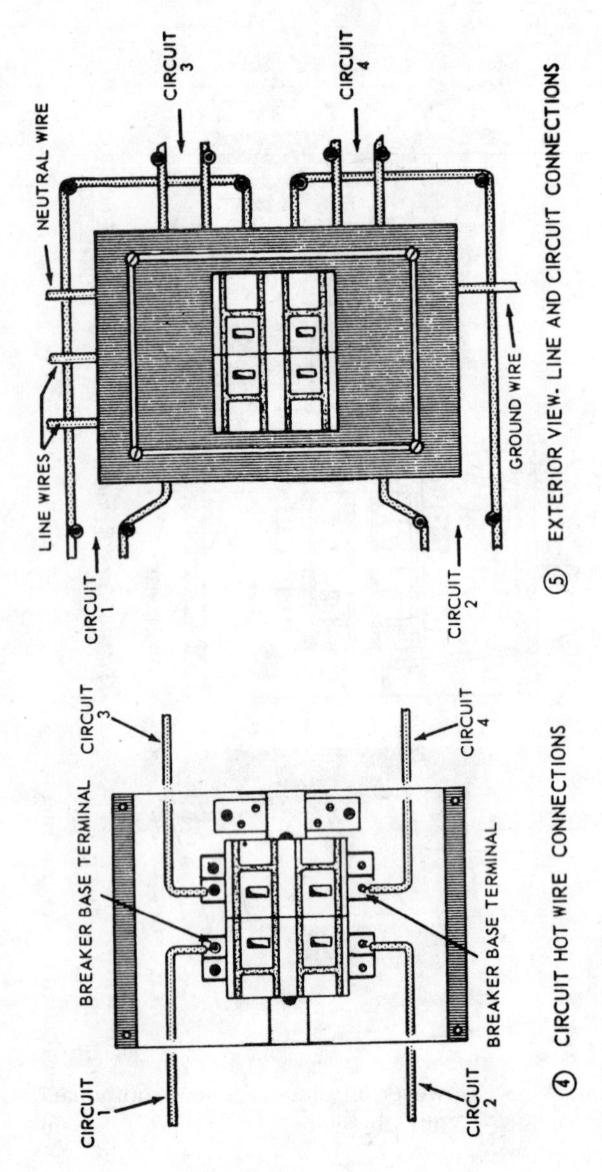

Fig. 15. (continued)

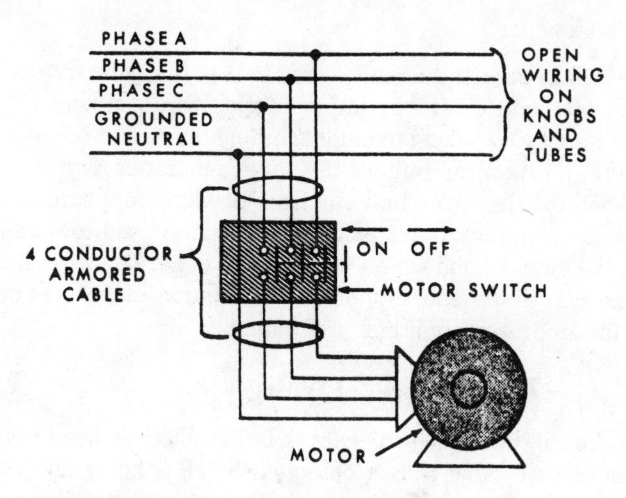

Fig. 16. Motor connection.

ADDITIONS TO EXISTING WIRING

Circuiting

Additions to existing circuits require analysis to determine whether additional circuit capacity is needed to handle the new load. These considerations are the same as those required for other types of installations and are outlined in Chapter 5, Design and Layout of Interior Wiring, in the section on additions to existing wiring.

Wire Connection

Where to connect. An open wiring system has a distinct advantage over the other wiring methods in that wires for new or additional outlets can be attached to the circuit runs by merely making tap splices in the wire runs, or by extending the circuit from an outlet box. However, in planning these additional outlets in the existing circuits, the electrician should be careful to have the shortest possible wire runs. This will result in attaining the lowest voltage drop.

How to Connect

First make sure the circuit is dead. This is a primary safety rule for all electricians working in existing wiring systems. This can be done by removing the fuse, tripping the circuit breaker to the OFF position, or pulling the service-entrance switch and disconnecting the entire building from power before commencing work. A voltage tester or test lamp is also used to double-check the circuit upon which work is to be done. The wires must then be connected and supported in the same manner as outlined for an original building installation.

Connections to Other Types of Wiring

Conduit and cable wiring cannot be installed with splices in the conduit or cable runs. Consequently, all splicing and connections must be made within the confines of an outlet, junction, or fuse box. Therefore, when open wiring is combined with one of the other wiring systems, the transition from one system to another must be made in one of these boxes. Since standard outlet, junction, or fuse boxes are used, open wiring must be encased in loom at the box entry. An example combining knob-and-tube wiring and conduit wiring is illustrated in Fig. 17.

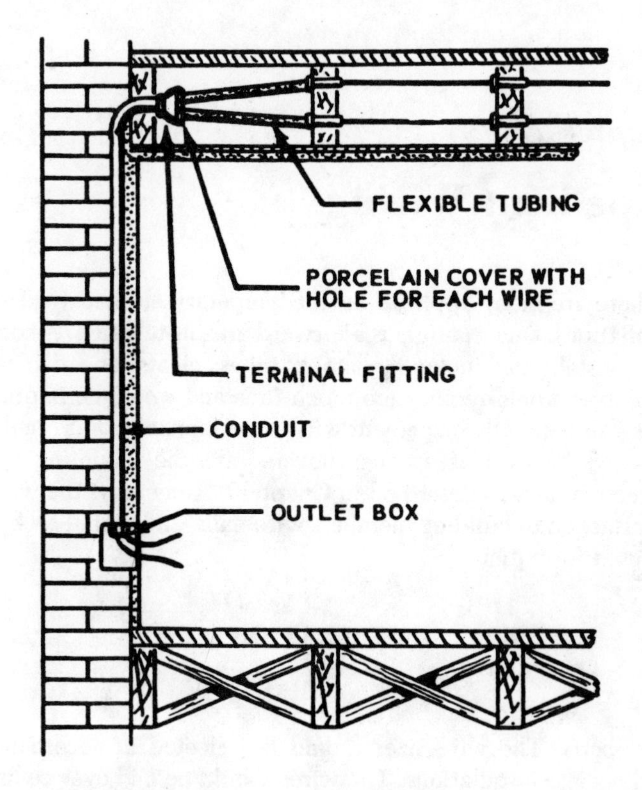

Fig. 17. Changing from knob-to-tube to conduit wiring.

Chapter 7

Expedient Wiring

There are many applications for temporary electrical wiring installations. One example is a forward area installation. A complete installation including knobs, tubes, cleats, and damage protection would require too much time and would be impractical. Consequently, expedient wiring used for temporary buildings and forward areas does not require the mounting and protective devices described in Chapter 6. Generally, the wires are attached to building members with nails, and pigtail sockets are used for outlets.

INSTALLATION

Wire

Supports. The wire sizes should be selected in accordance with normal installations. The wires should be laid over ceiling joists and fastened by nails driven into the joists and then bent over the wire as shown in Fig. 1. The nails should exert enough force to firmly grip the wire without injuring the insulation. If loom is available, it should be installed to protect the wire at the nail support. This is particularly important if the wooden joists are wet. If possible, expedient wiring installations should be fastened to joists or studs at least 7 feet above the floor. This will prevent accidental injury to the system or personnel which might result from the absence of damage protection.

Spacing. The spacing of wires should be the same as that outlined for exposed knob and tube wiring.

Joints, splices, and taps. Joints, splices, taps, and connections are made as outlined in Chapter 1, Fundamentals of Electricity, with the exception of the procedures outlined for soldering and

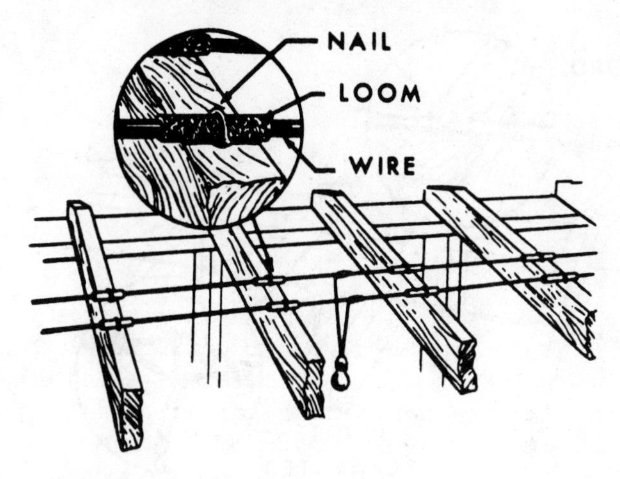

Fig. 1. Expedient wiring.

taping. In expedient wiring soldering is omitted, and only friction tape is used as a protective covering on the connections.

Fixture drops. Fixture drops, preferably pigtail sockets, are installed by tapping their leads to wires, as shown in Fig. 1, and then taping the taps. The sockets are supported by the tap wires.

Cord

Fig. 2 shows the application of a two-conductor cord in an expedient-wiring installation. The cord used should always be of the rubber-covered type and fastened securely to prevent the possibility of short circuits. The outer rubber sheathing should be removed at the point of fixture attachment and the fixture leads tapped into the conductor, purposely maintaining the separation between taps as shown. Each tap then should be individually taped.

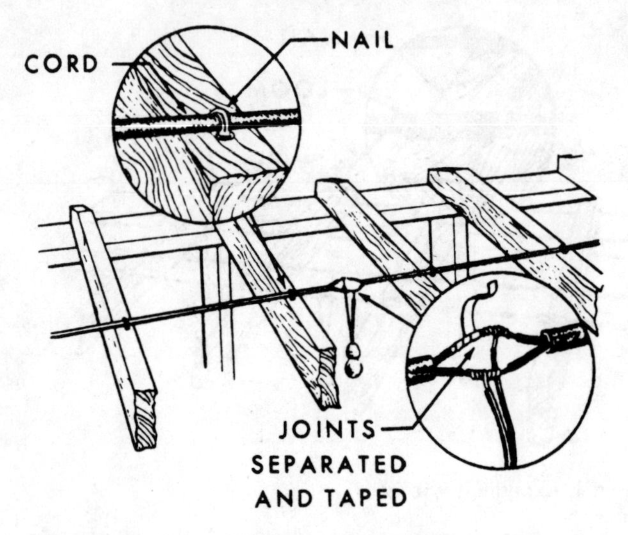

Fig. 2. Expedient-wiring cord installation.

Chapter 8

Cable Wiring

ARMORED CABLE WIRING

Armored cable wiring, commonly called BX, is permissible by Code for all interior installations, except where it is exposed to saturation by liquid or is in contact with acid fumes. In wet areas a lead-covered cable is required. From an economic viewpoint, the labor costs for armored cable installations compare favorably with the open or knob-and-tube wiring installation outlined in Chapter 6. The material requirements for armored cable wiring are greater, and thus overall cost is generally higher. This increased cost is often warranted because an armored conductor has greater mechanical damage protection, and thus eliminates the need for the porcelain insulators and loom which are required in open wiring.

MATERIALS

Cable

As outlined in Chapter 3, Wiring Materials, armored cable is constructed in two or three rubber- or thermoplastic-covered wire combinations encased in flexible steel armor. It is obtained from the manufacturer as Type AC without a lead sheath, and Type ACL with a lead sheath under the armor. Type AC cables have a copper or aluminum bonding strip. One of the conductors of armored cable is always white. Because of this color coding, the code allows a white wire in a switch installation (with armored cable and also nonmetallic sheathed cable) to be used as a hot wire and thereby allows its connection to a black wire. This is shown in the wiring of lampholder No. 3 and switch No. 4, in Fig. 1.

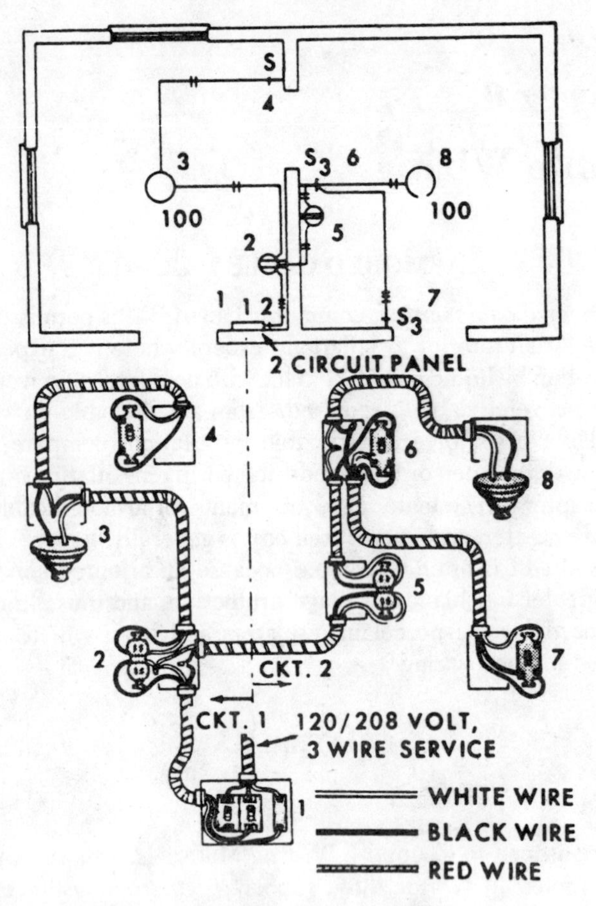

Fig. 1. Typical armored cable connections.

Three-Wire Armored Cable

Service. Three-wire armored cable is used to carry power from the service-entrance switch to the fuse panels or to local load centers if the system in a building is three-wire 120/240 or 120/208 volts. The minimum size of conductor recommended for this use is No. 10 AWG. In this type of service, the neutral wire can be of the same gage as the two hot wires because its maximum current will be no greater than the maximum for either of the other two wires. A three-wire cable connection is shown at the load center, No. 1, in Fig. 1.

Two-circuit. Sometimes in laying out and circuiting armored cable installations, several circuits feed out in the same general direction from the protection (fuse) panel. When the power distribution system uses three-wire 110 to 220 volts, and the circuits are not fed from a common hot-line wire, it is advantageous, both from voltage-drop and economic standpoints, to install three-wire cable as a two-circuit carrier as far as possible. An example of this use of three-wire cable is shown in Fig. 1, between the circuit breaker panel, No. 1, and the receptacle box, No. 2. Similarly, many installations are made wherein a switch and an outlet receptacle are connected from wires originating in an overhead light to be switch controlled. Rather than use two two-wire cables for circuiting these devices, a three-wire installation from the light to the switch (Fig. 2) is made.

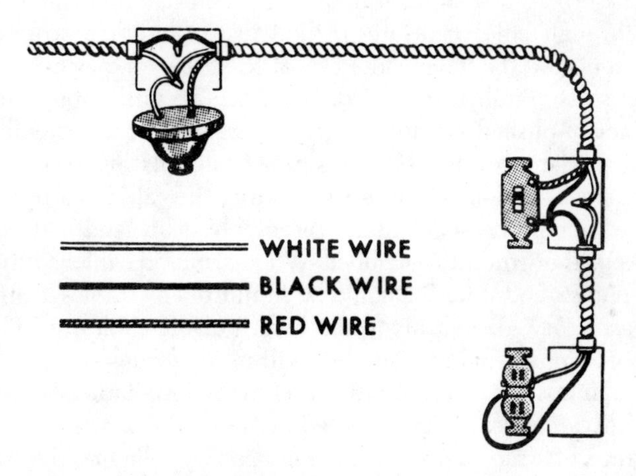

Fig. 2. Three-wire cable, two-circuit use.

Supports

Armored cable may be fastened to wooden building members with a one- or two-hole type mounting strap formed to fit the contour and size of the cable, or by staples made specifically for the armored cable used. (*See* Chapter 3, Wiring Materials, section on straps and staples.) The cable is normally supported at the box entry by integral BX clamps built into the boxes or by BX connectors.

Boxes and Devices

Boxes and devices recommended for use with armored cable wiring are described in Chapter 2, Electrician's Tools and Equipment. For quick installation, the electrical boxes with the integral cable clamps and attached mounting brackets are used.

INSTALLATION

Cable Support

Whenever possible, an armored cable installation should be run through holes centrally drilled in the building structural members, and the holes should be at least ⅛ inch oversize to facilitate easy "pull through" of the BX. The flush mounting of BX, accomplished by notching the joists and studs, should be avoided whenever possible. This type of installation exposes the BX to possible short circuits, by locating the cable in a position where it could be accidentally pierced by nails, and materially weakens the structural member. When armored cable is run between joists and studs, it should be supported by stables or straps at least every 4½ feet along the length of the cable run. These supports must also be installed within 12 inches of each box entry, unless the support interferes with installations that require extreme flexibility. This requirement also assures the continuance of a satisfactory box connection by relieving the strain on the splices and connection within the outlet box. Cable runs installed across the bottom of ceiling joists and studding faces at least 7 feet above the floor must be supported on each joist or stud. If preferred, they may also be installed on running boards similar to those used in open wiring installation.

Damage Protection

When armored cable is installed on the top of floor joists or studding in accessible locations (attics and temporary buildings) at a distance less than 7 feet from the floor, guard strips at least as high as the cable must be installed.

Armored Cable Bending

When installing armored cable, care must be taken to avoid bending or shaping the cable in a manner that damages the protective armor. This type of installation damage may occur in drilled holes for BX, in corner runs, or when locating boxes on studs and joists. To prevent this, the radius of the inner edge of any bend must not be less than five times the cable diameter. Fig. 3 shows an acceptable armored cable bend at box entry.

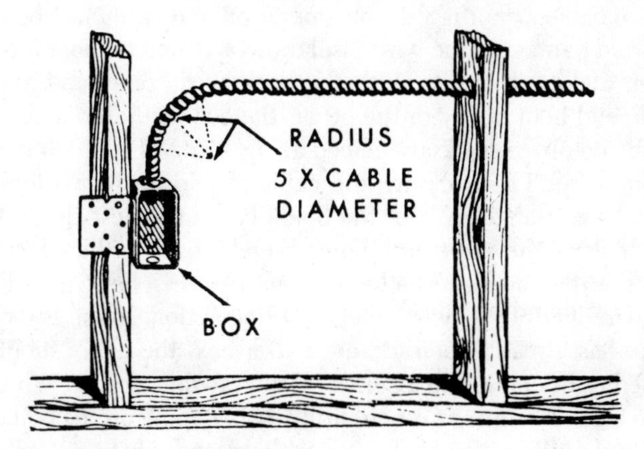

Fig. 3. Armored cable bend.

Box Connection

Procedure. Armored cable may be spliced or connected to devices only in standard junction or outlet boxes. All the cable used, therefore, must be cut long enough to run from box to box. To prevent cutting the cable too short, the BX should first be threaded through the mounting holes drilled in the joists or studs and attached to one box. The slack is taken out of the cable by using just enough force to maintain the proper bends. Keeping this tension, the cable is cut from the roll and connected to the box. The procedure to be followed in preparing and attaching the cable to a box is shown in Fig. 4.

Cable cutting. Though armored cable can be cut with a BX cutter specifically designed for the job, electricians generally use a hacksaw (Fig. 4). In making an outlet connection, the cable should first be cut completely through, about 8 inches longer on each end than required for the run. In removing the armored cable from the wire, the armor should be cut approximately 8 inches from the cable end so that ample wire will be inserted in the box for connecting to the outlet device. These lead lengths may be increased when the wire run terminates in a fuse or circuit breaker panel box and a longer cable is required. The cutting of cable armor is a simple operation, but care must be taken to avoid damaging the wire insulation or the metal bonding strip when making the cut. With the hacksaw in one hand and the cable end held firmly in the other, the cut should be made with the blade of the hacksaw placed at right angles to the lay of the armor strip. The hacksaw and cable should form two legs of a 60-degree triangle. When the blade has cut almost through the armor strip, the cable end should be bent back and forth several times until it breaks. The loose armor can then be stripped from the wire leads by a twisting and pulling action. If one end of the cable has already been attached to a box, the cable should be pulled tight enough to assure a steady sawing surface. When cut from a coil, the armor is held firmly by stepping on the coil cable end and pulling it tight. Rough or sharp ends of the cut are then smoothed with a file.

Unwrapping Paper

The fiber paper that is twisted around the conductors before the metallic armor is attached must be removed to allow free wire movement. Normally two or three turns of the paper are removed from under the armor by tearing the loose paper away

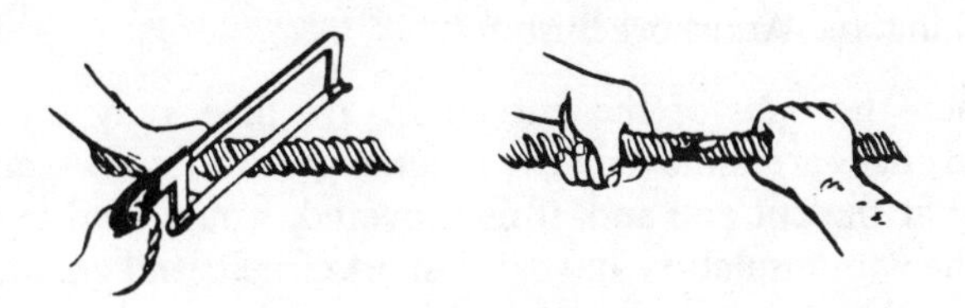

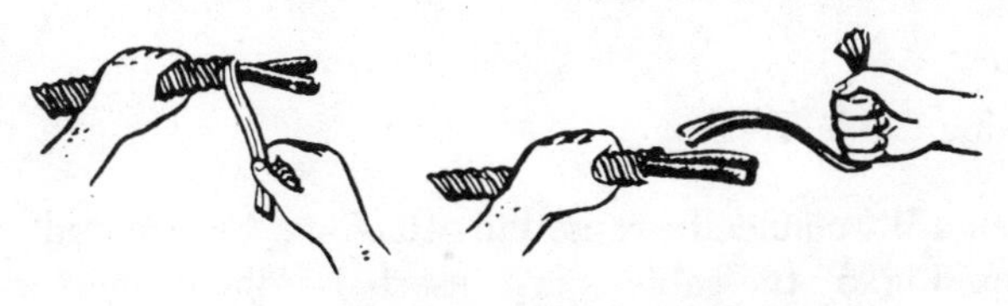

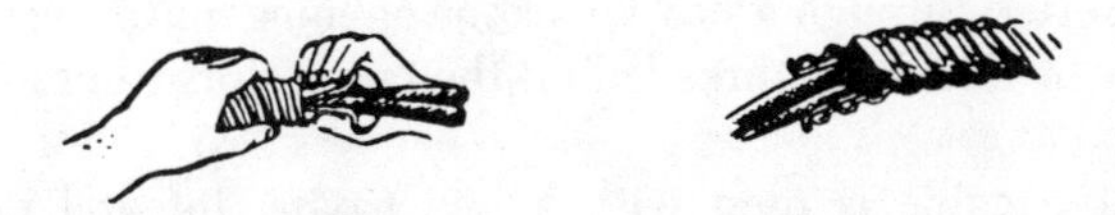

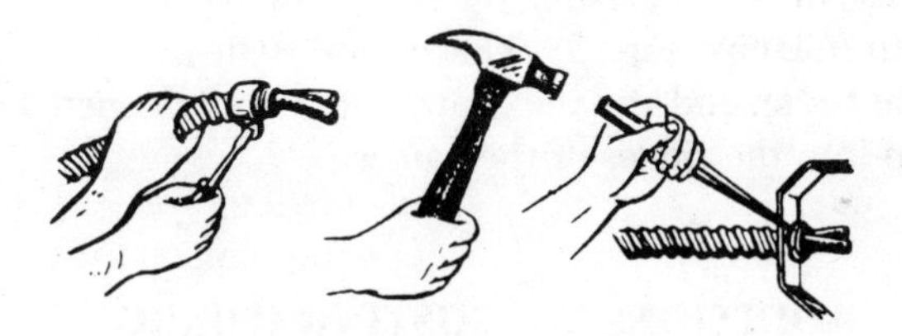

ATTACHING CONNECTOR TO
CABLE AND CABLE TO BOX

Fig. 4. Procedure for preparing and attaching cable.

from the wire at the armor end by a jerking action (Fig. 4). The free space between the armor and the wires facilitates mounting the antishort bushing.

Attaching and Antishort Bushing

When the ends of the cut armor are filed, only the outer burred edges are removed. The inner edges are always sharp and jagged at the cut end and, if not covered, would tend to puncture the wire insulation and cause short circuits and grounds. To prevent this, a tough fiber bushing, commonly called an antishort, must be inserted between the armor and the wire to protect the wire against damage (Fig. 4).

Attaching Cable To Box

When a BX connector is used in attaching the armored cable to a box (Fig. 5), the cable is first inserted in the connector and the holding screw or screws are tightened against the armor, securely connecting the cable and connectors. The BX connector is then inserted through a box knockout opening and is secured to the box by a locknut threaded on the connectors from inside the box.

When the cable is used with a box having integral cable clamps, the knockout at point of entry must first be pried out. Next the clamp-holder screw is loosened and the cable is inserted through the knockout opening and the leads threaded through the clamp (Fig. 5). The armor is then forced snugly against the clamp end and the clamp screw re-tightened, forcing the clamp into the ridges of the cables.

ADDITIONS TO EXISTING WIRING

Circuiting

Additions to existing armored cable layouts require analysis to determine whether additional circuit capacity is needed to handle the new load. These considerations are the same as those required for other types of installations and described in Chapter 5, Design and Layout of Interior Wiring, in the section on additions to existing wiring.

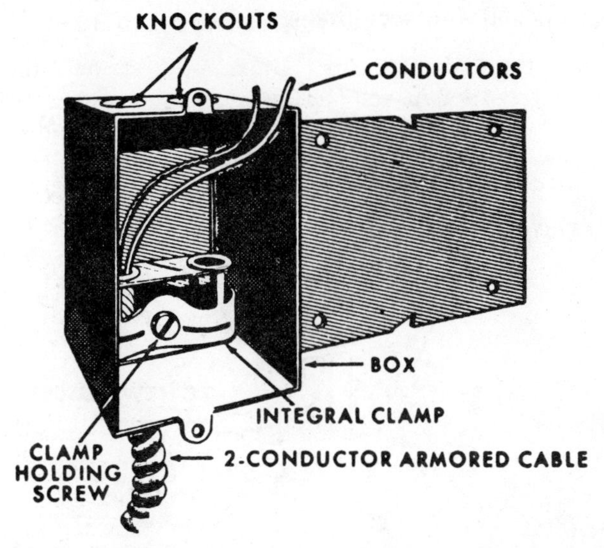

Fig. 5. Cable connection to box with integral clamps.

Cable Connection

Armored cable additions must always originate and terminate in electrical boxes. The junction box used for the addition should be located close enough to the desired outlets so that the voltage drop to the new device is within allowable limits. The box from which the additional outlet or outlets are to originate must also have both a neutral and a hot wire of the same circuit for the new load connection. This means that the conductors from an added outlet can be connected only to the conductors of an existing cable in an outlet box (white to white and black to black) if the existing conductors can be traced to the fuse or circuit breaker without interruption. Figure 6 illustrates two methods of connecting to existing conductors, one in a switch box and the other in a receptacle box.

Installation of Armored Cable Additions

Exposed. The installation of exposed armored cable additions to existing wiring must be patterned according to rules outlined

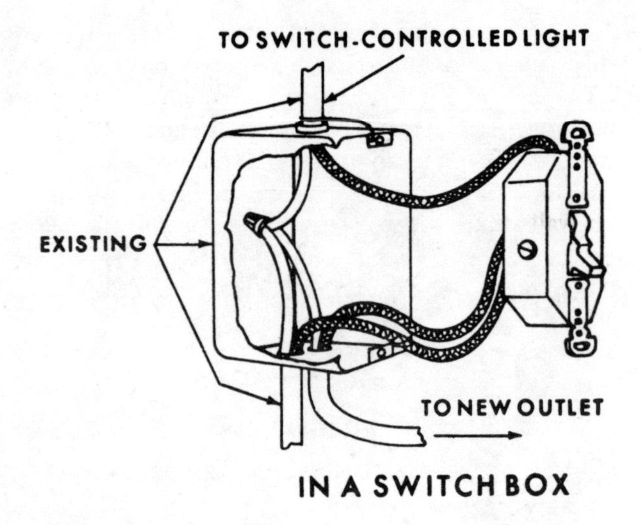

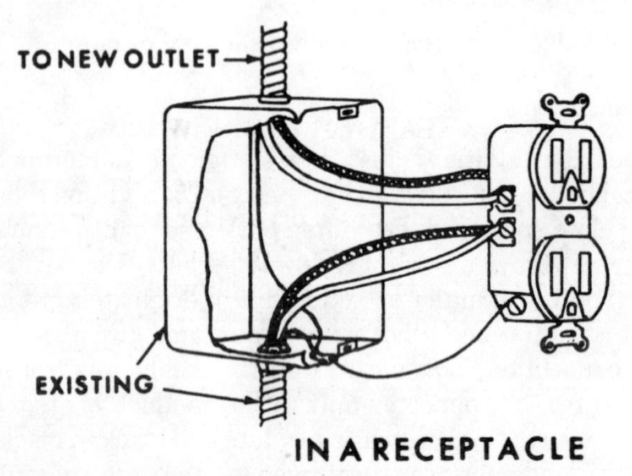

Fig. 6. Additions to existing wiring.

for original installations. If an armored cable installation is to be made into another type of wiring system, the changeover must be made in a junction box specifically installed for the purpose or in an existing outlet box, provided its conductor capacity will allow the entry of additional wires.

Concealed. Armored cable is preferred over all other types of wiring when additional outlets are required on completed buildings. The armor provides damage protection and, together with the bonding strip, adequate continuous ground to the metal outlet boxes. BX is also flexible enough to allow it to be fed through small openings from attic or basement areas to boxes mounted on walls and ceilings. The cable is usually pulled into the concealed box with a fish wire or drop chain. The fish wire is used when the cable is to be fed from below the box location, whereas the drop chain is used when the installation is to be made from above. In these cases, the junction box in which the power tap is to be made should be in a clear, readily accessible area, since the fishing and catching of a fish wire and drop chain become a tedious and time-consuming operation if the junction box is concealed. If it is difficult to feed into the power tap box because of building construction, the finished wall may have to be removed to allow entry. This will necessitate a replastering job after the addition has been installed.

NONMETALLIC SHEATHED CABLE WIRING

The conventional *nonmetallic sheathed cable* (NM) is approved for use in concealed or exposed dry indoor locations, and is recommended for use where a good system ground is not available. Since the cable is inexpensive, lightweight, and requires no special installation tools, it is suited for use in wiring systems. Because of its construction, it is not approved for imbedded installation in masonry, concrete, fill, or plaster. It should *not* be installed in potentially dangerous areas where wire damage may occur, such as commercial garages, theaters, storage-battery rooms, and hoistways. It is not used in humid or wet areas.

A newer nonmetallic sheathed cable (NMC) is the dual-purpose plastic sheathed cable with solid copper conductors. It needs no conduit, and its flat shape and gray or ivory color make it ideal for surface wiring. It resists moisture, acid, and corrosion, and can be run through masonry or between studding.

MATERIALS

Cable

Nonmetallic sheathed cable consists of rubber- or thermoplastic-covered wires in two- or three-wire combinations with a bare copper wire used for bonding. These are individually wrapped with a thick spiral paper tape for damage protection and covered with a woven fabric braid that has been saturated with a moisture-resistant and flame-retardant compound. The entire assembly is then coated with wax. The local codes in some areas also require the addition of a bare uninsulated conductor in the nonmetallic sheathed cable. This bare wire provides the same type of equipment ground or bonding at the outlet boxes as the armor in armored cable installation. The bare wire is attached to the outlet box by clamping it either under the connector locknut at box entry or under one of the screws of the cable clamp (Fig. 7).

Fig. 7. Three-wire nonmetallic sheathed ground.

Supports

Nonmetallic sheathed cable is generally mounted on wooden building members with one- or two-hole mounting straps formed to fit the contour of the cable. BX staples are not approved for this type of installation because of the danger of possible cable damage.

Boxes and Devices

The boxes and devices used in nonmetallic sheathed cable wiring are similar to those used with conduit. They are made of metal or nonmetallic materials such as porcelain or bakelite. They can be obtained with built-in clamps or knockout holes for the cable connectors. The knockouts in porcelain boxes, however, are designed for cable entry only. It is recommended that metal boxes with integral clamps be used whenever possible to assure a safe and efficient installation. Insulated switches, outlets, and lampholder devices may be used without boxes in exposed nonmetallic sheathed cable wiring. The cable-entry holes to these devices must clamp the cable securely, and the device must fully enclose the section of the cable from which the outer sheathing has been removed. No splicing can be done in these devices. Consequently, since all wires must be connected to terminals, use of these devices is limited to installation in rural or other areas wherein only a small number of outlets and switches are required.

INSTALLATION

Cable Support

Nonmetallic sheathed cable installation should be supported in a manner similar to that described for armored cable. As shown in Fig. 8, the cable can be installed either on running boards, in holes drilled in the center of the joists, or on the sides of joists and studs. When running boards or the sides of joists and studs are used, straps should support the cable at distances not greater than 4½ feet, and a cable strap should be attached within 12 inches of a box. When the cable run is to be made at an angle in an overhead installation and is supported on the edge of the joists, at least two No. 6 gage or three No. 8 gage wires must be used in the wire assembly. If smaller sized wires are used, they must be installed through holes bored in the joists or mounted on running boards.

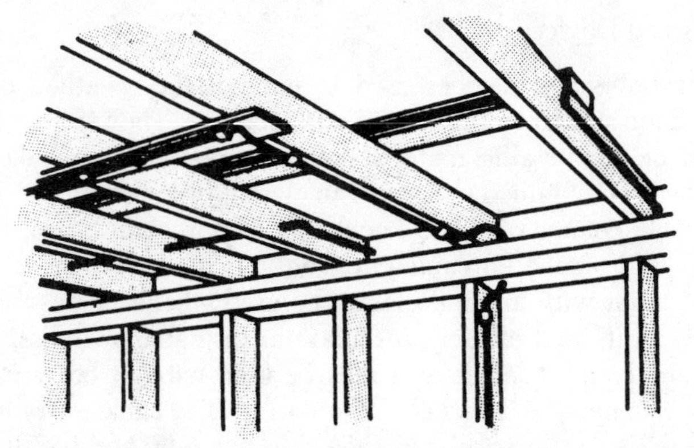

Fig. 8. Nonmetallic sheathed cable installation.

Damage Protection

If the cable is installed across the top of a floor or floor joists, it must be protected by guard strips at least as high as the cable. When the wire installation is made in a location not normally used, such as an attic or crawl space under a building, damage-protection devices such as guard strips are required only within 6 feet of the entrance. Concealed nonmetallic cable installations should not be installed near baseboards, door and window casings, or other possible locations of trim or equipment because of the possibility of damage from building nails. If thermal insulation is to be installed where nonmetallic sheathed cable is in place, only noncorrosive, noncombustible, nonconductive insulation should be used. During the installation of the insulation, care must be taken to prevent adding more strain on the cable, its supports, or its terminal connections. This is especially necessary if the NM installation includes porcelain outlet boxes. (NM is the code designation for nonmetallic sheathed cable.)

Nonmetallic Sheathed Cable Bending

To prevent accidental damage to the sheathing on nonmetallic sheathed cable, the minimum allowable radius of bends is five times the cable diameter. Though this bend limit is similar to the armored cable requirement, nonmetallic sheathed cable can be bent in a smaller arc. This is true because the cable diameters are smaller for the same wire gage combinations.

Box Connection

Cable runs must be continuous from outlet to outlet because wire splices are permitted only inside a box. NM cable is prepared for box connection in the same manner as outlined for armored cable. In removing the protective sheathing from the conductors for connection, an electrician's knife rather than a hacksaw is used. In removing the covering, a slit should be cut in the sheathing parallel to the wires without touching the individual wire insulation. A cut approximately 8 inches long for cable entry to ordinary boxes is satisfactory, but can be increased to suit entry to panels. The knife is then used to remove the slit sheathing. The moisture-preventive paper should also be removed from the wires. Figure 9 illustrates the slitting of a cable end. Figure 10 shows a special tool called a cable stripper which can be used instead of a knife to remove the sheathing from NM cable, lead-covered cable, and portable cords. In operation the

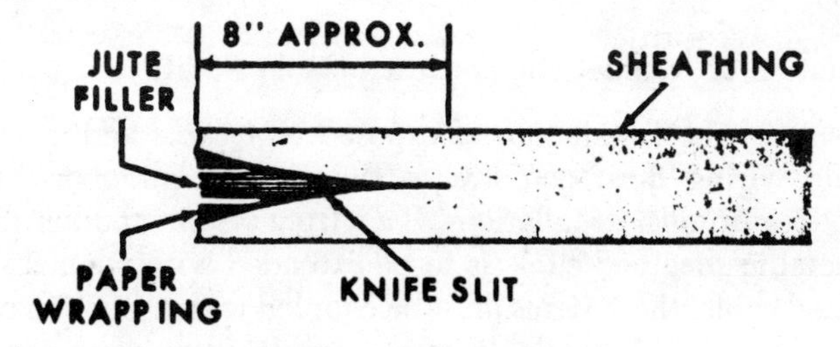

Fig. 9. Slitting cable.

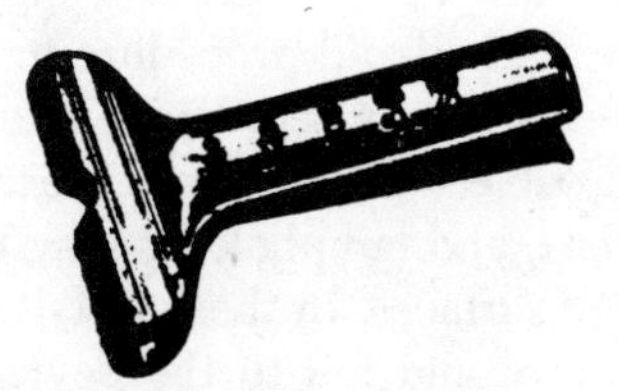

Fig. 10. Cable stripper.

stripper is inserted over the cable, squeezed together, and then pulled off the conductor. This action rips off the outer braid quickly and efficiently. The use of a stripper instead of a knife for outer braid removal is recommended, since it cannot damage the wire insulation.

ADDITIONS TO EXISTING WIRING

Circuiting

The factors pertinent for additions to existing wiring systems outlined in Chapter 5, Design and Layout of Interior Wiring, in the section on additions in existing wiring, are the same as those which should be considered for NM wiring.

Cable Connections

Connection additions for nonmetallic sheathed cable are the same as the connection additions previously described in this chapter in the section on additions to existing wiring for armored cable.

Installation of Nonmetallic Sheathed Cable Additions

Exposed. The installation of exposed NM cable additions to existing wiring must conform to the same requirements described for original installations. If a wiring system of other than nonmetallic sheathed cable is to be extended with nonmetallic sheathed cable, the systems must be coupled with a junction box. Existing boxes with available spare conductor capacity can be used.

Concealed. Additions to concealed nonmetallic sheathed cable are similar in method and procedure to those outlined in this chapter previously in the section on additions to existing wiring for armored cable wiring, with the exception that insulated switches, outlets, and lampholders may be installed without boxes on the wall surfaces, In these installations the cable is fished through the wall and fed to the device at the point of entry.

Chapter 9

Conduit Wiring

RIGID CONDUIT INSTALLATION

Either black enameled or galvanized rigid metal conduit is approved for use under all conditions and locations, except that metal conduit and fittings protected from corrosion only by enamel may be used only indoors and in areas not subject to severe corrosion. Though it is generally the most expensive type of wiring installation, its inherent strength permits installation without running boards and damage protection. Its conductor capacity allows it to carry more conductors in one run than any other system, and its rigidity permits installation with fewer supports than the other types of wiring systems. Moreover, the sizes of conduit used in the system's installation generally provide for the possible addition of several more conductors in the conduit when additional circuits and outlets are required in the run.

MATERIALS

Though the materials used in rigid conduit wiring have been previously described in detail, we will review the advantages of these standard materials as well as their limitations in this chapter.

Rigid Conduit

Rigid conduit (Fig. 1) has the same size designations as water pipe. Under the Code limitations, no conduit smaller than ½ inch may be used except in finished buildings where extensions are to be made under plaster. In these installations $^{5}/_{16}$-inch conduit or tubing is permitted. The size of conduit is determined by the inside diameter—for example, ½ inch (0.622). Standard conduit sizes used in interior wiring are ½, ¾, 1, 1¼, 1½, 2, and 2½ inches. Larger sizes, up to six inches, are available for use in certain commercial and factory installations. Though conduit is made in dimensions similar to water pipe, it differs from water pipe in a number of ways. It is softer than water pipe and thus can be bent fairly easily. In addition, the inner surface is smooth, to prevent damage to wires being pulled through it. Moreover, the finish is rust-resistant. Black enamel conduit is used for dry and indoor installations, and exterior galvanized conduit is used in outside installations to provide moisture protection for the conductors. For wiring installations in corrosive atmospheres, aluminum, copper alloy, or plastic-jacketed conduit is available.

Conductors

Rubber-covered insulated-type R or RH wire is used with conduit in most interior wiring installations, although the thermoplastic insulation types T or TW are gaining favor because of their superior insulating characteristics. Underground or wet installations require the insertion of lead-covered cables in rigid galvanized conduit for permanent protection.

Supports

The conduit straps illustrated and described in Chapter 3, Wiring Materials, in the section on straps and staples, are preferred for use in mounting conduit in interior wiring systems. According to Code requirements, the conduit should be supported on spacings as shown in Table 23, Appendix 2.

Fittings

There are two types of fittings: the standard ordinary-size outlet box and the small junction or pull boxes called condulets. The standard outlet box fittings are classified as type F and are used normally in exposed installations to house receptacles or switches where high-quality installation is desired. The junction or pull box fittings (Fig. 1) are used to provide either intermediate points in long conduit runs for pull-through of wire or junctions for several concealed installations where they will not be accessible. They are classified by the manufacturers as follows:

1. Service entrance, type SE.
2. Elbow or turn fittings, type L.
3. Through fittings, type C.
4. Through fittings with 90-degree take-off, type T.

Boxes and Connectors

Steel or cast-iron outlet boxes are used in rigid conduit installations. Boxes normally used have knockouts which are removable for conduit insertion. Bushings and locknuts are provided for attachment of the conduit to the boxes, as shown in Fig. 1.

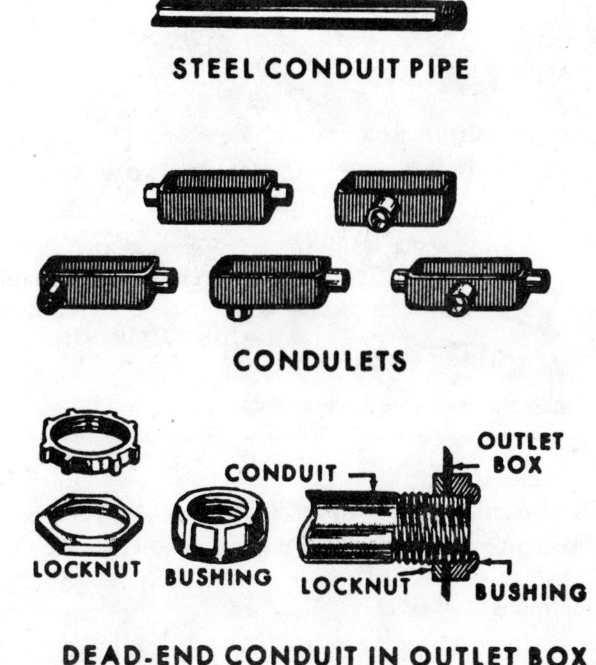

Fig. 1. Rigid conduit and fittings.

Boxes to be used in wet or hazardous locations must have threaded hubs into which the conduit is screwed.

Devices

The devices used in conduit installations are all box-mounted units and are covered in Chapter 3, Wiring Materials, in the section on electrical boxes.

Conduit Accessories

Threaded couplings. A threaded conduit coupling (Fig. 2) is furnished with each length of rigid conduit.

Threadless couplings. Rigid conduit may be installed using threadless couplings (Fig. 2), provided the couplings are installed tightly.

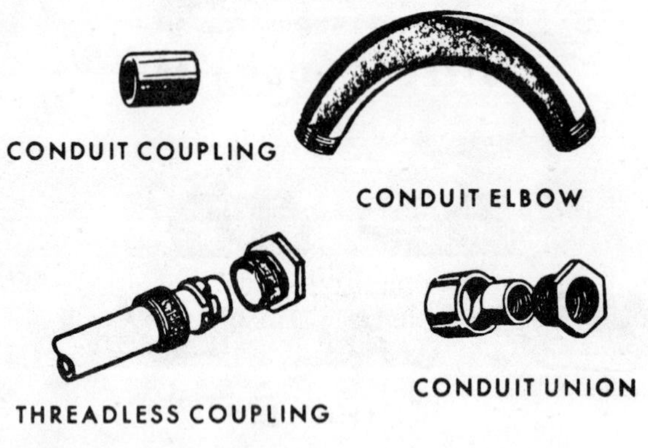

Fig. 2. Rigid conduit accessories.

Elbows

Standard conduit elbows (Fig. 2) are manufactured for use where 90-degree bends are required.

Conduit Unions

To permit the opening of a conduit at any point without sawing or breaking the conduit run, conduit unions (Fig. 2) are installed. With the use of unions, conduit may be started from two outlets and joined together at any convenient place in the run.

INSTALLATION

Bends

Bends of rigid conduit must be made without collapsing the conduit wall or reducing the internal diameter of the conduit at the bend.

Most bends are made on the job by the electrician as an integral part of the installation procedure. These are called field bends. The radius of the curve of the inner edge of any field bend must be at least 6 times the internal diameter of the conduit for rubber-, braid-, or thermoplastic-covered conductors, and not less than 10 times the internal diameter of the conduit for lead-covered conductors. Table 24, Appendix 2, shows the minimum radii for field bends. The maximum number of quarter bends for a conduit run between two openings is four. Moreover, a 10-foot length of conduit should have no more than three quarter bends. Factory-made bends may be used instead of bending conduit on the job. However, they are not generally used, since they increase the wiring cost; more conduit cutting and threading is required and additional couplings must be used.

Conduit up to and including ¾ inch is usually bent with a hand conduit bender called a hickey, as shown in Fig. 3. This can be slipped over the conduit. Conduit bending forms are also available as built-in units of pipe-vise stands. If neither of these tools is available, bends can be made using the lever advantage between two fixed posts or building members.

Fig. 3. Factory-made head for hickey.

The following procedure, illustrated in Fig. 4, is recommended as one method of making a right-angle bend in a length of ½-inch conduit. If a 90-degree bend is to be made in a length of conduit at a distance of 20 inches from one end, you must:

1. Mark off 20 inches from the end of the conduit.
2. Place the conduit hickey 2 inches in front of the 20-inch mark and bend the conduit about 25 degrees.
3. Move the bender to the 20-inch mark and bring the bend up to 45 degrees.
4. Move the bender about 1 inch behind the 20-inch mark and bring the conduit up to 70 degrees.
5. Move the hickey back about 2 inches behind the 20-inch mark and bring the bend up to 90 degrees.

Miscellaneous conduit bends (offset bends, Fig. 4) can be made more accurately if the contour of the bend is drawn with chalk on the floor and the bend in the pipe is matched with the chalk diagram as the bend is formed. Conduit in excess of one inch is usually bent by a hydraulic bender.

Cutting Conduit

Conduit can be cut with either a hacksaw or a standard pipe cutter. When a hand hacksaw is used, the conduit should be held in a vise, and care should be taken to make the cut at right angles to the length of the pipe. If a large number of conduits are to be cut, a power hacksaw is recommended. Cutting oil should always be used when cutting pipe.

Reaming Conduit

Irrespective of the cutting method used, a sharp edge always remains inside the conduit after cutting. Consequently, to avoid conductor damage, this edge must be removed before the conduit is installed. Pipe reamers or files, as illustrated in Chapter 2, Electrician's Tools and Equipment, Figs. 8 and 10, are generally used for the reaming operation.

Cutting Threads

Since the outside and inside diameters of rigid conduit are the same as those of gas, water, or steam pipes, the standard thread

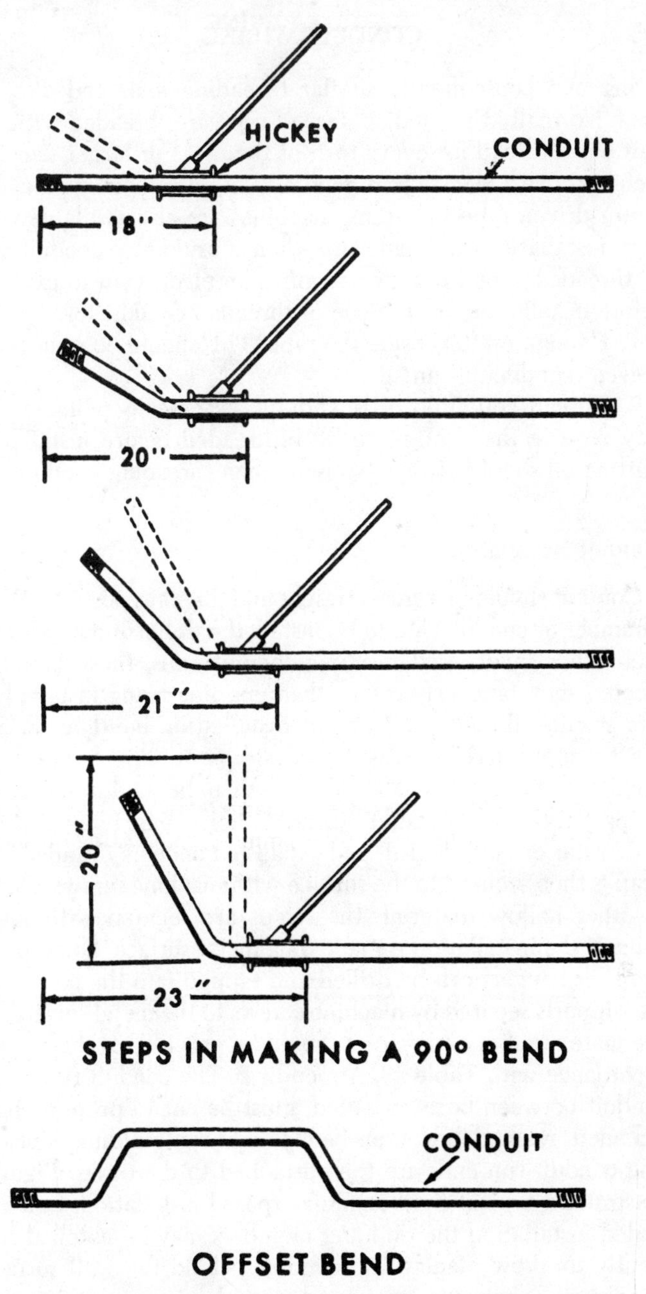

Fig. 4. Bending rigid conduit.

forms, and consequently similar threading tools and dies, are used. Normally the smaller sizes of pipe are threaded with dies that cut a thread for every turn of the die. For larger sizes (1½ inch and over) electricians generally use a ratchet type cutter. Motor-driven pipe-threading machines are also available when large installations are made and when considerable conduit must be threaded. Good practice requires an electrician to examine, before installation, each piece of threaded conduit for:

1. Foreign matter inside the pipe. This should be removed to prevent conductor damage.
2. Thread condition. Mishandling, extraneous paint, or dirt may require the conduit to be rethreaded before installation. Cutting oil should always be used when threading conduit.

Conduit Installation

Conduit should be run as straight and direct as possible. When a number of conduits are to be installed parallel and adjacent to each other in exposed multiple-conduit runs, they should be erected simultaneously rather than one line being installed before starting the others. Conduit installed on building surfaces can be supported by either pipe straps or pipe hangers. On wooden surfaces, nails or wood screws can be used to secure the straps. On brick or concrete surfaces, holes must be drilled first with a star or carbide drill and expansion anchors installed. The strap is then secured to the surface with machine screws. On tile or other hollow material, the straps are secured with toggle bolts. If the installation is made on a metal surface, holes for the straps or hangers can be drilled and tapped into the metal, and the supports secured by machine screws to the metal surface. An adequate number of supports should be provided at distances in accordance with Table 23, Appendix 2. The conduit run, as the conduit between boxes is called, must be cut to proper length, threaded, reamed, and then bent to suit the building contours. The conduit-run ends are then attached to the boxes. Figure 5 illustrates a typical rigid conduit exposed installation. In a concealed installation the building members may be notched sufficiently to allow placing the conduit behind the wall surfaces. Care must be taken to avoid undue weakening of the structure.

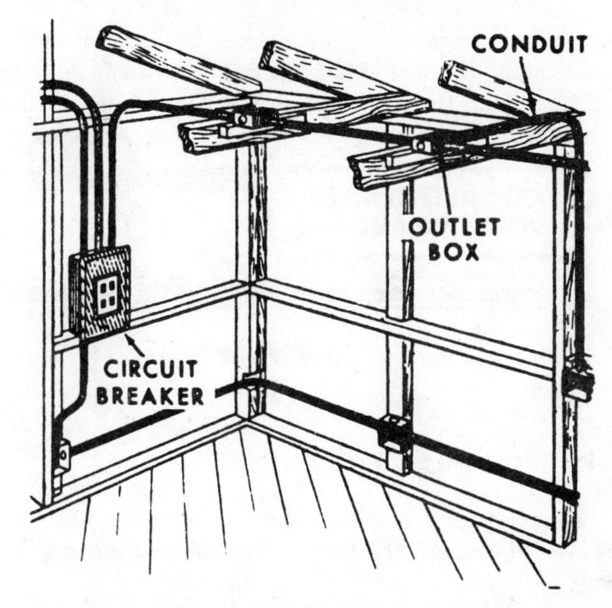

Fig. 5. Typical installation of conduit wiring.

Box Connection

When the boxes are of threaded-hub construction, the conduit ends are screwed into the box hubs and the conduit runs are connected at some midpoint by a coupling.

If the boxes are of knockout-type construction, they should be loosely located in the required position on studs and joists. A bowed locknut having teeth on one side should then be screwed onto the threads at the run ends of the conduit with the teeth of the locknut adjacent to the box. The conduit ends should next be inserted into the knockout openings. Bushings, which have smooth surfaces on their inside diameter to insure damage-free conductor installation, should then be screwed tightly onto the conduit ends in the boxes. Finally, the locknuts should be tightened against the boxes so that the teeth will dig into the metal sides of the box. This operation can be accomplished by driving a drive punch against one of the locknut lugs and forcing the locknut to move on the threaded conduit against the box. Figure 6 shows a standard box connection for conduit using locknuts

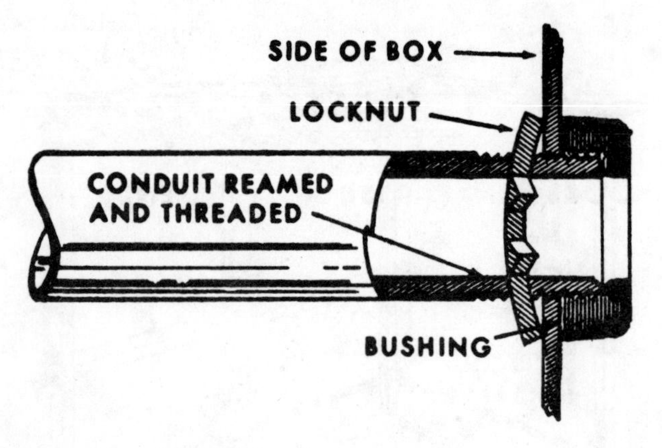

Fig. 6. Conduit and box connection.

and bushings. After this connection and all other box connections have been made, the box can be fastened securely to the building.

Wire Pulling

Upon installation of the boxes and conduit runs, the conductor wires should be pulled into the conduit. For short runs with few wires, conductors can be paired and pushed through the conduit run from box to box. When the conduit run has several bends and more than two conductors, a fish wire must be used in pulling wire. For normal runs the fish wire (or tape) is pushed through the conduit run from one end to the other. Occasionally, on long conduit runs, separate fish wires are used from either end of the conduit, as shown in Fig. 7. After the conductor ends are bared of insulation, they are wrapped around the fish wire and taped (Fig. 7) for pulling through the conduit. Taping of the fish wire and conductor junction is required to prevent damage to the conduit interior and existing conductors in the conduit. In taping, the joint is also compacted and strengthened, thus insuring easier pulling. For efficient and safe operation, wire pulling is generally a two-person operation. One electrician is required to pull the conductors through the conduit while the other feeds

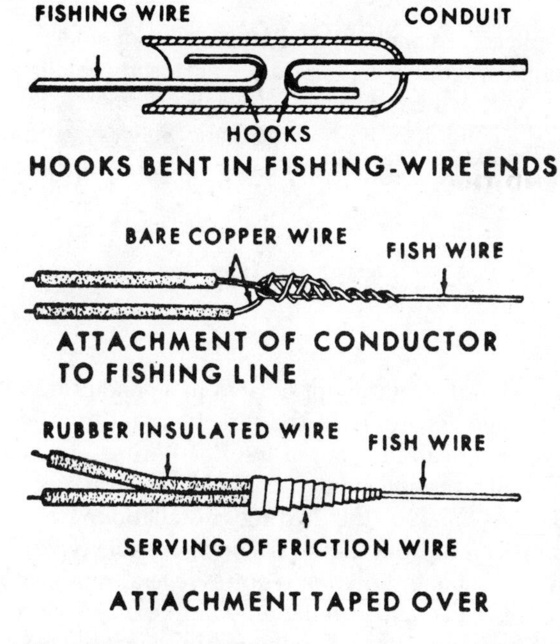

Fig. 7. Fish-wire pulling.

the conductors into the conduit. In this operation, care must be used in feeding and pulling the wires so that they maintain their same relative position in the conduit throughout the run length, thus avoiding insulation injury. For ease of operation, a wire lubricant such as powdered soapstone may be rubbed on the conductors or blown into the conduit. In intricate runs, wire pulling may be performed in sections between boxes. This procedure requires a large amount of additional splicing to be made in the boxes and requires that more time be taken in wiring. The preferred practice in wire pulling is to pull the conductors from the source through to the last box in the conductor run. Loops that extend about 8 inches from the box openings are made for each conductor that is to be tapped or connected to a device in the box. Conductors which are not to be tapped are pulled directly through the box to their connection.

Splices

Wire splices in conduit installations are not under tension, and a simple pigtail splice, carefully made to obtain a good electrical joint, can be used. No wire splices which will be concealed in the conduit runs are to be made. Splices would reduce the pulling area in a conduit and could easily be sources of electrical failure.

CIRCUITING

Layout

The layout and circuiting of devices in a conduit installation should be made according to the directions and procedures in Chapter 5, Design and Layout of Interior Wiring. The availability of different sizes of conduit along with their varying conductor capacities makes the wiring installation for conduit somewhat different from that of the open or cable types. For example, where cable installation requires several runs in a particular location, a conduit installation would use a single conduit with multiple conductors. Consequently, conduit layouts and runs should be planned to use the minimum amount of conduit possible and also keep the conductor runs to each outlet short enough to maintain a low voltage drop. Figure 8 shows a typical wiring layout in conduit.

Conductor Connection

No exceptions to the standard color coding of wires as outlined in the other systems are permitted in conduit wiring. All load utilization devices (fixtures, receptacles) operating at line to neutral voltage in a grounded neutral system must be connected to both a white and a black (or substitute color) wire. The white wire is always the grounded neutral wire. Black wires are the hot leads which are fused and connected to the switch when controlling power to a lampholder or outlet. Red-, blue-, and orange-colored insulation wire can be substituted for black wire when wires are combined in a conduit or circuit. The white wire

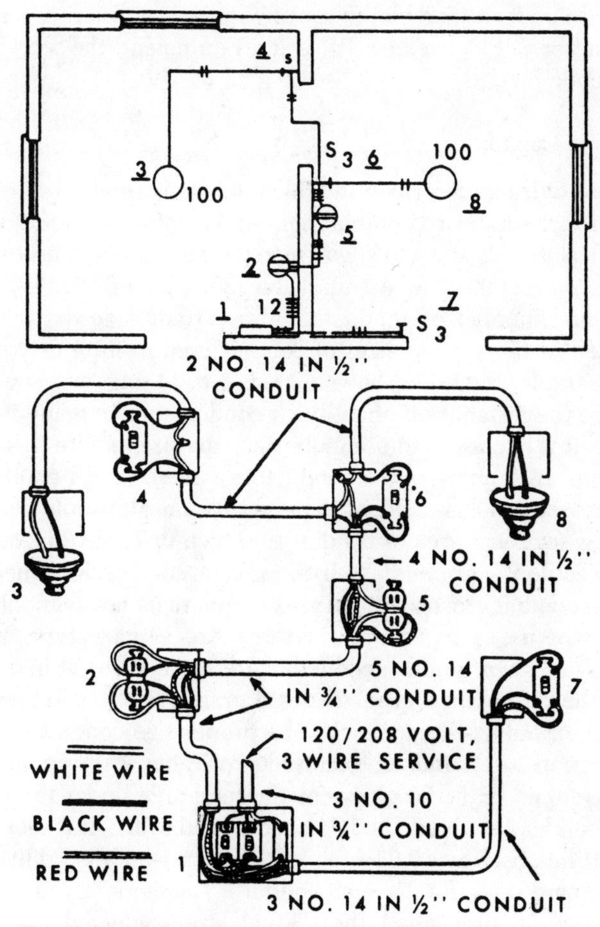

Fig. 8. Typical wiring layout in conduit.

must never be connected to a black or substitute color wire. The white wire must not be fused or switched except in a multipole device opening all conductors of the circuit simultaneously. A green-colored insulated conductor denotes a wire used to provide an auxiliary equipment ground. As an expedient measure, the ends of the wire insulation may be painted to obtain proper color coding when the colored insulation is not available. They may also be identified by the use of wire code markers described in Chapter 2, Electrician's Tools and Equipment, the section on wire code markers.

Conduit Capacity

Cable wiring, described in Chapter 8, is normally limited to two or three standard combinations of wire sizes. Conduit, however, has the capacity to accommodate many more than two or three conductors in one run. Table 25, Appendix 2, lists the maximum number of conductors of a certain gage that can be inserted in the various sizes of conduit used in interior wiring. For example, the table shows that six No. 14 gage wires would require the installation of a ¾-inch conduit run. In many installations it is necessary to use more than one size of wire in a conduit run. In such cases the conductors cannot have a combined or cross-sectional area equal to more than the allowable percent of cross-sectional area of conduit as shown in Table 26, Appendix 2. Table 27, Appendix 2, lists, for each size conduit, the percent of conduit cross-sectional area in square inches available for conductor use. For example, if three No. 10 gage type-R and four No. 8 gage type-R conductors are to be inserted in a conduit, their combined cross-sectional area is $3 \times 0.0460 + 4 \times 0.0760$, or 0.4420 square inch. The proper size conduit for this installation is 1¼ inches, according to Table 27, Appendix 2. This is found by first looking for the total area under the headings "not lead covered" and "4 conduit and over." It is seen that 0.4420 lies between 0.34 square inch (for 1-inch conduit) and 0.60 square inch (for 1¼-inch conduit). Consequently, as the 1-inch conduit is too small, the 1¼-inch size is selected.

Circuit Wiring

A fundamental law of electricity generation can be restated for wiring purposes as follows: When a conductor carrying current changes position or the current reverses direction in the conductor, it induces a current in an iron or steel conduit carrying the conductor. Consequently, if this conductor were isolated in an iron or steel conduit, the conduit would be heated by the induced current. This would result in considerable power loss. In an alternating current system, both wires of a circuit are encased in a single conduit, thereby causing the induced current of each to balance and cancel each other. To eliminate any possibility of induced heating of iron or steel conduit, both the wires of a circuit must travel in the same conduit. If the conductors in the circuit (Fig. 9) are run separately in this conduit, induced current (indicated by arrows) will flow through the conduit.

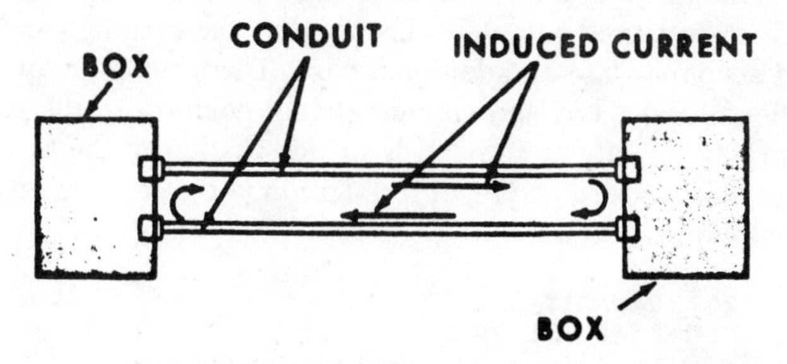

Fig. 9. Circulating current in conduit, showing induced-current flow.

ADDITIONS TO EXISTING WIRING

Increase of Circuit Amperage

A standard conduit installation has enough flexibility to accommodate a normal increase in circuit load even if an increase in circuit amperage is required. For example, a ½-inch conduit in a standard conduit-wiring installation generally carries two No. 14 gage conductors, which have a 15-ampere capacity. From Table 25, Appendix 2, it is evident that the ½-inch conduit can also accommodate two No. 12 gage conductors having a 20-ampere capacity. Consequently, if the load in an existing circuit must be increased, the No. 14 gage wire can be replaced by two No. 12 gage wires. Hence, when all the wires in the circuit are replaced, the amperage for the fuse or circuit breaker in the central fuse panel for the circuit can be safely increased from 15 to 20 amperes to accommodate the additional load.

Addition of New Circuit

When a new load is added to an existing building with conduit wiring and the circuit analysis indicates the need for a new circuit, the existing conduit in many cases can be used to carry the new circuit most of its distance. Figure 10 illustrates this principle. The new circuit is installed by pulling in an additional wire (red) from the circuit breaker panel to the existing outlet, and then adding the required outlet box beyond this location. The new load is connected to the additional circuit. Table 25, Appendix 2, was used to determine whether the existing conduit could accommodate an additional wire. The installation of the additional outlet box and conduit should conform to the rules and practices outlined previously in this chapter in the section on installation. In this type of installation, a common grounded neutral wire is used.

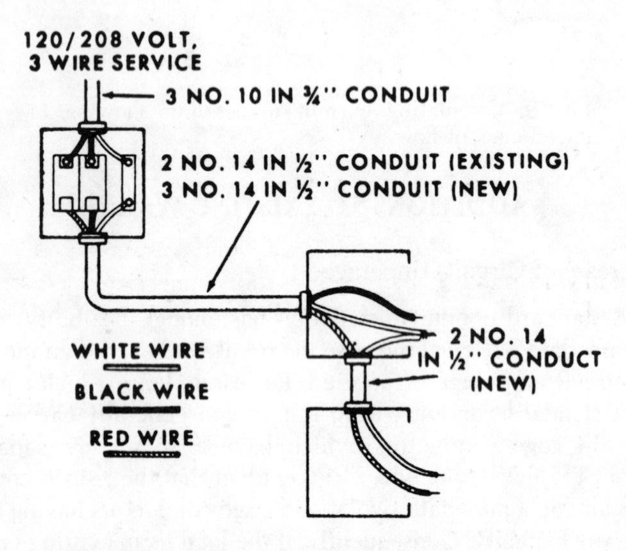

Fig. 10. Circuit addition in existing conduit.

THIN-WALL CONDUIT WIRING

Thin-wall conduit is a metallic tubing which can be used for either exposed or concealed electrical installations. Its use should be confined to dry interior locations. This is necessary because it has a very thin plating which does not protect it from rusting when exposed to the elements or humid conditions. It is less expensive than rigid conduit and much easier to install. The process of bending requires less effort, and it is not necessary for the ends to be threaded. In comparison with the other systems of wiring, it ranks behind rigid conduit, but ahead of the other types of wiring, when considering the quality and durability of the installation. For this reason, and because of the decreased cost in materials and labor, it is most generally specified for home building construction. It is installed in the same manner as rigid conduit, except that pressure-type couplings and connectors are used instead of threaded units.

MATERIALS

Thin-Wall Conduit and Fittings

Electrical metallic tubing (EMT), commonly called "thin-wall conduit," is more easily installed than rigid conduit. This conduit, as its name implies, has a thinner wall than rigid conduit but has the same interior diameter and cross-sectional area. EMT is available in sizes from ⅜ to 2 inches. The ⅜-inch size is used only for underplaster extensions. The inside surface is enameled to protect the wire insulation and minimize the friction in wire pulling. All couplings and connections to boxes are threadless and are of either the clamp or the compression type. Figure 11 illustrates thin-wall conduit and the fittings commonly used. Some fittings are similar to sleeves, and are secured to the conduit by an impinger tool that pinches circular indentations in the fitting. (*See* Chapter 2, Fig. 15.) This holds it firmly against the conduit. Others have threaded bushings which, when tightened, force a tapered sleeve firmly against the tubing. Figure 11 shows a box connector and coupling fitting of this type.

Wire Conductors

The same type, capacity, and maximum number of conductors per size of conduit previously given in Tables 23 through 27, Appendix 2, for rigid conduit also apply to thin-wall installations.

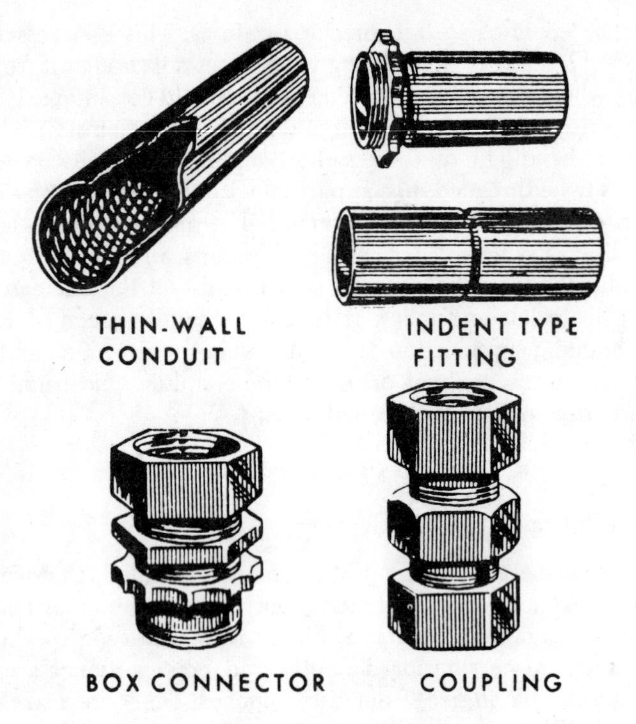

Fig. 11. Thin-wall conduit and fittings.

BENDING

Extreme care must be used when bending metallic tubing to avoid kinking the pipe or reducing its inside area. The radius of the curve of the inner edge of any field bend must not be less than 6 times the internal diameter of the tubing when braid-covered conductors are used, and not less than 10 times the interior diameter of the tubing when lead-covered conductors are used. This is the same rule that applies to rigid conduit. Table 24, Appendix 2, shows the minimum radii for field bends.

Construction

The thin-wall conduit bender (Fig. 12) has a cast steel head which is attached to a steel-pipe handle approximately 4 feet

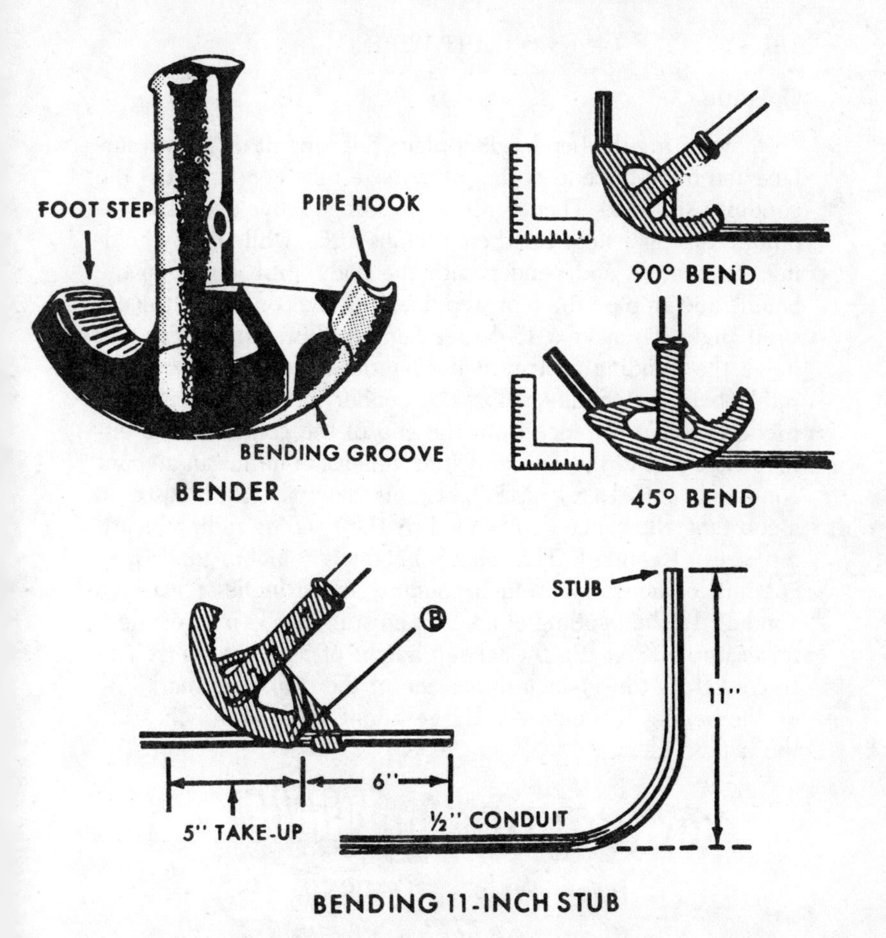

Fig. 12. Bending thin-wall conduit.

long. It is used in the field to form thin-wall conduit into standard and offset bends. Benders are made for each size of conduit and must be used only on those sizes for which they are designed. Each size bends the conduit to the recommended safe radius. The projection on the head of the bender, sometimes called a "foot step," is used to steady the bender in operation, and reduces the pressure required on the handle for bending. The numbers cast on the bender shaft are inch measurements and are used to check the depths of offset bends.

Operation

In operating the bender, first place the conduit on a level surface and hook the end of the proper size tube bender under the conduit's stub end. Then, with the bending groove over the conduit and using a steady and continuous force, while firmly holding the conduit and bender with the body, push down on the handle and step on the foot step, bending the conduit to the desired angle. To make a 45-degree bend in this manner (Fig. 12), move the bending tool until the handle is vertical. For accurately bending conduit stubs, the bender must be placed at a predetermined distance from the end of the conduit. This distance is equal to the required stub dimension minus an amount commonly called a takeup height. This takeup height is based on a constant allowance determined by the bending radii for various sizes of conduit. The takeup height is 5 inches for ½-inch conduit, 6 inches for ¾-inch conduit, and 8 inches for 1-inch conduit. In the bending of an 11-inch stub in a ½-inch conduit, for example (Fig. 12), the takeup height of 5 inches is first subtracted from the 11-inch dimension of the stub. The mark "B" on the bender is then set at the resultant value of 6 inches and the bends made.

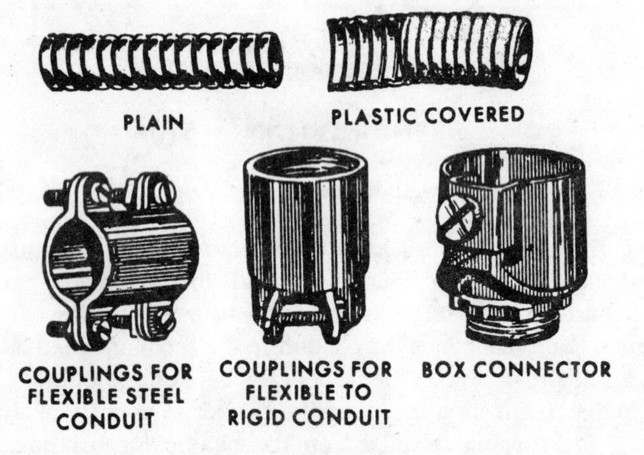

Fig. 13. Greenfield flexible conduit and fittings.

INSTALLATION

Thin-wall conduit may be cut either with a hacksaw or with a special thin-wall cutter. As with rigid conduit, the sharp edge in thin-wall tubing should be reamed after cutting to prevent premature wire damage. Exposed thin-wall conduit is supported in a similar manner and with the same type of supports as used with rigid conduit. Since there is no positive link between the couplings, box connectors, and thin-wall conduit, care must be taken during the installation to make sure each conduit joint is electrically and mechanically secure. All conduit ends must be inserted into the fittings until they touch the inner limiting edges. The fittings should then be tightened firmly, securely gripping the conduit walls. Care is also necessary to prevent the loosening of the conduit from the fittings, which could cause a loose connection, short circuit, or electrical fire at the point of wire and conduit contact. A mechanically loose conduit joint will not maintain the ground continuity required in an electrical wiring installation.

FLEXIBLE CONDUITS

Materials

Flexible metal conduit is generally called "Greenfield." It resembles armored cable in appearance, but it is more adaptable than cable, since various sizes and numbers of wires can be pulled into it after it is installed. Plastic-covered Greenfield may be used where the internal conductors are exposed to oil, gasoline, or other materials that have a deteriorative effect on the wire insulation. This metal conduit has a thermoplastic sheath covering similar to that used on type-T wire, the characteristics and uses of which are detailed in Table 8, Appendix 2. Figure 13 shows Greenfield conduit and the various fittings available.

Installation

Flexible conduit installation is similar to that covered previously in this chapter in the section for thin-wall conduit wiring, except that Greenfield must be supported more frequently, as an armored cable. Its high cost limits its use to connections between rigid wiring systems and movable or vibrating equipment such as motors or fans. It may also be installed where the construction requires a conduit bend which is difficult or impossible to make otherwise.

Chapter 10

Foreign Wiring

Most foreign wiring systems have not been installed according to prescribed standards, and consequently do not conform to the rigid limitations and practices outlined by the United States National Electrical Code. This fact may be attributed largely to the material shortages in most foreign countries, which have dictated the use and employment of materials at hand. In many instances these would be considered below standard or expedient substitutes in the United States. These limitations have been advantageous in some areas, such as the Scandinavian countries, for they have provided an incentive for electrical development which has resulted in more rapid advancement.

This chapter describes the major differences to be found and precautions needed when making wiring installations or using equipment from a foreign country. To properly illustrate and discuss the major differences in foreign wiring, United States standards are reviewed when necessary.

WIRING INSTALLATIONS

The wiring system generally used in most foreign countries is similar to open wiring installations. (*See* Chapter 6, Open Wiring, Knobs, and Tubes.) This is because of low material requirements. Other types of installations, outlined in the preceding chapters, are being installed in the more urban areas.

DISTRIBUTION

Voltages

Domestic. The United States uses nominal voltages which range from 120 to 240 volts for single-phase alternating current and 208 to 600 volts for three-phase alternating current, low-voltage distribution. Though these are considered to be standard voltage ratings because of their prevalent use, some locations and areas throughout the country still have direct-current systems or use alternating-current systems with nonstandard voltages.

Foreign. Since a considerable number of countries employ other voltages than those we accept as standard, the electrical equipment in use has to be converted, modified, or operated inefficiently when powered by foreign electrical installations.

Frequency

Domestic. The standard frequency of alternating-current distribution in the United States is 60 cycles.

Foreign. In most foreign lands, 50-cycle frequency generation is common, but the electrician may also encounter such frequencies as 25, 40, 42, and 100 cycles.

MATERIALS

Background

The wiring materials commonly used in foreign countries are usually peculiar to the territory's manufacture. In recent years the export of electrical goods from Germany, England, and the United States has been increasingly reflected in the established wiring patterns.

Wire

The United States employs the American Wire Gage system denoted by the symbol AWG, which is peculiar to our installations. Table 28 gives the metric diameter of AWG standard sizes.

Devices

The receptacles, switches, and plugs used in foreign wiring systems are also peculiar to the installations found in a particular area and normally cannot be mated or used with similar types of receptacles manufactured in the United States.

Section III

Maintenance, Troubleshooting, and Miscellaneous Equipment Maintenance

The rules and routines outlined for a maintenance program for any electrical wiring system are primarily determined by the selection, location, and installation of the original equipment installed. In a well-planned system, maintenance is merely a system of routines designed to keep the equipment in satisfactory operating condition through periodic inspection, cleaning, testing, tightening, adjusting, and lubricating. These basic maintenance operations should be set down in the above-listed order and the various duties performed to prevent operating breakdowns.

Chapter 11

Preventive Maintenance and Miscellaneous Equipment Maintenance

This chapter reviews and outlines the various procedures and recommended practices necessary to perform maintenance operation duties efficiently.

INSULATION

The insulation materials designed to shield or protect the conductors from accidental contact with other conducting substances are built into the conductor during manufacture, or may be installed as part of the system's installation. Since it is important to maintain these protective coatings or shields on the wire conductors, preventive maintenance should include periodic tests and checks to expose potential trouble locations where the wire insulation has become frayed or where protective devices have been damaged. These wire areas should be taped, repaired, or replaced as required.

Conductor shielding installed in the field, such as loom, anti-short bushings, and damage protection, should always be maintained, and should be replaced when dislodged or damaged.

Conductor or conductor-enclosure supports should also be inspected and maintained periodically to insure a trouble-free operation system. In case of damage, they should be replaced.

LOOSE FITTINGS

To avoid the possibility of short circuits, the maintenance organization responsible for power distribution in an electrical system should periodically spot-check the electrical fittings. These fittings include such items as conduit couplings, connectors, and box-entry devices. The fittings should be checked for looseness or separation and should be tightened or the conduit reclamped when necessary.

CONDUCTOR CONNECTION

The conductor connections made to electrical devices or other conductors should also be included in the periodic maintenance checks to determine the condition of solder splices, wire taps, or terminal connections. Loose, partly contacting, or partly broken connections at the screw terminals or splices of an electrical device can cause short circuits, arcing and burning, or radio interference. This may result in the rapid oxidation of the connecting materials. This may also result in a dangerous short circuit if the free wire contacts other metallic components which are grounded. Moreover, the increased resistance resulting from a loose or poor connection increases the voltage drop in the circuit, causing inefficient operation of the devices on the system. If this increased resistance in the wire or terminal connections is high enough, the heat resulting from the resistance in an electrical connection may reach a temperature which will ignite the surrounding materials and cause a fire. Consequently, all electrical connections should be kept tight.

DEVICES

The importance of periodically inspecting all operating devices for defects as a preventive maintenance function to forestall more serious difficulties must be stressed for all of the wiring systems discussed. As outlined in Chapter 5, Design and Layout of Interior Wiring, the inspection checks include normal operation and operation under rated load. Any devices that fail these tests, or that are broken or loosely supported in their mountings, should be replaced or repaired to prevent breakdown or potential danger to you or others nearby. If breakage repeatedly occurs in specific locations, the electrical devices should be replaced with items able to withstand the use intended, or the outlets or switches should be relocated.

MISCELLANEOUS EQUIPMENT MAINTENANCE

Housekeeping

Rotating equipment. All electrical rotating equipment is manufactured to operate at a particular temperature rated in degrees above ambient temperature. This term limits the maximum operating temperature of the equipment, which is derived by adding this rating to the atmospheric temperature of the operating location. For example, if a motor was rated at 30° F. above ambient, and the temperature of the surrounding area was 80° F. the maximum operating temperature of the motor would be 110° F. To help maintain the operating temperatures of the rotating equipment below a danger point, it is necessary to keep the equipment clean and dry. An excessive amount of dust or moisture on the equipment surface acts as an insulator, preventing the heat from being dissipated to the atmosphere through the housings of the equipment. Poor housekeeping conditions in a wiring area or wiring installation increase the possibility of short circuits.

Lighting. The efficiency of a lighting installation is also reduced when poor housekeeping conditions prevail. When dirt collects on the reflectors, lamps, walls, or ceilings, the initial or designed foot-candlepower of the installation drops. Though original installations are usually planned with an expected drop of 10 to 15 percent in candlepower, sporadic or unplanned cleaning may lower the lighting output as much as 50 percent. For proper maintenance under normal conditions, all fixtures and lamps should be cleaned at least once every three months. This period is shortened when necessary.

Lubrication

All rotating equipment, such as motors and fans, rotate in their housings on either ball, roller, or sleeve bearings. To insure maximum operating performance of this equipment, maintenance routines should include definite periodic lubrications according to the lubrication orders for the equipment. The data on the type of lubricant and lubricating period, as well as the points of lubrication on the equipment, are often attached to the equipment as an extra data plate.

Though it is important to lubricate the equipment at regular intervals, it is of equal importance not to overlubricate by using too much oil or grease, or by shortening the lubrication intervals. This can cause as much malfunctioning in the equipment as not lubricating at all. For instance, when oil or grease comes into contact with insulated conductors, their deterioration is hastened. Overlubrication also results in overheating and grease leakage. In addition, oily or greasy surfaces collect dust and abrasive materials in the air and, if not cleaned promptly, can cause wear on the bearing ends and eventual breakdown of the equipment.

Chapter 12

Troubleshooting

OPEN CIRCUITS

An open circuit occurs in a wiring system when one or more conductors in the circuit are broken, burned out, or otherwise separated. During operation, an open circuit is identified by the failure to operate of a part or all of an electrical circuit even though the fuses are not blown. The maintenance procedure for locating the source of the trouble is as follows:

1. Initially, a visual check should be made for a broken or loose connection at the first "dead" (nonoperating) outlet in the circuit. If a defective connection is found, the connection should be tightened or repaired. If necessary, a new wire may be spliced into the circuit at this point.

2. If the trouble or "open" is not found by a visual check, a test lamp should be used to determine whether the circuit is "alive" (operating) up to the point of the outlet. For this operation, two 115-volt lamps connected in series or one 230-volt lamp can be used.

3. When the circuit is open between the outlet and the fuse block in the power panel, a test lamp can be used to determine which wire is open. If the test prods are placed between a phase or hot wire and the conduit, cable sheathing, or ground connection, and the lamp does not glow, the phase or hot wire is open. If the phase or hot wire is alive up to the outlet, the neutral or ground lead may be open. If the neutral wire is believed to be open, a similar test cannot be made between the neutral wire and the conduit, cable sheathing, or ground, since they are at the same potential, and the test lamp would not light. In this case, if the neutral wire is in a two-wire circuit and is insulated, the

leads of the phase and neutral wire can be temporarily interchanged at the fuse block and the test outlined above for an "open" in the hot wire circuit is applicable. If this method is used, care must be exercised to make sure the neutral lead is returned to its former position in the fuse panel as soon as the test is over.

SHORT CIRCUIT OR GROUND

A short circuit results when two bare conductors of different potential come in contact with each other. If a conductor inadvertently contacts a metallic part of a wiring system such as a motor frame or conduit, as illustrated in Fig. 1, the system is sometimes said to be grounded instead of having a short circuit. Grounds or short circuits can be solid, partial, or floating.

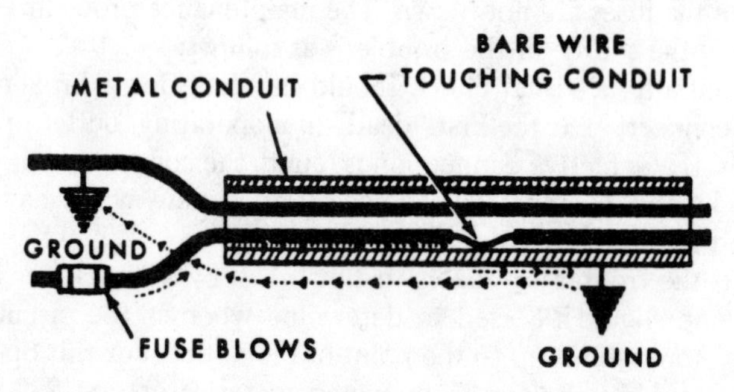

Fig. 1. Short circuit or ground.

Solid

A *solid ground* or short circuit is one in which a full voltage can be measured across the terminals of a blown fuse when the load is disconnected from the circuit. The circuit resistance in this case is very low and the current very high, so that the fuse blows or the circuit breaker trips immediately.

Partial

A *partial ground* or short is one in which the resistance in the short circuit path is partially lowered, but remains high enough to prevent enough current flow to blow the fuse or trip the circuit breaker. Grounds of this type are generally more difficult to locate than solid grounds. A partial ground or short in a light circuit operated from a two- or three-phase source can be detected by the dimness of some lights while others are operating normally.

Floating

A *floating ground* is a condition in which the resistance of the defect in the system varies from time to time. Grounds of this type may be present in an electrical system for some time before their existence becomes known. A floating ground is indicated when fuses are blown on the phase side of a circuit a number of times, and a circuit test shows no defects in the system. In grounds of this type, fuse trouble may not occur for several days. Then the ground reappears and the fuses are blown again.

LOCATING MALFUNCTION

Troubleshooting should be undertaken only when a malfunction occurs or an electrical circuit experiences repeated blowing of fuses. To determine the cause and location of the malfunction in a building-wiring circuit, proceed as follows:

1. Open the main switch on the fuse panel. Remove the blown fuse and screw a 25-watt light bulb (no larger) into the fuse receptacle on the phase or hot side of the circuit. Open the neutral lead on the defective circuit. Disconnect or turn off all load sources on the defective circuit. Close the main switch on the fuse panel. If the light in the fuse receptacle burns brightly, the phase or hot side of the circuit is grounded. Repair. If the light does not burn, the trouble has been isolated in one of the switch circuits or one of the disconnected loads.

2. Prior to troubleshooting the switch circuits or disconnected loads, make one additional test. Open the main switch. Reconnect the neutral wire. Close the main switch on the fuse panel. If the light in the fuse receptacle burns brightly, the phase or hot side of the circuit is shorted to the neutral. While

this malfunction is rare, it is a possibility. Repair. If the light in the fuse receptacle does not burn, this indicates the malfunction is in the switch circuits or in one of the disconnected loads.

3. Since all portable equipment has been disconnected and all switches turned off, troubleshooting time can be reduced by giving prime consideration to the portable equipment. Approximately 90 percent of the short circuits on an interior wiring system occur in motors or in the flexible cord feeding the fixtures or electrical devices. In many cases a physical check of the flexible cord will indicate the trouble area. Repair.

4. Assuming that the inspection to this point has not indicated the source of the trouble, proceed as follows:

Open the main switch on the fuse panel. Remove the 25-watt bulb. Check to make sure the hot side and neutral leads are properly connected. Insert a new fuse. Close the main switch on the fuse box. The fuse should not blow. A fuse is used during this portion of the test to insure circuit protection and sufficient motor starting current.

Starting with the portable equipment nearest the fuse box, reconnect it to the circuit. If the equipment functions properly, the circuit receptacle and equipment are good. Disconnect the equipment under test and reconnect a different piece of equipment. Continue this process until all equipment has been checked out or until such time as a piece of equipment does not function. Immediately disconnect the equipment and check the fuse. A blown fuse would indicate that the equipment or receptacle circuit is bad. Repair.

Assuming that all equipment has been checked out and found to function properly, the malfunction must be in one of the switch or lighting circuits. Proceed as follows:

1. Make a quick check to insure that all equipment is still disconnected from the circuit.

2. Open the main switch on the fuse panel. Replace the fuse with a 25-watt light bulb and close the main switch.

3. Taking each switch-controlled lighting circuit in turn, remove the lamps and all other energy-consuming devices. Place the switch in the ON position. This completes the electrical circuit from the fuse holder to the hot side of the lighting outlet or

switch leg. If the lamp in the fuse box lights, the malfunction has been discovered. Repair. If the lamp does not light, the circuit can be assumed to be good and the test is continued with other switch circuits.

4. If several fixtures are connected to a circuit and not controlled by remote switches, each individual fixture is normally controlled by a pull chain. Follow the same testing procedure for this type of fixture as was followed for the circuit controlled by the remote switch.

During this troubleshooting procedure, the individual load or a lamp in the fuse box has been used as an indicator device. On occasion, it may not be possible to use these devices to troubleshoot a circuit. When this is the case, a self-contained power bell–buzzer arrangement or test lamp may be used.

Chapter 13

Homeowner Tips on Electrical Wiring

Many electrical maintenance and repair jobs around a home do not justify calling an electrician, but can be done by the homeowner who will take time to learn what is involved.

Circuit Fuses and Breakers

Probably the most baffling part of the wiring system to many people is the fuse box, and it is likely that the fuse box will need attention sooner or later in most homes.

Electric current enters the home through the main fuses (or circuit breakers), then goes through the individual circuit fuses or breakers to the circuits. The *purpose* of fuses or breakers is to limit the current flow, in case of a fault, and prevent overheating of the wiring.

To save time when trouble strikes, it is a good idea to label all the circuits in a fuse box as to what they serve. This can be determined by disconnecting circuits one at a time (by removing the fuse or switching the circuit breaker off), and checking to see what does not work by operating the switches and plugging in a portable lamp around the house.

If there is no room to put labels directly on the box, you can place a number by each fuse and fasten a listing on a sheet of paper inside the door of the fuse box. Then, when power is off in part of the wiring system, you can quickly find which fuse or circuit breaker is involved.

The circuit protection you find in the fuse box may be cartridge fuses, screw-in fuses, or circuit breakers. Fuses must be replaced when they blow, but when a circuit breaker trips you need only push it all the way to the OFF position, then back ON, to restore service. It is wise to disconnect all appliances from the circuit before doing this, or it will probably trip the breaker or blow the fuse again. If all equipment is disconnected, and it happens again, the trouble is likely to be in the wiring, and you probably should call an electrician.

There are several kinds of *screw-in fuses.* Besides the ordinary screw-in fuse with brass threads, there is the "non-temperable" fuse (usually labeled "fustat"), which has porcelain threads and must be replaced with one of the same ampere rating (the wrong size will not fit). Amperage is stamped on the top of the fustat, and different sizes are color-coded so that if you replace with the same color fustat, it should work.

When the fustat is missing, you can look into the bottom of the hole and read the amperage needed. If you read "SA20," it uses a 20-amp fustat. Fuses and circuit breakers protect electric wires from overloads and short circuits. *Never* replace a fuse with one of a larger size (Fig. 1).

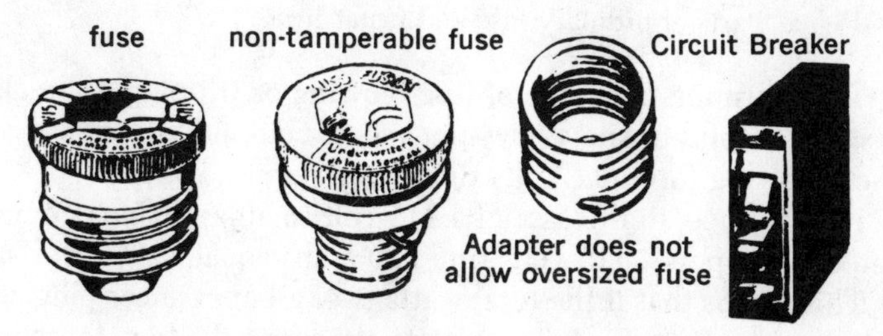

Fig. 1. Fuses and circuit breaker.

Fustats perform better on circuits that serve electric motors, since they will carry several times their rated current for a few seconds and will handle the high starting current of motors. A fustat will also tell you what caused it to blow.

Inside a fustat are two elements. One element is like an ordinary fuse link, and the other, called a thermal element, looks like a drop of solder. There is a small spring pulling against the top of the conductor where it enters the thermal element.

If the fuse link is partly missing, or if the transparent window is black, this tells you that a short circuit caused it to blow. If the spring has contracted, separating at the thermal element, it means that the cause was an overload. Usually information and instructions are printed on the box in which you buy fustats. Parts of an ordinary fuse and the newer fustat fuse (also called Type S fuse) are shown in Fig. 2. The fustat will carry temporary overloads (such as motor starting currents) better than an ordinary fuse. It cannot be replaced by one of higher amp rating.

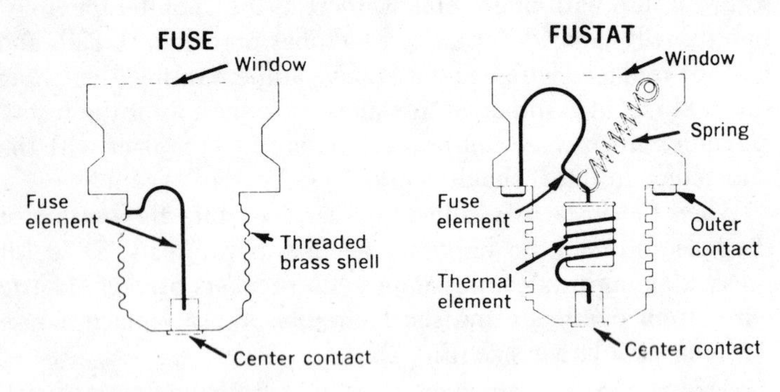

Fig. 2. Parts of ordinary fuse and fustat fuse.

To determine the cause of fuse-blowing or tripping of breakers, it is helpful to know how much load a circuit will carry. Use the following formula: Volts times amps equals watts.

For instance, in the case of a 115-volt small-appliance circuit rated 20 amperes, 115 volts times 20 amps equals 2,300 watts.

This means that if the total wattage of all appliances plugged into this circuit exceeds 2,300 watts, the fuse will blow. Most appliances have the watts stamped on them somewhere. Motors may be labeled with amps. In the case of our 20-amp circuit, it would work as follows:

Coffee maker	600 watts
Toaster	1,000 watts
Frypan	1,100 watts
Total	2,700 watts

If all three appliances are operated at the same time, you can expect the fuse to blow. To operate the three at one time, you will need two circuits.

Plugs and Cords

Probably more electrical problems originate in plugs and cords than anywhere else.

Cords should be checked periodically to be sure they are in good condition. To check a cord, after disconnecting it, pull the length of the cord around your finger, watching for cracks, worn spots, or a point where it bends too easily, indicating a broken wire.

Cords that have part of the insulation worn off can be restored by wrapping with plastic electrician's tape, but cords that are broken or cut should not be spliced. A good splice takes quite a bit of time and few people will do it right. Rather, put a plug on one cord and a cord-end receptacle on the other. Make certain that you do not get a plug on both ends of a section of cord.

When replacing cords, be sure you get the type of cord that is suitable for the intended use, and of the correct wire size.

TABLE 6

ALLOWABLE CURRENT-CARRYING CAPACITIES AND WATTAGES

Wire Size	*Amps*	*Watts Load on 115 Volts*	*Approximate Thickness*
No. 18	5 amps	575 watts	
No. 26	7 amps	805 watts	
No. 14	15 amps	1,725 watts	One Penny
No. 12	20 amps	2,300 watts	One nickel
No. 10	30 amps	3,450 watts	Two dimes

Capacities shown above are for ordinary rubber-covered cords. For special heater cords, capacities are higher. Wire sizes are usually stamped on the cord itself or on the reel on which it is purchased by the store.

There are a number of types of cords, and a replacement should be of the type it is replacing, if the original was correct. Here are some cord types and uses:

Parallel cord—types SP and SPT—is used on lamps, radios, and other light loads. *Junior hard service cord*—types SJ, SJT, and SJO—is a round, jacketed cord, used on appliances such as washing machines, drills, and trouble lamps. The SJ has a rubber jacket, the SJT a thermoplastic jacket, and the SJO an oil-resistant jacket material. Another type, *hard-service cord,* types S, ST, and SO, has similar uses, but for rougher service.

Heater cords—types HPN and HSJ—are used on heating appliances. Type HPN has a neoprene insulation, and HSJ has asbestos insulation with a rubber outer jacket. *Asbestos-insulated* cord with a cloth outer braid is also used. The type with gold thread will stand the most flexing, the one with red thread less flexing, and white thread the least flexing.

There are also *special cords,* as for a range or dryer, usually made with a molded cap to fit the proper receptacle.

Christmas tree lighting strings and cords are made for either indoor or outdoor use. Outdoor lights can be used inside, but indoor lights must not be used outside, because of moisture.

When replacing the cord in a plug, be sure to run the wires around the prongs and clockwise under the screws. Wires stay under the screws better for tightening, and running wires around the prongs helps relieve the stress on the individual strands where they are held down by the screw (Fig. 3). Heavy-duty plugs, which clamp to the cord and will stand more pulling, are available.

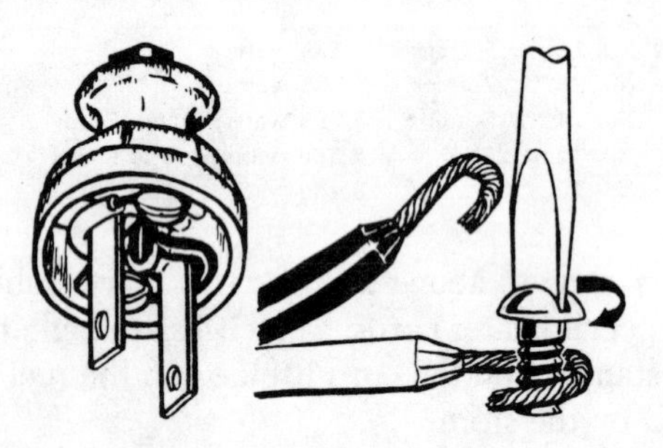

Fig. 3. Wiring a plug.

If the plug or cord you are replacing is of the three-wire grounding type, you must be very careful to replace with a three-wire cord and plug, and put the green-insulated wire under the green screw in the plug. The other end of the green-insulated wire fastens to the frame of the appliance, so it is very important that it be connected correctly. If connected wrong, it could energize the appliance and produce a serious shock. Conventional color coding calls for connecting the red or black wire to the brass-colored screw and the white wire to the silver-colored screw.

When purchasing appliance cords, remember that it may not be necessary to have a full-length cord where the appliance is used on the kitchen counter. You can buy coiled cords that will retract like a telephone cord, and short cords only about two feet long. This reduces cord clutter on the kitchen counter.

Switches

Switches (Fig. 4) wear out eventually, and can be replaced if care is taken to put things together the same way that they came apart. Switches are available with a pilot light to be seen in the dark, and you can get switches that operate silently. Also available are "dimmer" switches that allow turning lights up and down as you wish. These fit into a standard switch box, and are efficient in that they are solid-state devices that do not waste power, as would a rheostat.

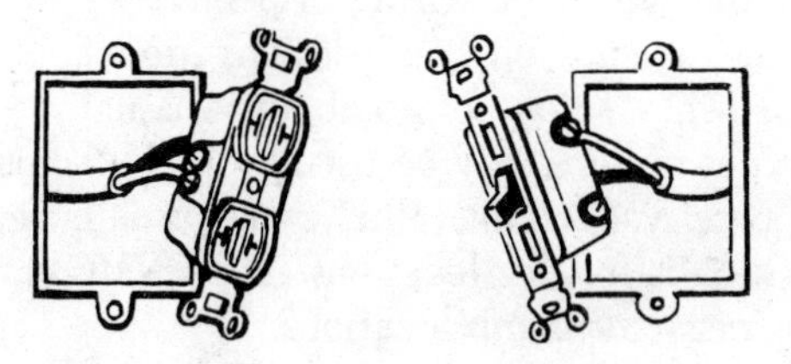

Fig. 4. Switches and receptacles.

Receptacles

Receptacles (Fig. 4) also wear out, though this is not always realized. Bending the prongs of a loose-fitting plug is only a temporary solution at best, and on a heating device could cause overheating of the plug. What usually happens is that the contacts inside the receptacle no longer exert pressure on both sides of the prong. A new receptacle is the answer.

If the receptacle is the three-prong grounding type, be sure the bare or green grounding wire is connected under the green, hexagon-shaped screw. The white wire goes to the silver screw, and the black or red wire to the brass-colored screw.

When purchasing electrical parts, be sure they carry the label of Underwriters' Laboratories. The UL label means that the cord or device has met minimum safety requirements for the purpose intended. It could still be unsafe if improperly used. For instance, a brass-shelled socket is acceptable for a lamp for use only in dry places, not for use in a basement or outdoors. The cardboard separator can become wet and conduct electricity to the shell.

The most common receptacle is the *parallel-blade type*, which is intended for use on 115 volts. There are several types of receptacles for 230-volt use, depending on the size of the electrical load, ranging from 15 amps up to 50 amps.

Unfortunately, in many homes more circuits are needed. Adding a receptacle may increase the convenience of using electricity, but it will not add anything to the capacity of that circuit.

Check with your electric power supplier on your legal status in doing wiring. In most states you can do your own, but the law may require that you have wiring inspected.

Safe wiring depends on good design, proper materials, and good workmanship, so do not go about it casually.

Several types of wire may be used around a home. The most common is type *NM* (nonmetallic) cable, usually referred to as "romex," which is for dry locations. Type *NMC* is nonmetallic cable for corrosive or damp locations.

Type *AC* (armored cable) has a metal armor spiralled around the wires, and is required for home wiring in some areas. Type *UF* (underground feeder) is meant for burial underground, and usually is also labeled for NMC use.

Cable comes stamped with the type number, number of wires, and size of wire stamped on the outside every foot or two. For instance, "NM 12/2 w.g." would mean type NM cable, wire size 12, two insulated wires with one bare wire for grounding. Sometimes wiring is run inside electrical conduit (pipe) where mechanical protection is needed.

Receptacles on the outside of buildings require a weatherproof box and cover. They usually have spring-loaded covers with rubber gaskets to keep out moisture.

Ground-Fault Interrupter

The National Electrical Code for new outdoor receptacles in residential locations requires that a "Ground-Fault Interrupter" (referred to as GFI) be used. (The National Electrical Code is a voluntary standard for the electrical and insurance industries, which has been made compulsory in many states.)

The GFI (Fig. 5) is a device which will sense the fact that current is going to ground (such as through a person receiving a shock) and shut off the power before the person is injured. These are available as portable plug-in units, as separate devices in a metal box, and as part of a circuit breaker for mounting in a breaker box.

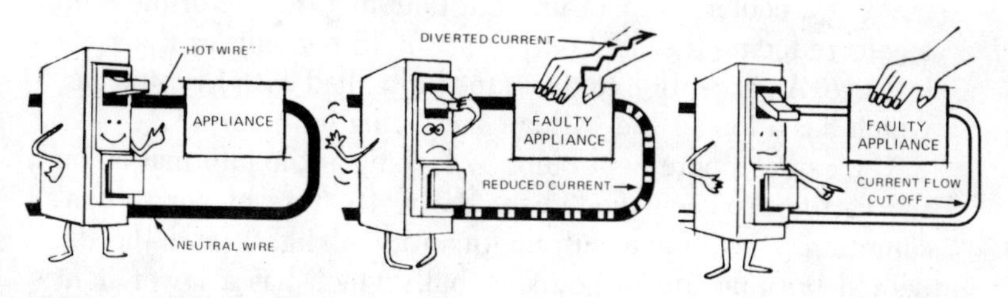

Fig. 5. Ground-fault interrupter.

It must be mentioned that a GFI does not protect from all kinds of shock, such as from grasping one wire in each hand, but since shocks usually involve current flow to ground, it will protect against most of them.

Each GFI has a test button. When pressed, it simulates a current flow to ground, so you can check periodically to be sure the GFI is working.

Outdoor Electrical Cords and Fittings

Electrical cords and fittings used *outdoors* should be of the nonmetallic type, or weatherproofed and properly grounded.

Generally, outdoor lights require porcelain-base sockets, and cords should be heavy-duty rubber. Permanent outdoor weatherproof receptacles can be installed for plugging in outdoor lighting. If fixtures are intended for outdoor use, they will carry the UL label and a statement as to their purpose.

Light Bulbs and Fluorescent Tubes

A great variety of light bulbs and fluorescent tubes is available for replacement.

Light bulbs may be clear, frosted inside, or all-white. For enclosing in a diffuser, the clear bulb is acceptable. Where the bulb will be visible, the all-white is preferred, since its "hot spot" brightness is less. Also, the all-white bulb does not darken as it is used.

Most light bulbs are designed to burn from 750 to 1,000 hours. It is possible to build a bulb to last longer, by operating the filament at a cooler temperature, but doubling the life of the bulb means reducing the light output about 15 percent for the same wattage. An exception to this is the bulb filled with Krypton gas, which has a longer life without sacrificing light output.

You can compare light bulbs by looking at the information on the carton. You will find there the watts (rate of power consumption), the lumen output (quantity of light), and the expected burning life in hours. A bulb which has a lower light output but a longer life may be a logical choice for a hard-to-reach place, such as a yard light.

Fluorescent lamps are available in color ranges from "cool-white" to "warm-white." The "warmer" the lamp, the more pleasing are the colors from skin and food. The "cool" lamp, though efficient, makes people and food look less attractive. For most home uses, a "warm white" or "Deluxe Warm White" lamp is advised.

Section IV

Batteries and Transformers

Selecting a *battery* can be as simple as buying a cell for a pen light—or as complicated as specifying a source of stored energy for a satellite transmitter. Although the many kinds of batteries may seem to make a proper choice difficult, the problem can be somewhat simplified by first outlining the application requirements and then matching a battery to the job.

In alternating current circuits it is possible to change the value of the voltage by means of *transformers.* A transformer is a device consisting of a magnetic iron core about which are wound electrical coils for input (primary) and output (secondary) voltages. The number of windings on the input coil is in the same ratio to those of the output coils as the input voltage is to the output (or voltage to be used). This property of transformers makes possible efficient utilization of power for plant power, lighting systems, and so on.

Section V

Batteries and

Chapter 14

Batteries

Batteries are widely used as sources of direct-current electrical energy in automobiles, boats, aircraft, ships, portable electric/electronic equipment, and lighting equipment. In some instances, they are used as the only source of power, while in others they are used as a secondary or standby power source.

A battery consists of a number of cells assembled in a common container and connected together to function as a source of electrical power.

CELL

A *cell* is a device that transforms chemical energy into electrical energy. The simplest cell, known as either a galvanic or a voltaic cell, is shown in Fig. 1. It consists of a piece of carbon (C) and a piece of zinc (Zn) suspended in a jar that contains a solution of water (H_2O) and sulfuric acid (H_2SO_4).

The cell is the fundamental unit of the battery. A simple cell consists of two strips, or electrodes, placed in a container that holds the electrolyte.

Electrodes

The *electrodes* are the conductors by which the current leaves or returns to the electrolyte. In the simple cell, they are carbon and zinc strips that are placed in the electrolyte, while in the dry cell (Fig. 2), they are the carbon rod in the center and the zinc container in which the cell is assembled.

Electrolyte

The *electrolyte* is the solution that acts upon the electrodes which are placed in it. The electrolyte may be a salt, an acid, or

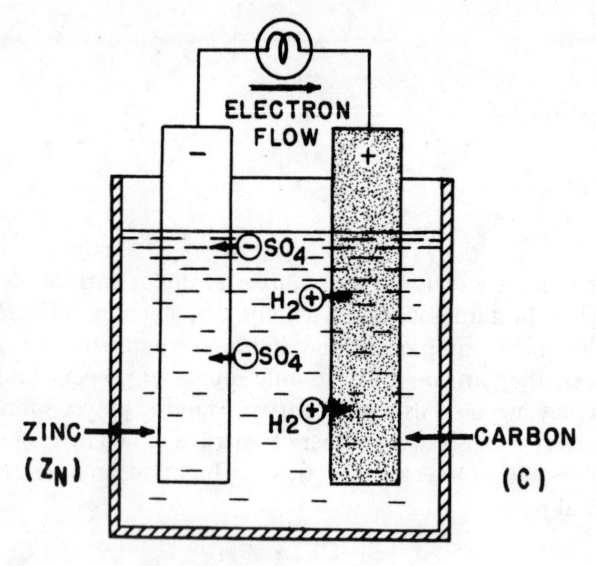

Fig. 1. Simple voltaic cell.

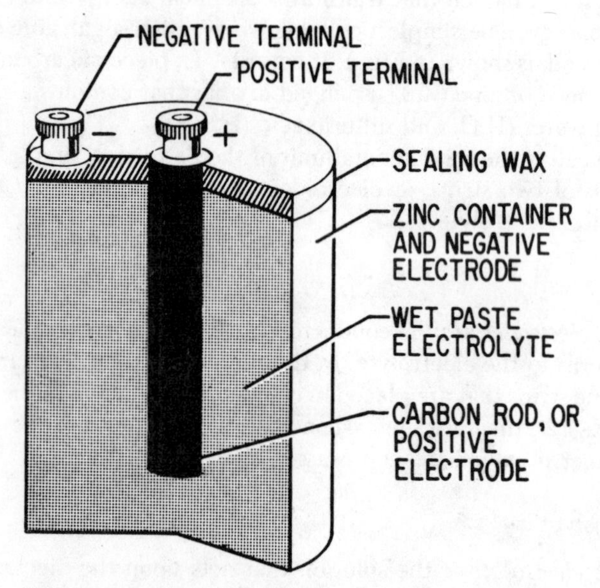

Fig. 2. Dry cell, cross-sectional view.

an alkaline solution. In the simple galvanic cell and in the automobile storage battery, the electrolyte is in a liquid form, while in the dry cell, the electrolyte is a paste.

Primary Cell

A *primary cell* is one in which the chemical action eats away one of the electrodes, usually the negative. When this happens, the electrode must be replaced or the cell must be discarded. In the galvanic cell, the zinc electrode and the liquid solution are usually replaced when this happens. In the case of the dry cell, it is usually cheaper to buy a new cell. Some primary cells have been developed to the state where they can be recharged.

Secondary Cell

A *secondary cell* is one in which the electrodes and the electrolyte are altered by the chemical action that takes place when the cell delivers current. These cells may be restored to their original condition by forcing an electric current through them in the opposite direction to that of discharge. The automobile storage battery is a common example of the secondary cell.

BATTERY

As previously mentioned, a battery consists of two or more cells placed in a common container. The cells are connected in series, in parallel, or in some combination of series and parallel, depending upon the amount of voltage and current required of the battery. The connection of cells in a battery is discussed in more detail later in this chapter.

Battery Chemistry

If a conductor is connected externally to the electrodes of a cell, electrons will flow under the influence of a difference in potential across the electrodes from the zinc (negative) through the external conductor to the carbon (positive), returning within the solution to the zinc. After a short period of time, the zinc will begin to waste away because of the "burning" action of the acid. If zinc is surrounded by oxygen, it will burn (become oxidized) as a fuel. In this respect, the cell is like a chemical furnace in which energy released by the zinc is transformed into electrical energy rather than heat energy.

The voltage across the electrodes depends upon the materials from which the electrodes are made and the composition of the solution. The difference of potential between carbon and zinc electrodes in a dilute solution of sulfuric acid and water is about 1.5 volts.

The current that a primary cell may deliver depends upon the resistance of the entire circuit, including that of the cell itself. The internal resistance of the primary cell depends upon the size of the electrodes, the distance between them in the solution, and the resistance of the solution. The larger the electrodes and the closer together they are in solution (without touching), the lower the internal resistance of the primary cell and the more current it can supply to a load.

When current flows through a cell, the zinc gradually dissolves in the solution and the acid is neutralized. A chemical equation is sometimes used to show the chemical action that takes place. The symbols in the equation represent the different materials that are used. The symbol for carbon is C, and for zinc, Zn. The equation is quantitative and equates the number of parts of the materials used before and after the zinc is oxidized. All matter is composed of atoms and molecules, with the atom being the smallest part of an element and the molecule the smallest part of a compound.

A compound is a chemical combination of two or more elements in which the physical properties of the compound are different from those of the elements comprising it. For instance, a molecule of water (H_2O) is composed of two atoms of hydrogen (H_2) and one atom of oxygen (O). Ordinarily, hydrogen and oxygen are gases; but when combined, as just mentioned, they form water, which normally is a liquid. However, sulfuric acid (H_2SO_4) and water (H_2O) form a mixture (not a compound), because the identity of both liquids is preserved when they are in solution together.

When a current flows through a primary cell having carbon and zinc electrodes and a dilute solution of sulfuric acid and water, the chemical reaction that occurs can be expressed as

$$Zn + H_2SO_4 + H_2O \xrightarrow[\text{discharge}]{} ZnSO_4 + H_2O + H_2\uparrow$$

The expression indicates that as current flows, a molecule of zinc combines with a molecule of sulfuric acid to form a molecule of zinc sulfate ($ZnSO_4$) and a molecule of hydrogen (H_2). The zinc sulfate dissolves in the solution and the hydrogen appears as gas bubbles around the carbon electrode. (A gas is designated by the arrow pointing upward in the equation.) As current continues to flow, the zinc is gradually consumed and the solution changes to zinc sulfate and water. The carbon electrode does not enter into the chemical changes taking place but simply provides a return path for the current.

In the process of oxidizing the zinc, the solution breaks up into positive and negative ions that move in opposite directions through the solution (Fig. 1). The positive ions are hydrogen ions that appear around the carbon electrode (positive terminal). They are attracted to it by the free electrons from the zinc that are returning to the cell by way of the external load and the positive carbon terminal. The negative ions are SO_4 ions that appear around the zinc electrode. Positive zinc ions enter the solution around the zinc electrode and combine with the negative SO_4 ions to form zinc sulfate ($ZnSO_4$), a grayish-white substance that dissolves in water. At the same time that the positive and negative ions are moving in opposite directions in the solution, electrons are moving through the external circuit from the negative zinc terminal, through the load, and back to the positive carbon terminal. When the zinc is used up, the voltage of the cell is reduced to zero. There is no appreciable difference in potential between zinc sulfate and carbon in a solution of zinc sulfate and carbon in a solution of zinc sulfate and water.

Polarization. The chemical action that occurs in the cell (Fig. 1) while the current is flowing causes hydrogen bubbles to form on the surface of the positive carbon electrode in great numbers until the entire surface is surrounded. This action is called *polarization.* Some of these bubbles rise to the surface of the solution and escape into the air. However, many of the bubbles remain until there is no room for any more to be formed.

The hydrogen tends to set up an electromotive force in the opposite direction to that of the cell, thus increasing the effective internal resistance, reducing the output current, and lowering the terminal voltage.

A cell that is heavily polarized has no useful output. There are several ways to prevent polarization from occurring or to overcome it after it has occurred. The simplest method might be to remove the carbon electrode and wipe off the hydrogen bubbles. When the electrode is replaced in the electrolyte, the emf and current are again normal. However, this method is not practicable, because polarization occurs rapidly and continuously in the simple voltaic cell. A commercial form of voltaic cell, known as the dry cell, employs a substance rich in oxygen as part of the positive carbon electrode, which will combine chemically with the hydrogen to form water, H_2O. One of the best depolarizing agents used is manganese dioxide (MnO_2), which supplies enough free oxygen to combine with all of the hydrogen so that the cell is practically free from polarization.

The chemical action that occurs may be expressed as

$$2MnO_2 + H_2 \rightarrow Mn_2O_3 + H_2O$$

The manganese dioxide combines with the hydrogen to form water and a lower oxide of manganese. Thus the counter electromotive force (emf) of polarization does not exist in the cell, and the terminal voltage and output current are maintained at normal.

Local action. When the external circuit is opened, the current ceases to flow, and theoretically all chemical action within the cell stops. However, commercial zinc contains many impurities, such as iron, carbon, lead, and arsenic. These impurities form many small cells within the zinc electrode in which current flows between the zinc and its impurities. Thus the zinc is oxidized, even though the cell itself is an open circuit. This wasting away of the zinc on open circuit is called *local action.* For example, a small local cell exists on a zinc plate containing impurities of iron, as shown in Fig. 3. Electrons flow between the zinc and iron, and the solution around the impurity becomes ionized. The negative SO_4 ions combine with the positive Zn ions to form $ZnSO_4$. Thus the acid is depleted in solution and the zinc consumed.

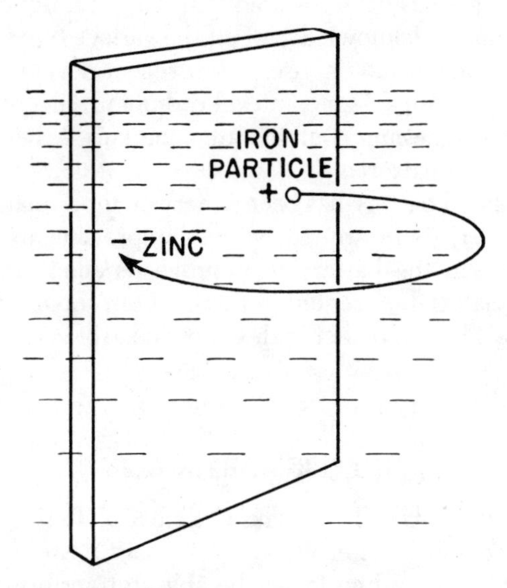

Fig. 3. Local action on zinc electrode.

Local action may be prevented by using pure zinc (which is not practical), by coating the zinc with mercury, or by adding a small percentage of mercury to the zinc during the manufacturing process. The treatment of the zinc with mercury is called amalgamating (mixing) the zinc. Since mercury is 13.5 times as heavy as an equal volume of water, small particles of impurities having a lower relative weight than that of mercury will rise (float) to the surface of the mercury. The removal of these impurities from the zinc prevents local action. The mercury is not readily acted upon by the acid; and even when the cell is delivering current to a load, the mercury continues to act on the impurities in the zinc, causing them to leave the surface of the zinc electrode and float to the surface of the mercury. This process greatly increases the life of the primary cell.

TYPES OF BATTERIES

The development of new and different types of batteries in the past decade has been so rapid that it is virtually impossible to have a complete knowledge of all the various types currently being developed or now in use. A few recent developments are the silver-zinc, nickel-zinc, nickel-cadmium, silver-cadmium, magnesium-magnesium perchlorate, mercury, thermal, and water-activated batteries.

The lead-acid battery has been in service for a relatively long time; however, there are still various improvements being incorporated into the battery to improve its efficiency and life span. The material presented in this chapter, though not all-inclusive, provides the reader with a knowledge of various types of batteries.

DRY (PRIMARY) CELL

The *dry cell* is so called because its electrolyte is not in a liquid state. Actually, the electrolyte is a moist paste. If it should become dry, it would no longer be able to transform chemical energy to electrical energy. The name "dry cell," therefore, is not strictly correct in a technical sense.

Construction of the Dry Cell

The construction of a common type of dry cell is shown in Fig. 4. The internal parts of the cell are located in a cylindrical zinc container. This zinc container serves as the negative electrode of the cell. The container is lined with a nonconducting material, such as blotting paper, to insulate the zinc from the paste. A carbon electrode is located in the center, and it serves as the positive terminal of the cell. The paste is a mixture of several substances. Its composition may vary depending on its manufacturer. Generally, however, the paste will contain some

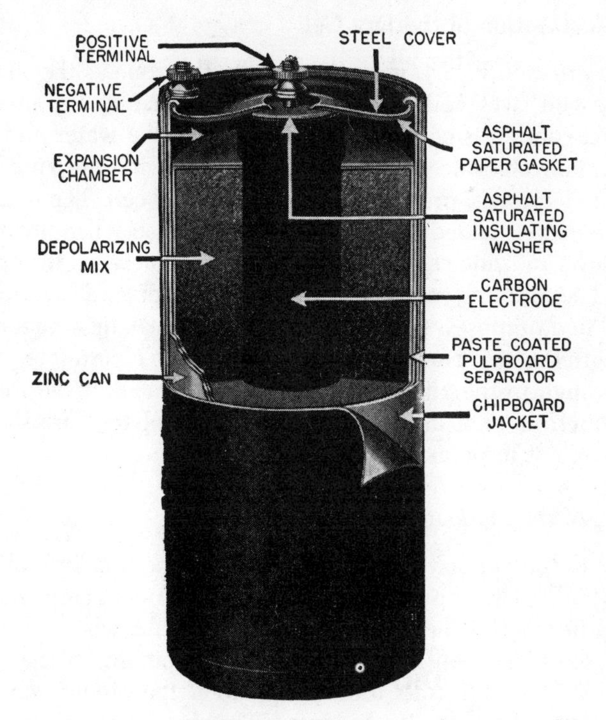

Fig. 4. Cutaway view of the general-purpose dry cell.

combination of the following substances: ammonium chloride (sal ammoniac), powdered coke, ground carbon, manganese dioxide, zinc chloride, graphite, and water.

This paste, which is packed in the space between the carbon and the blotting paper, also serves to hold the carbon electrode rigid in the center of the cell. When the paste is packed in the cell, a small expansion space is left at the top. The cell is then sealed with asphalt-saturated cardboard.

Binding posts are attached to the electrodes so that wires may be conveniently connected to the cell.

Since the zinc container is one of the electrodes, it must be protected with some insulating material. Therefore, it is common practice for the manufacturer to enclose the cells in cardboard containers.

Chemical Action of the Dry Cell

The *dry cell* (Fig. 4) is fundamentally the same as the simple voltaic cell (wet cell) described earlier, as far as its internal chemical action is concerned. The action of the water and the ammonium chloride in the paste, together with the zinc and carbon electrodes, produces the voltage of the cell. The manganese dioxide is added to reduce the polarization when line current flows and zinc chloride reduces local action when the cell is idle. The blotting paper (paste-coated pulpboard separator) serves two purposes: to keep the paste from making actual contact with the zinc container, and to permit the electrolyte to filter through to the zinc slowly. The cell is sealed at the top to keep air from entering and drying the electrolyte. Care should be taken not to break this seal.

Rating of the Standard-Size Cell

One of the popular sizes in general use is the standard, or No. 6, dry cell. It is approximately 2½ inches in diameter and six inches in length. The voltage is about 1½ volts when new but decreases as the cell ages. When the open-circuit voltage falls below 0.75 to 1.2 volts (depending upon the circuit requirements), the cell is usually discarded. The amount of current that the cell can deliver and still give satisfactory service depends upon the length of time that the current flows. For instance, if a No. 6 cell is to be used in a portable radio, it is likely to supply current constantly for several hours. Under these conditions, the current should not exceed ⅛ ampere, the rated constant-current capacity of a No. 6 cell. If the same cell is required to supply current only occasionally, for only short periods of time, it could supply currents of several amperes without undue injury to the cell. As the time duration of each discharge decreases, the interval of time between discharges increases, the allowable amount of current available for each discharge becomes higher, up to the amount that the cell will deliver on short circuit.

The short-circuit current test is another means of evaluating the condition of a dry cell. A new cell, when short circuited through an ammeter, should supply not less than 25 amperes. A cell that has been in service should supply at least 10 amperes if it is to remain in service.

Rating of the Unit-Size Cell

Another popular size of dry cell, the size D, is 1⅜ inches in diameter and 2¾ inches in length. It is also known as the unit cell. The size D cell voltage is 1.5 volts when new. A discharged cell may expand, allowing the electrolyte to leak and cause corrosion. Some manufacturers place a steel jacket around the zinc container to prevent this action.

Shelf Life

A cell that is not being used (sitting on the shelf) will gradually deteriorate because of slow internal chemical actions (local action) and changes in moisture content. However, this deterioration is usually very slow if cells are properly stored. High-grade cells of the larger sizes should have a shelf life of a year or more. Smaller-size cells have a proportionately shorter shelf life, ranging down to a few months for the very small sizes. If unused cells are stored in a cool place, their shelf life will be greatly increased; therefore, to minimize deterioration, they should be stored in refrigerated spaces (10° F. to 35° F.) that are not dehumidified.

MERCURY CELLS

With the advent of the space program and the development of small transceivers and miniaturized equipment, a power source of miniature size was needed. Such equipment requires a small battery which is capable of delivering maximum electrical capacity per unit volume while operating in varying temperatures and at a constant discharge voltage. The mercury battery, which is one of the smallest batteries, meets these requirements.

Present mercury batteries are manufactured in three basic structures. The wound anode type (Fig. 5) has an anode composed of a corrugated zinc strip with absorbent paper wound in an offset manner so that it protrudes at one end. The zinc is

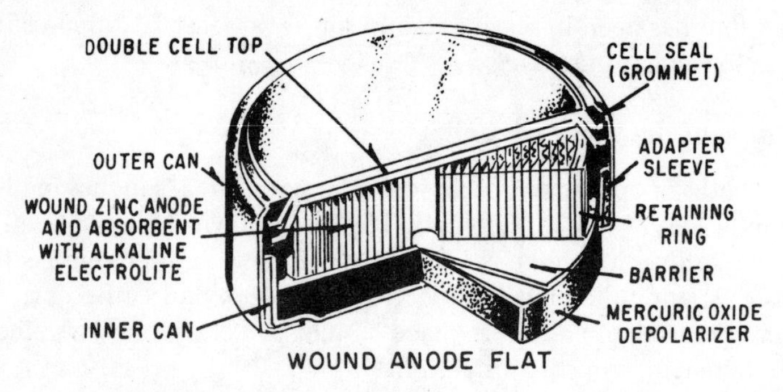

Fig. 5. Wound anode mercury cells.

amalgamated (mixed) with mercury (10 percent), and the paper is impregnated with the electrolyte, which causes it to swell and produce a positive contact pressure.

In the pressed-powder cells (Fig. 6 and Fig. 7), the zinc powder is preamalgamated before it is pressed into shape; its porosity allows electrolyte impregnation with oxidation in depth when current is discharged. A double can structure is used in the larger-sized cells. The space between the inner and outer containers provides passage for any gas generated by an improper chemical balance or impurities present within the cell. The construction is such that, if excessive gas pressures are experienced, the compression of the upper part of the grommet by internal pressure allows the gas to escape into the space between the two cans. A paper tube surrounds the inner can so that any liquid

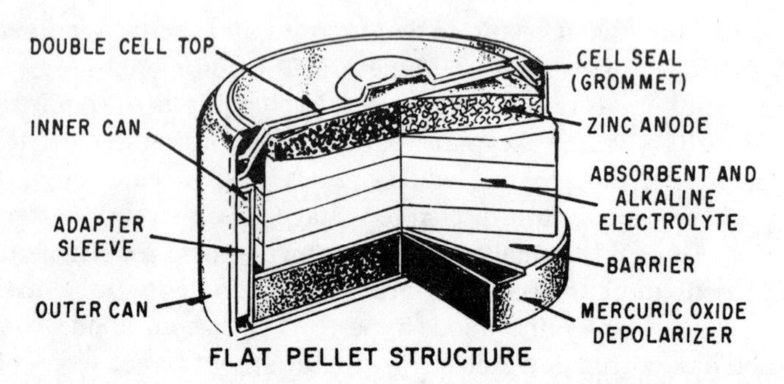

Fig. 6. Pressed powder cells.

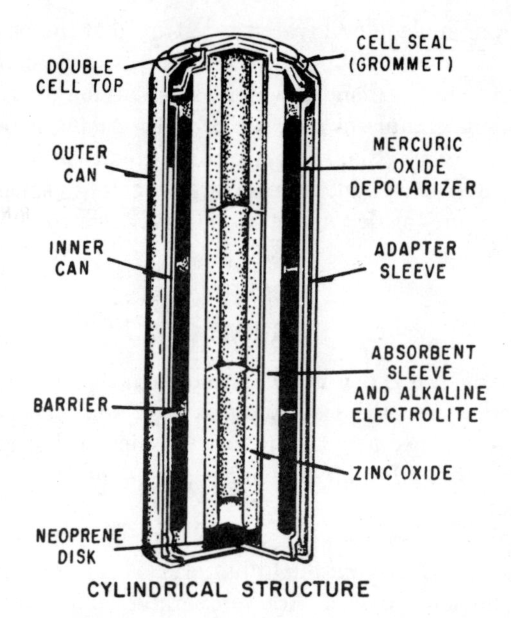

Fig. 7. Pressed powder cells.

carried by discharging gas will be absorbed, maintaining leak-resistant structure. Release of excessive gas pressure automatically reseals the cell.

Note: Mercury batteries have been known to explode with considerable force when shorted. *Caution* should be exercised to insure that the battery is not accidentally shorted.

The overall chemical action by which the mercury cell produces electricity is given by the following chemical formula:

$$Zn + H_2O + HgO \rightarrow ZnO + H_2O + Hg$$

This action, the same as in other types of cells, is a process of oxidation. The alkaline electrolyte is in contact with the zinc electrode. The zinc oxidizes (Zn changes to ZnO), thus taking atoms of oxygen from water molecules in the electrolyte. This leaves positive hydrogen ions, which move toward the mercuric oxide pellet, causing polarization. These hydrogen ions take oxygen from the mercuric oxide (thus changing HgO to Hg).

Where one molecule of water is destroyed at the negative electrode, one molecule is produced at the positive electrode, maintaining the net amount of water. By absorbing oxygen, the zinc electrode accumulates excess electrons, making it negative. By giving up oxygen, the mercuric oxide electrode loses electrons, making it positive. In the discharged state, the negative electrode is zinc oxide, and the positive electrode is ordinary mercury.

RESERVE CELL

A *reserve cell* is one in which the elements are kept dry until the time of use; the electrolyte is then admitted and the cell starts producing current. In theory, this means that a reserve cell may be stored for an indefinite period of time before it is activated.

One new reserve cell (Fig. 8) is the alkaline manganese cell of the standard D size (flashlight battery). This reserve cell exhibits a high efficiency over a wide temperature range and is capable

Fig. 8. Reserve cell.

of momentary high-current pulses in the range of 12 to 15 amperes.

The reserve cell is manufactured in a dry state, the electrolyte being contained in a plastic vial within the cell. When stored in this manner, the cell has a shelf life of over 10 years. To activate the cell, the activating mechanism is rotated 35 degrees in either direction. This releases a spring-loaded plunger which breaks the plastic vial of electrolyte. Continued rotation permits the activating mechanism to be removed and discarded, resulting in a D-size cell. A safety device is incorporated to prevent accidental activation during handling and transit.

Activation time is approximately two seconds when the cell is not under load. When under a 4-ohm load, the activation time (to reach a 1.35-volt level) is less than five seconds at 70° F. and less than 30 seconds at 30° F.

The cell has been designed so that it is not position-sensitive during either the activation or the discharge period, and after activation can be handled and used as a standard D cell. After activation, shelf life of a reserve cell is approximately two years less than that of the standard alkaline manganese cell.

Reserve cells are used for emergency lighting and communications equipment, and in any other situation where long storage ability is of prime importance.

COMBINING CELLS

In many cases, a battery-powered device may require more electrical energy than one cell can provide. The device may require either higher voltage or more current, and in some cases both. Under such conditions it is necessary to combine, or interconnect, a sufficient number of cells to meet the higher requirements. Cells connected in series provide a higher voltage, while cells connected in parallel provide a higher current capacity. To provide adequate power when both voltage and current requirements are greater than the capacity of one cell, a combination series-parallel network of cells must be interconnected.

Series-Connected Cells

Assume that a load requires a power supply with a potential of six volts and a current capacity of ⅛ ampere. Since a single cell normally supplies a potential of only 1.5 volts, more than one cell is obviously needed. To obtain the higher potential, the cells are connected in series, as shown in Fig. 9, A.

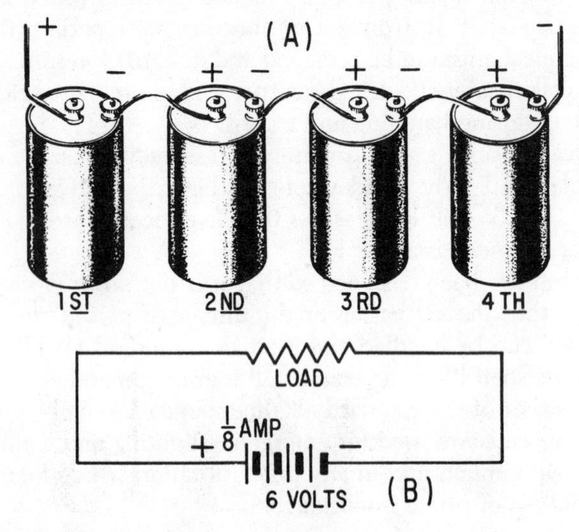

Fig. 9. A, Pictorial view of series connected cells; B, schematic of series connections.

In a series hookup, the negative electrode of the first cell is connected to the positive electrode of the second cell, the negative electrode of the second to the positive of the third, and so on. The positive electrode of the first cell and negative electrode of the last cell then serve as the power takeoff terminals of the battery. In this way, the potential is boosted 1.5 volts by each cell in the series line. There are four cells, so the output terminal voltage is $1.5 \times 4 = 6$ volts. When the cells are connected to the load, ⅛ ampere flows through the load and each cell of the battery. This is within the capacity of each cell. Therefore, only four series-connected cells are needed to supply this particular load.

Parallel-Connected Cells

In this case, assume an electrical load requires only 1.5 volts, but will draw ½ ampere of current. (Assume that a cell will supply only ⅛ ampere.) To meet this requirement, the cells are connected in parallel, as shown in Fig. 10, A. In a parallel connection, all positive cell electrodes are connected to one line, and all negative electrodes are connected to the other. No more than one cell is connected between the lines at any one point, so the potential between the lines is the same as that of one cell, or 1.5 volts. However, each cell may contribute its maximum allowable current of ⅛ ampere to the line. There are four cells, so the total line current is ⅛ × 4 = ½ ampere. Hence, four cells in parallel have the capacity to supply a load requiring ½ ampere at 1.5 volts.

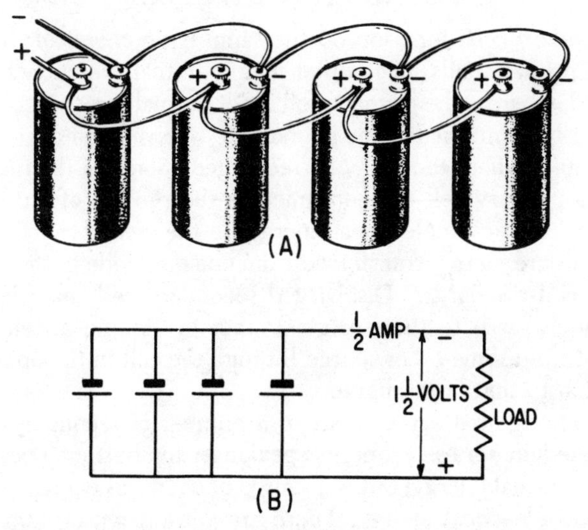

Fig. 10. A, Pictorial view of parallel-connected cells; B, schematic of parallel connection.

Series-Parallel–Connected Cells

Figure 11 depicts a battery network supplying power to a load requiring both a voltage and a current greater than one cell

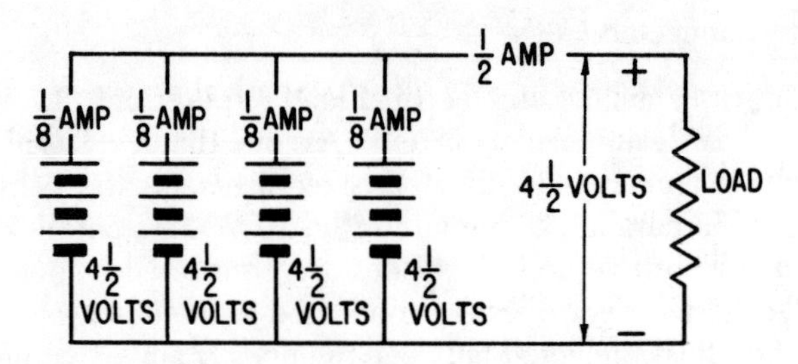

Fig. 11. Series-parallel connected cells.

can provide. To provide the required ½ ampere of current, four series groups are connected in parallel, each supplying ⅛ ampere.

SECONDARY (WET) CELLS

Secondary cells function on the same basic chemical principles as primary cells. They differ mainly in that they may be recharged, whereas the primary cell is not normally recharged. (As previously mentioned, some primary cells have been developed to the state where they may be recharged.) Some of the materials of a primary cell are consumed in the process of changing chemical energy to electrical energy. In the secondary cell, the materials are merely transferred from one electrode to the other as the cell discharges. Discharged secondary cells may be restored (charged) to their original state by forcing an electric current from some other source through the cell in the opposite direction to that of discharge.

The storage battery consists of a number of secondary cells connected in series. Properly speaking, this battery does not store electrical energy, but is a source of chemical energy which produces electrical energy. There are various types of storage batteries: the lead-acid type, which has an emf of 2.2 volts per cell; the nickel-iron alkali type; the nickel-cadmium alkali type, with an emf of 1.2 volts per cell; and the silver-zinc type, which has an emf of 1.5 volts per cell. Of these types, the lead-acid type is the most widely used.

LEAD-ACID BATTERY

The *lead-acid battery* is an electrochemical device for storing chemical energy until it is released as electrical energy. Active materials within the battery react chemically to produce a flow of direct current whenever current-consuming devices are connected to the battery terminal posts. This current is produced by chemical reaction between the active material of the plates (electrodes) and the electrolyte (sulfuric acid). The lead-acid battery is used extensively throughout the world. The parts of a lead-acid battery are illustrated in Fig. 12 and are discussed in the following paragraphs.

Battery Construction

A *lead-acid battery* consists of a number of cells connected together, the number depending upon the voltage desired. Each cell produces approximately two volts.

A cell consists of a hard rubber, plastic, or bituminous-material compartment, into which is placed the cell element, consisting of two types of lead plates, known as positive and negative plates. (*See* Fig. 13.) These plates are insulated from each other by suitable separators (usually made of plastic, rubber, or glass) and submerged in a sulfuric acid solution (electrolyte).

There are a variety of plates used in the lead-acid battery: pasted plates, spun-lead (Plante) plates, Gould plates, and iron-clad plates. Each kind of plate is designed to fulfill a specific purpose. Pasted plates are most commonly used.

The pasted plates are formed by applying lead-oxide pastes to a grid (Fig. 14) made of lead-antimony alloy. The grid is designed to give the plates mechanical strength, hold the active material in place, and provide adequate conductivity for the electric current created by the chemical action. The active material (lead oxide) is applied to the grids in paste form and allowed to dry. The plates are then put through an electrochemical process that converts the active material of the positive plates into lead peroxide, and that of the negative plates into

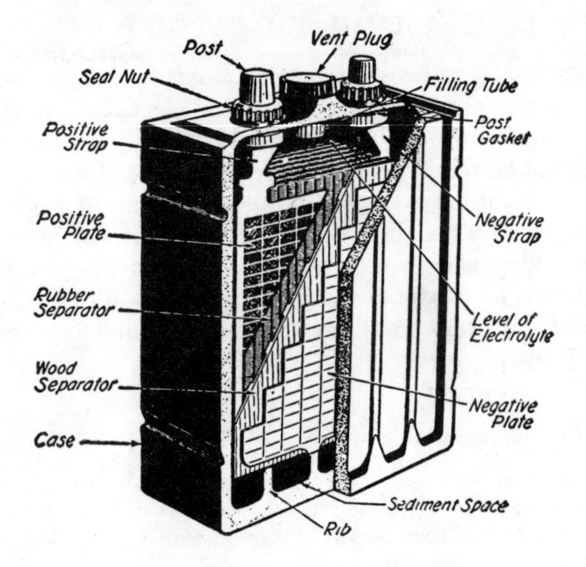

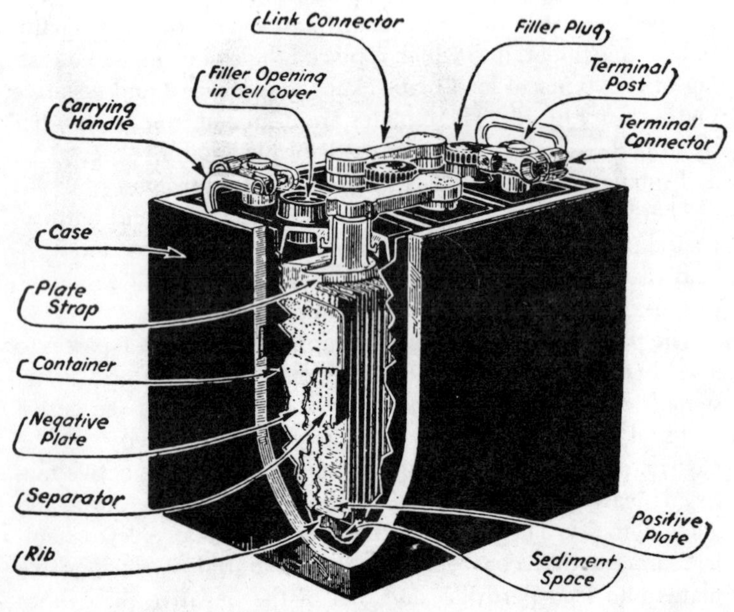

Fig. 12. Lead-acid battery construction.

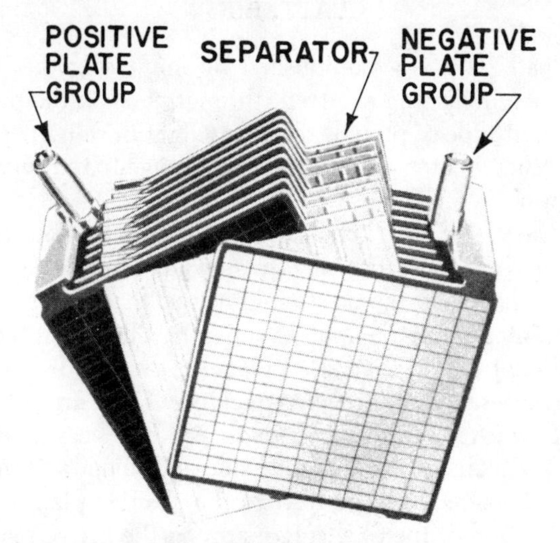

Fig. 13. Plate arrangement.

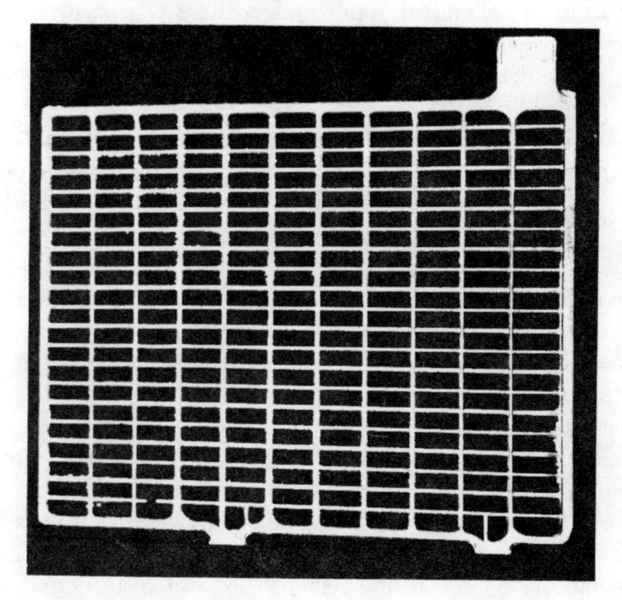

Fig. 14. Grid structure.

sponge lead. This is accomplished by immersing the plates in electrolyte and passing a current through them in the proper direction. This type of plate is relatively light in weight compared to the other plates that are more rugged and durable in construction.

After the plates have been formed, they are built into positive and negative groups. The negative group of plates always has one more plate than the positive group, so that both sides of the positive plates are acted upon chemically. This keeps the expansion and contraction that takes place in the positive plates the same on both sides and prevents buckling. These groups are then assembled with separators to become cell elements. The separators are vertically grooved on one side and smooth on the other. The grooved side is placed next to the positive plate to permit free circulation of the electrolyte around the active material.

The positive plates, which are lead peroxide, and the negative plates, which are spongy lead, are referred to as the active material of the battery. However, these materials alone in a container will cause no chemical action unless there is a path for interaction between them. To provide this path for interaction and to carry the electric current within the battery are the functions of the electrolyte.

A battery container is the receptacle for the cells that make up the battery. Most containers are made from hard rubber, plastic, or bituminous composition that is resistant to acid and mechanical shock and is able to withstand extreme weather conditions. Most batteries are assembled in a one-piece container with a compartment for each cell. The bottom of the container has ribs molded into it to provide support for the elements and a sediment space for the flakes of active material that drop off the plates during the life of the battery.

The battery or cell covers and the battery container are usually made of the same material. The cell covers provide openings for the two-element terminals and a vent plug.

Cell connectors are used to connect the cells of a battery in series. The element in each cell is placed so that the negative terminal of one cell is adjacent to the positive terminal of the next cell; they are connected both physically and electrically by a cell connector. Connectors must be of sufficient size to carry the current demands of the battery without overheating.

Vent plugs are of various designs to function in conjunction with the cover vent openings, permitting the escape of gases that form within the cells while preventing leakage or loss of the electrolyte. A typical vent plug used in an automobile battery is depicted in Fig. 12.

Some batteries utilize a nonspill type of vent plug that makes it possible to place the battery in any position without loss of the electrolyte. (*See* Fig. 15.) This type of vent plug has found wide use in aircraft.

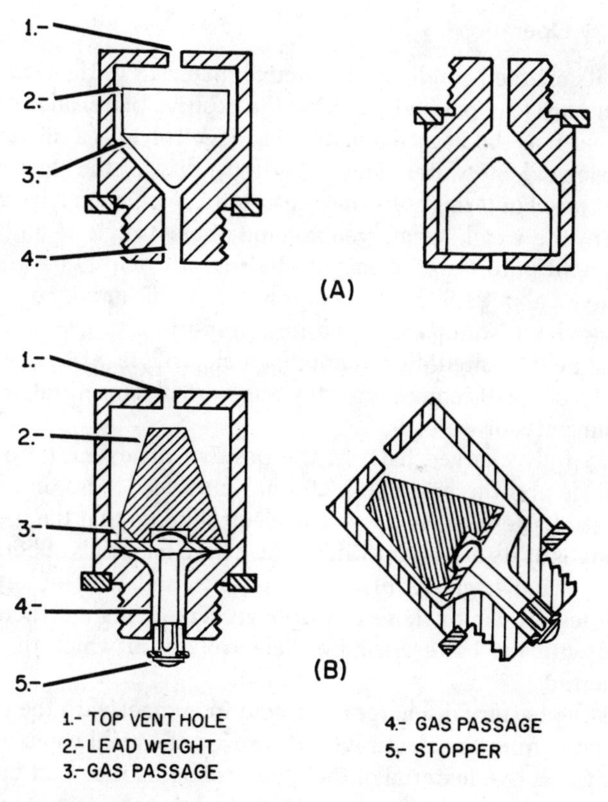

Fig. 15. Nonspill vent plugs.

Sealing compound, generally made of a bituminous substance, is used to form a seal between the cell cover and the container. The compound is an acid-resistant material that must conform to rigid vibration and heat standards. This insures that the sealing compound does not melt or flow at summer temperatures and does not crack at winter temperatures. Batteries with a polystyrene jar use a polystyrene cement as a sealer.

The terminals of a lead-acid battery are normally distinguishable from one another by their physical size and the marking by the manufacturer. The positive terminal marked (+) is slightly larger than the negative terminal marked (−).

Battery Operation

In its charged condition, the active materials in the lead-acid battery are lead peroxide (used as the positive plate) and sponge lead (used as the negative plate). The electrolyte is a mixture of sulfuric acid and water. The strength (acidity) of the electrolyte is measured in terms of its specific gravity. Specific gravity is the ratio of the weight of a given volume of electrolyte to an equal volume of pure water. Concentrated sulfuric acid has a specific gravity of about 1,830; pure water has a specific gravity of 1,000. The acid and water are mixed in a proportion to give the specific gravity desired. For example, an electrolyte with a specific gravity of 1,210 requires roughly one part of concentrated acid to four parts of water.

In a fully charged battery, the positive plates are pure lead peroxide and the negative plates are pure lead. Also, in a fully charged battery, all acid is in the electrolyte, so that the specific gravity is at its maximum value. The active materials of both the positive and negative plates are porous, and have absorptive qualities similar to those of a sponge. The pores are therefore filled with the battery solution (electrolyte) in which they are immersed.

As the battery discharges, the acid in contact with the plates separates from the electrolyte. It forms a chemical combination with the active material of the plate, changing it to lead sulfate. Thus, as the discharge continues, lead sulfate forms on the plates, and more acid is taken from the electrolyte. The water content of the electrolyte becomes progressively higher; that is, the ratio of water to acid increases. As a result, the specific gravity of the electrolyte will gradually decrease during discharge.

When the battery is being charged, the reverse takes place. The acid held in the sulfated plate material is driven back into the electrolyte; further charging cannot raise its specific gravity any higher. When fully charged, the material of the positive plates is again pure lead peroxide and that of the negative plates is pure lead.

Electrical energy is derived from a cell when the plates react with the electrolyte. As a molecule of sulfuric acid separates, part of it combines with the negative sponge-lead plates. Thus, it makes the sponge-lead plates negative, and at the same time forms lead sulfate. The remainder of the sulfuric acid molecule, lacking electrons, has thus become a positive ion. The positive ions migrate through the electrolyte to the opposite (lead peroxide) plates, and take electrons from them. This action neutralizes the positive ions, forming ordinary water. It also makes the lead peroxide plates positive, by taking electrons from them. Again, lead sulfate is formed in the process.

The action just described is represented in more detail by the following chemical equation:

$$Pb + PbO_2 + 2H_2SO_4 \overset{\text{discharging}}{\rightleftarrows} 2PbSO_4 + 2H_2O$$

The left side of the expression represents the cell in the charged condition, and the right side represents the cell in the discharged condition.

In the charged condition, the positive plate contains lead peroxide, PbO_2; the negative plate is composed of sponge lead, Pb; and the solution contains sulfuric acid, H_2SO_4. In the discharged condition, both plates contain lead sulfate, $PbSO_4$, and the solution contains water, H_2O. As the discharge progresses, the acid content of the electrolyte becomes less and less because it is used in forming lead sulfate, and the specific gravity of the electrolyte decreases. A point is reached where so much of the active material has been converted into lead sulfate that the cell can no longer produce sufficient current to be of practical value. At this point the cell is said to be discharged (Fig. 16, C). Since the amount of sulfuric acid combining with the plates at any time during discharge is in direct proportion to the ampere-hours (product of current in amperes and time in hours) of discharge,

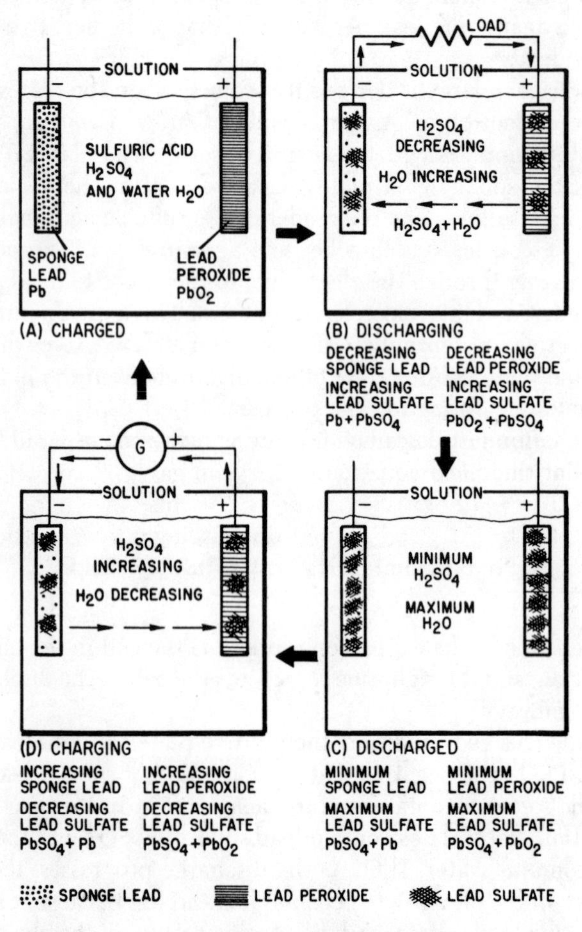

Fig. 16. Chemical action in lead-acid cell.

the specific gravity of the electrolyte is a guide in determining the state of discharge of the lead-acid cell.

If the discharged cell is properly connected to a direct-current charging source (the voltage of which is slightly higher than that of the cell), current will flow through the cell, in the opposite direction to that of discharge, and the cell is said to be charging (Fig. 16, D). The effect of the current will be to change the lead sulfate on both the positive and negative plates back to its original active form of lead peroxide and sponge lead, respectively. At the same time, the sulfate is restored to the electrolyte with the result that the specific gravity of the electrolyte increases. When all the sulfate has been restored to the electrolyte, the specific gravity will be maximum. The cell is then fully charged and is ready to be discharged again.

It should always be remembered that the addition of sulfuric acid to a discharged lead-acid cell does not recharge the cell. Adding acid only increases the specific gravity of the electrolyte and does not convert the lead sulfate on the plates back into active material (sponge lead and lead peroxide), and consequently does not bring the cell back to a charged condition. A charging current must be passed through the cell to do this.

As a cell charge nears completion, hydrogen gas (H_2) is liberated at the negative plate and oxygen gas (O_2) is liberated at the positive plate. This action occurs because the charging current is greater than the amount that is necessary to reduce the small remaining amount of lead sulfate on the plates. Thus, the excess current ionizes the water in the electrolyte. This action is necessary to assure full charge to the cell.

Specific Gravity

The ratio of the weight of a certain volume of liquid to the weight of the same volume of water is called the specific gravity of the liquid. The specific gravity of pure water is 1.000. Sulfuric acid has a specific gravity of 1.830; thus sulfuric acid is 1.830 times as heavy as water. The specific gravity of a mixture of sulfuric acid and water varies with the strength of the solution from 1.000 to 1.830.

As a storage battery discharges, the sulfuric acid is depleted and the electrolyte is gradually converted into water. This action provides a guide in determining the state of discharge of the lead-acid cell. The electrolyte that is usually placed in a lead-acid battery has a specific gravity of 1.350 or less. Generally, the specific gravity of the electrolyte in standard storage batteries is adjusted between 1.210 and 1.220.

Hydrometer. The specific gravity of the electrolyte is measured with a hydrometer. In the syringe hydrometer (Fig. 17), part of the battery electrolyte is drawn up into a glass tube by means of a rubber bulb at the top.

The hydrometer float consists of a hollow glass tube weighted at one end and sealed at both ends. A scale calibrated in specific gravity is laid off axially along the body (stem) of the tube. The hydrometer float is placed inside the glass syringe and the electrolyte to be tested is drawn up into the syringe, thus immersing the hydrometer float in the solution. When the syringe is held approximately in a vertical position, the hydrometer float will sink to a certain level in the electrolyte. The extent to which the hydrometer stem protudes above the level of the liquid depends upon the specific gravity of the solution. The reading on the stem at the surface of the liquid is the specific gravity of the electrolyte in the syringe.

Caution: Hydrometers should be flushed daily with fresh water to prevent inaccurate readings. Storage battery hydrometers must not be used for any other purpose.

Corrections. The specific gravity of the electrolyte is affected by its temperature. The electrolyte expands and becomes less dense when heated and its specific gravity reading is lowered. On the other hand, the electrolyte contracts and becomes denser when cooled and its specific gravity reading is raised. In both cases the electrolyte may be from the same fully charged storage cell. Thus, the effect of temperature is to distort the readings.

Most standard storage batteries use 80°F. as the normal temperature to which specific gravity readings are corrected. To correct the specific gravity reading of a storage battery, add four points to the reading for each 10°F. above 80° and subtract four

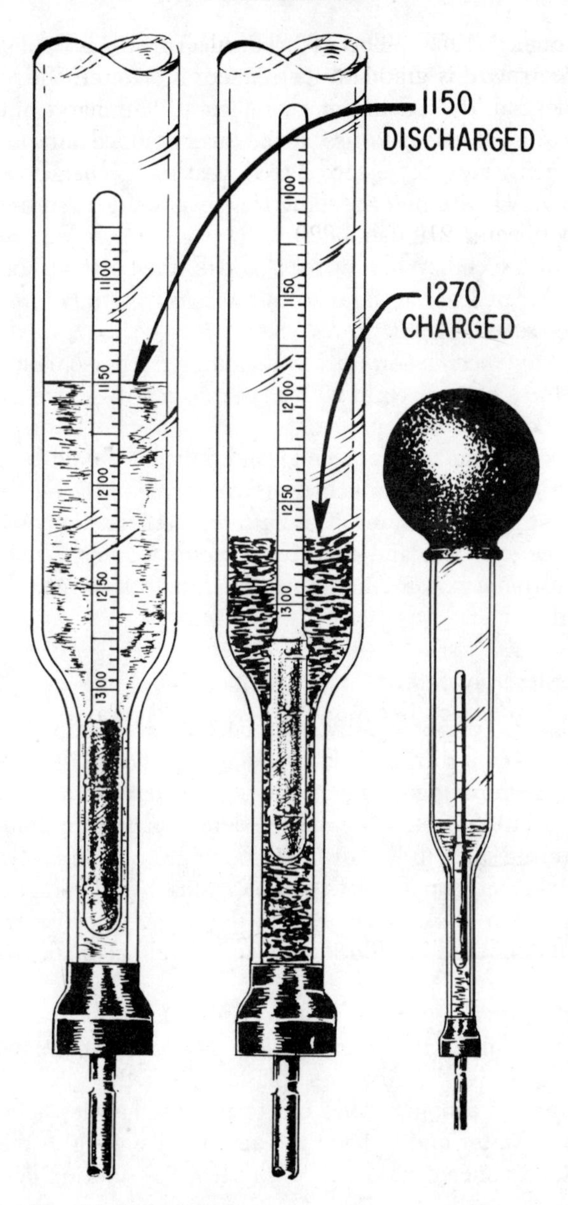

Fig. 17. Type-B hydrometer.

points for each 10°F. below 80°. The electrolyte in a cell should be at the normal level when the reading is taken. If the level is below normal, there will not be sufficient fluid drawn into the tube to cause the float to rise. If the level is above normal there is too much water, the electrolyte is weakened, and the reading is too low. A hydrometer reading is inaccurate if taken immediately after water is added, because the water tends to remain at the top of the cell. When water is added, the battery should be charged for at least an hour to mix to electrolyte before a hydrometer reading is taken.

Adjusting Specific Gravity. Only authorized personnel should add acid to a battery. Acid with a specific gravity above 1.350 is never added to a battery.

If the specific gravity of a cell is more than it should be, it can be reduced to normal limits by removing some of the electrolyte and adding distilled water. The battery is charged for one hour to mix the solution, and then hydrometer readings are taken. The adjustment is continued until the desired true readings are obtained.

Mixing Electrolytes

The electrolyte of a fully charged battery usually contains about 38 percent sulfuric acid by weight, or about 27 percent by volume. In preparing the electrolyte, distilled water and sulfuric acid are used. New batteries may be delivered with containers of concentrated sulfuric acid of 1.830 specific gravity or electrolyte of 1.400 specific gravity, both of which must be diluted with distilled water to make electrolyte of the proper specific gravity. The container used for diluting the acid should be made of glass, earthenware, rubber, or lead.

When mixing electrolyte, *always pour acid into water*—never pour water into acid. Pour the acid slowly and cautiously to prevent excessive heating and splashing. Stir the solution continuously with a nonmetallic rod to mix the heavier acid with the lighter water and to keep the acid from sinking to the bottom. When concentrated acid is diluted, the solution becomes very hot.

Treatment of Acid Burns

If acid or electrolyte from a lead-acid battery comes into contact with the skin, the affected area should be washed as soon as possible with large quantities of fresh water, after which a salve such as petrolatum, boric acid, or zinc ointment should be applied. If none of these salves are available, clean lubricating oil will suffice. When washing, large amounts of water should be used, since a small amount of water might do more harm than good in spreading the acid burn.

Acid spilled on clothing may be neutralized with dilute ammonia or a solution of baking soda and water.

Capacity

The capacity of a battery is measured in ampere-hours. As previously mentioned, the ampere-hour capacity is equal to the product of the current in amperes and the time in hours during which the battery is supplying this current. The ampere-hour capacity varies inversely with the discharge current. The size of a cell is determined generally by its ampere-hour capacity. The capacity of a cell depends upon many factors, the most important of which are as follows:

1. The area of the plates in contact with the electrolyte.
2. The quantity and specific gravity of the electrolyte.
3. The type of separators.
4. The general condition of the battery (degree of sulfating, plates buckled, separators warped, sediment in bottom of cells, and the like).
5. The final limiting voltage.

Rating

Storage batteries are rated according to their rate of discharge and ampere-hour capacity. Most batteries, except those in aircraft and some used for radio and sound systems, are rated according to a 20-hour rate of discharge—that is, if a fully charged battery is completely discharged during a 20-hour period, it is discharged at the 20-hour rate. Thus if a battery can deliver 20 amperes continuously for 20 hours, the battery has a rating of 20 × 20, or 400 ampere-hours. The 20-hour rating is thus equal to the average current that a battery is capable of supplying without interruption for an interval of 20 hours.

All standard batteries deliver 100 percent of their available capacity if discharged in 20 hours or more, but they will deliver less than their available capacity if discharged at a faster rate. The faster they discharge, the less ampere-hour capacity they have.

The low-voltage limit, as specified by the manufacturer, is the limit beyond which very little useful energy can be obtained from a battery. For example, at the conclusion of a 20-hour discharge test on a battery, the closed-circuit voltmeter reading is about 1.75 volts per cell and the specific gravity of the electrolyte is about 1.060. At the end of a charge, its closed-circuit voltmeter reading, while the battery is being charged at the finishing rate, is between 2.4 and 2.6 volts per cell. The specific gravity of the electrolyte corrected to 80°F. is between 1.210 and 1.220. In climates of 40°F. and below, authority may be granted to increase the specific gravity to 1.280. Other batteries, of higher normal specific gravity, may also be increased.

Test Discharge

The *test discharge* is the best method of determining the capacity of a battery. Most battery switchboards are provided with the necessary equipment for giving test discharges. To determine the battery capacity, a battery is normally given a test discharge once every six months. Test discharges are also given whenever any cell of a battery after charge cannot be brought within 10 points of full charge, or when one or more cells are found to have less than normal voltage after an equalizing charge.

A test discharge must always be preceded by an equalizing charge. Immediately after the equalizing charge, the battery is discharged at its 20-hour rate until either the total battery voltage drops to a value equal to 1.75 times the number of cells in series or the voltage of any individual cell drops to 1.65 volts, whichever occurs first. The rate of discharge should be kept constant throughout the test discharge. Because standard batteries are rated at the 20-hour capacity, the discharge rate for a 200-ampere-hour battery is 200/20, or 10 amperes. If the temperature of the electrolyte at the beginning of the charge is not exactly 80°F., the time duration of the discharge must be corrected for the actual temperature of the battery.

A battery of 100-percent capacity discharges at its 20-hour rate for 20 hours before reaching its low-voltage limit. If the battery or one of its cells reaches the low-voltage limit before the 20-hour period has elapsed, the discharge is discontinued immediately and the percentage of capacity is determined from the equation

$$C = \frac{H_a}{H_t} \times 100$$

where C is the percentage of ampere-hour capacity available, H_a the total hours of discharge, and H_t the total hours for 100-percent capacity. The date for each test discharge should be recorded on the storage battery record sheet.

For example, a 200-ampere-hour 6-volt battery delivers an average current of 10 amperes for 20 hours. At the end of this period the battery voltage is 5.25 volts. On a later test the same battery delivers an average current of 10 amperes for only 14 hours. The discharge was stopped at the end of this time because the voltage of the middle cell was found to be only 1.65 volts. The percentage of capacity of the battery is now 14/20 × 100, or 70 percent. Thus, the ampere-hour capacity of this battery is reduced to 0.7 × 200 = 140 ampere hours.

State of Charge

After a battery is discharged completely from full charge at the 20-hour rate, the specific gravity has dropped about 150 points, to about 1.060. The number of points that the specific gravity drops per ampere-hour can be determined for each type of battery. For each ampere-hour taken out of a battery, a definite amount of acid is removed from the electrolyte and combined with plates.

For example, if a battery is discharged from full charge to the low-voltage limit at the 20-hour rate and if 100 ampere-hours are obtained with a specific gravity drop of 150 points, there is a drop of 150/100, or 1.5 points per ampere-hour delivered. If the reduction in specific gravity per ampere-hour is known, the drop in specific gravity for this battery may be predicted for any number of ampere-hours delivered to a load. For example, if 70 ampere-hours are delivered by the battery at the 20-hour rate or any other rate or collection of rates, the drop in specific gravity is 70 × 1.5, or 105 points.

Conversely, if the drop in specific gravity per ampere-hour and the total drop in specific gravity are known, the ampere-hours delivered by a battery may be determined. For example, if the specific gravity of the previously considered battery is 1.210 when the battery is fully charged and 1.150 when it is partly discharged, the drop in specific gravity is 1.210 − 1.150, or 60 points, and the number of ampere-hours taken out of the battery is 60/1.5, or 40 ampere-hours. Thus, the number of ampere-hours expended in any battery discharge can be determined from the following items:

1. The specific gravity when the battery is fully charged.
2. The specific gravity after the battery has been discharged.
3. The reduction in specific gravity per ampere-hour.

Voltage alone is not a reliable indication of the state of charge of a battery except when the voltage is near the low-voltage limit on discharge. During discharge the voltage falls. The higher the rate of discharge, the lower will be the terminal voltage. Open-circuit voltage is of little value because the variation between full charge and complete discharge is so small—only about 0.1 volt per cell. However, abnormally low voltage does indicate injurious sulfation or some other serious deterioration of the plates.

Types of Charges

The following types of charges may be given to a storage battery, depending upon the condition of the battery.

1. Initial charge.
2. Normal charge.
3. Equalizing charge.
4. Floating charge.
5. Fast charge.

Initial Charge. When a new battery is shipped dry, the plates are in an uncharged condition. After the electrolyte has been added, it is necessary to convert the plates into the charged condition. This is accomplished by giving the battery a long low-rate initial charge. The charge is given in accordance with the manufacturer's instructions, which are shipped with each battery. If the manufacturer's instructions are not available, refer to the detailed instruction in current directives.

Normal Charge. A normal charge is a routine charge that is given in accordance with the nameplate data during the ordinary cycle of operation to restore the battery to its charged condition. The following steps should be observed.

1. Determine the starting and finishing rate from the nameplate data.
2. Add water, as necessary, to each cell.
3. Connect the battery to the charging panel and make sure the connections are clean and tight.
4. Turn on the charging circuit and set the current through the battery at the value given as the starting rate.
5. Check the temperature and specific gravity of pilot cells hourly.
6. When the battery begins to gas freely, reduce the charging current to the finishing rate.

A normal charge is complete when the specific gravity of the pilot cell, corrected for temperature, is within five points (0.005) of the specific gravity obtained on the previous equalizing charge.

Equalizing Charge. An equalizing charge is an extended normal charge at the finishing rate. It is given periodically to insure that all the sulfate is driven from the plates and that all the cells are restored to a maximum specific gravity. The equalizing charge is continued until the specific gravity of all cells, corrected for temperature, shows no change for a four-hour period. Readings of all cells are taken every half hour.

Floating Charge. A battery may be maintained at full charge by connecting it across a charging source that has a voltage maintained within the limits of from 2.13 to 2.17 volts per cell of the battery. In a floating charge, the charging rate is determined by the battery voltage rather than by a definite current value. The voltage is maintained between 2.13 and 2.17 volts per cell, with an average as close to 2.15 volts as possible.

Fast Charge. A fast charge is used when a battery must be recharged in the shortest possible time. The charge starts at a much higher rate than is normally used for charging. It should be used only in an emergency, as this type charge may be harmful to the battery.

Charging Rate

Normally, the *charging rate* of storage batteries is given on the battery nameplate. If the available charging equipment does not have the desired charging rates, the nearest available rates should be used. However, the rate should never be so high that violent gassing occurs. *Never allow the temperature of the electrolyte in any cell to rise above 125°F.*

Charging Time

The charge must be continued until the battery is fully charged. Frequent readings of specific gravity should be taken during the charge. These readings should be corrected to 80°F. and compared with the reading taken before the battery was placed on charge. If the rise in specific gravity in points per ampere-hour is known, the approximate time in hours required to complete the charge is as follows:

$$\frac{\text{rise in specific gravity in points to complete charge}}{\text{rise in specific gravity in points per ampere-hour} \times \text{charging rate in amperes}}$$

Gassing

When a battery is being charged, a portion of the energy is dissipated in the electrolysis of the water in the electrolyte. Thus, hydrogen is released at the negative plates and oxygen at the positive plates. These gases bubble up through the electrolyte and collect in the air space at the top of the cell. If violent gassing occurs when the battery is first placed on charge, the charging rate is too high. If the rate is not too high, steady gassing, which develops as the charging proceeds, indicates that the battery is nearing a fully charged condition. A mixture of hydrogen and air can be dangerously explosive. No smoking, electric sparks, or open flames should be permitted near charging batteries.

NICKEL-CADMIUM BATTERIES

The *nickel-cadmium batteries* are far superior to the lead-acid type. Some are physically and electrically interchangeable with the lead-acid type, while some are sealed units which use standard plug and receptacle connections which are used on other electrical components. These batteries generally require less maintenance than lead-acid batteries throughout their service life in regard to the adding of electrolyte or water.

The nickel-cadmium and lead-acid batteries have capacities that are comparable at normal discharge rates, but at high discharge rates the nickel-cadmium battery can:

1. Be charged in a short time.
2. Deliver a large amount of power.
3. Stay idle in any state of charge for an indefinite time and keep a full charge when stored for a long time.

4. Be charged and discharged any number of times without any appreciable damage.

5. The individual cells may be replaced if a cell wears out; the rest of the cells do not have to be replaced.

Due to their superior capabilities, nickel-cadmium batteries are being used extensively in many applications that require a battery with a high discharge rate. A prime example is the aircraft storage battery.

Some lead-acid batteries are equipped with the same quick-disconnect receptacle and plug used on nickel-cadmium batteries. In distinguishing a lead-acid battery from a nickel-cadmium battery or a silver-zinc battery, the nameplate of each battery should be checked, since the physical appearance could be the same. (*See* Fig. 18.)

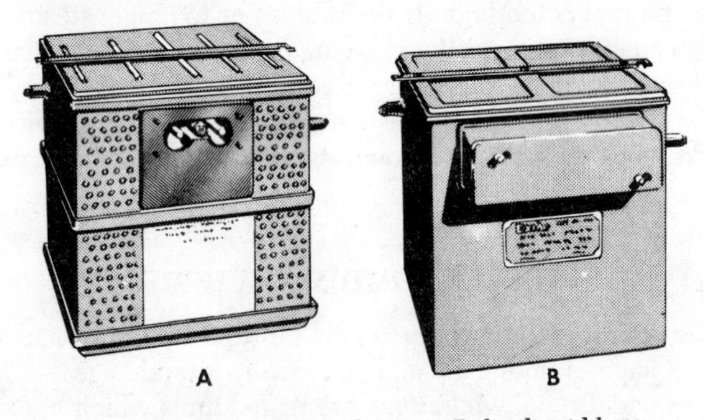

Fig. 18. A, Nickel-cadmium battery; B, lead-acid battery.

The nickel-cadmium battery plates are constructed of nickel powder sintered to a nickel-wire screen. The active materials (nickel-hydroxide on the positive plate and cadmium-hydroxide on the negative plate) are electrically bonded to the basic plate structure. The separators are constructed of plastic, nylon cloth, or a special type of cellophane, and assembled as a cell core with plates. (*See* Fig. 19.)

The construction of the sintered-plate cell is accomplished by a powder metallurgy process. Carbonyl nickel powder is lightly compressed in a mold and then is subjected either to a temperature of about 1,600°F. in a sintering furnace or to a sudden

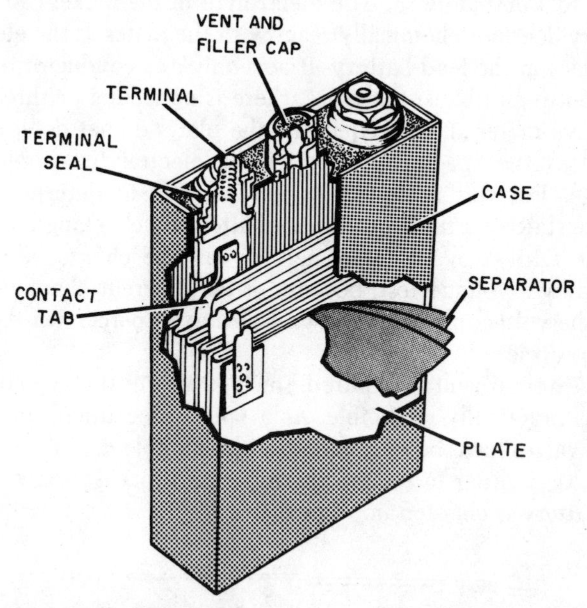

Fig. 19. Nickel-cadmium cell.

heavy electric current. Either process causes the individual grains of nickel to weld at their points of contact, producing a porous plaque that is approximately 80 percent open holes and 20 percent solid nickel. The plaques are then impregnated with active materials and soaked in a solution of nickel salts to make the positive plates and in a solution of cadmium salts to make negative plates. The bath is repeated until the plaques contain the amount of active material necessary to give them the desired capacity. When the plaques are impregnated, they are classified as plates.

The electrolyte used in a nickel-cadmium battery is a 30-percent-by-weight solution of potassium hydroxide in distilled water. Chemically speaking, this is just about opposite to the diluted sulfuric acid used in the lead battery. As with lead-acid batteries, there are limitations on the concentration of electrolyte solution that can be used in nickel-cadmium cells. The specific gravity of the solution should not be outside the range of

1.240 to 1.300 at 70°F. The electrolyte in the nickel-cadmium battery does not chemically react with the plates as the electrolyte does in the lead battery. It acts only as a conductor of current between plates; therefore, there is no flaking or shredding of active material. Consequently, the plates do not deteriorate, nor does the specific gravity of the electrolyte appreciably change. For this reason, it is not possible to determine the charge state of a nickel-cadmium battery by checking the electrolyte with a hydrometer; neither can the charge be determined by a voltage test, because of the inherent characteristic that the voltage remains constant during 90 percent of the discharge cycle.

No external vent is required, since gassing of this type of battery is practically negligible. As a safety precaution, however, relief valves have been installed in the fill hole cap of each cell (Fig. 20) in order to release any excess gas that is formed when the battery is charged improperly.

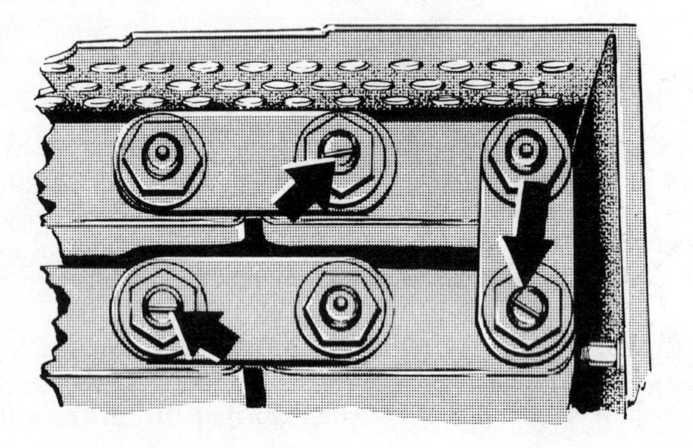

Fig. 20. Relief valves in negative post of cells.

Charge Determination

Two methods may be used for determining the state of charge of a nickel-cadmium battery; namely, the constant potential method and the discharge method. If the current falls to three amperes or less within five minutes, the battery is charged.

The *discharge method* consists of placing a 15-ampere load across the battery for five minutes. If the voltage does not drop below 22 volts during the discharge period, the battery may be returned to service after being recharged.

The available ampere-hour capacity cannot be accurately determined. Therefore, it is recommended that any battery whose charge is unknown or subject to doubt be discharged to or beyond the manufacturer's set end point of 1.0 or 1.1 volts, and then be recharged in accordance with the appropriate instructions. This process will prevent possible damage to the cells from overcharge.

Charging

Nickel-cadmium batteries should be charged at an ambient temperature of 70° to 80°F. Never allow a battery on charge to exceed 100°F., as this may cause overcharging and gassing. In the battery shop, a thermometer should be placed between the central cells in such a manner that the bulb of the thermometer is located below the top of the cell. Whenever the temperature of the battery is 100°F. or higher, the battery should not be charged.

The rate of charging of a nickel-cadmium battery is dependent upon two factors, the first being the charging voltage and the second being the temperature of the battery. In hot weather, where the ground air temperature approaches 90°F. or higher, the battery can be adequately charged at 27 volts. In mild ground air temperatures ranging from 35° to 85°F., the battery can be satisfactorily charged at 27.5 volts. In subfreezing weather, the battery requires a charging voltage of 28.5 volts.

The nickel-cadmium battery was designed and constructed to operate without gassing of the cells. The charging voltage should be maintained below the gassing voltage (approximately 29.4 volts at 80°F.) so that the life of the battery is prolonged. Therefore, on constant-potential charging in the battery shop, the voltage should be set at 28 volts or less. Under no circumstances should it exceed 28.5 volts.

If the battery has never been placed in service, follow the manufacturer's instructions accompanying the battery for the initial charge. If possible, the battery should be charged by the constant potential method.

For constant-potential charging, maintain the battery at 28 volts for four hours, or until the current drops below three amperes. Do not allow battery temperature to exceed 100°F.

For constant-current charging, start the charge at 10 to 15 amperes and continue until the voltage reaches 28.5 volts—then reduce the current to four amperes and continue charging until the battery voltage reaches 28.5 volts, or until the battery temperature exceeds 100°F. and the voltage begins to decline.

Never add electrolyte unless the battery is fully charged. Allow the fully charged battery to stand for a period of three or four hours before distilled water is added to bring the electrolyte to the proper level. A hydrometer or syringe can be used for introduction of the distilled water—just enough to cover the top of the plates. The battery solution is then recycled to stir the water and prevent it from freezing during cold-weather operation.

Safety Precautions

The electrolyte used in nickel-cadmium batteries is potassium hydroxide (KOH). This is a high corrosive alkaline solution, and should be handled with the same degree of caution as sulfuric acid (H_2SO_4). You should always wear rubber gloves, a rubber apron, and protective goggles when handling and servicing these batteries. If the electrolyte is spilled on the skin or clothing, the exposed area should be rinsed immediately with water, or, if available, vinegar, lemon juice, or boric acid solution. If the face or eyes are affected, treat as above and immediately get a medical examination and treatment.

SILVER-ZINC BATTERIES

Silver-zinc batteries are used largely in some industrial applications where their unique characteristics are sufficiently important to justify their comparatively high cost.

The silver-zinc battery was developed for two purposes. The major purpose was to secure a large quantity of electrical power for emergency operations. The secondary purpose was to permit a design that would save weight in new batteries. A lightweight silver-zinc battery provides as much electrical capacity as a much larger lead-acid or nickel-cadmium battery.

Operational silver-zinc batteries have a nominal operating voltage of 24 volts, obtained with sixteen 1.5-volt cells. Cell electrolytic levels should be monitored and adjusted periodically. The other required operations that might be considered maintenance are the normal recharging of the battery and keeping the top surfaces of the cells reasonably clean.

Characteristics

Because of its extremely low internal resistance, the silver-zinc battery is capable of discharge rates of up to 30 times its ampere-hour rating. The low internal resistance (as low as 0.0003 ohms per cell) is due primarily to the excellent conductivity of its plates, the close plate spacing (possible because small amounts of electrolyte may be used effectively), and the fact that the composition (and therefore the conductivity) of the electrolyte does not change during discharge. The internal conductivity of the battery increases during discharge as the positive plates are changed from oxides of silver (fair conductors) to metallic silver.

The high electrical capacity per unit of space and weight is a result of the close plate spacing, the large degree to which the active plate materials are utilized, and the absence of heavy supporting grids in the plate. Silver-zinc batteries are capable of producing as much as six times more energy per unit of weight and volume than other types. Silver-zinc cells have been built with capacities ranging from tenths of ampere-hours to thousands of ampere-hours.

Good voltage regulation is provided by the relatively constant voltage discharge characteristic of the silver-zinc battery. Terminal voltage is essentially constant throughout most of the discharge when discharged at higher than the two- or three-hour rate.

Silver-zinc batteries have a maximum service cycle life which is less than that of other types, but their life expectancy compares favorably with that of other types of batteries that are designed for maximum capacity per unit of space and weight, such as nickel-cadmium batteries.

Operation

The construction and electrochemical reactions of the silver-zinc battery are somewhat similar to those of the nickel-cadmium type. In the fully charged condition, the positive plates are composed of silver oxide and the negative plates of zinc. As the battery discharges, the positive plates are reduced to metallic silver and the negative plates are oxidized. Thus, when the battery is discharging, electrons are flowing out of the cathode (negative plates) and into the anode (positive plates) by way of the external circuit.

The electrolyte, potassium hydroxide in aqueous solution, exists as potassium (K) and hydroxide (OH) ions, which serve only to conduct the electric charge between the plates. Thus, the electronic or metallic conduction in the external circuit is balanced by the ionic or electrolytic conduction through the electrolyte, maintaining an equal net charge transfer into and out of each electrode.

As with other types of alkaline cells, and unlike lead-acid cells, the electrolyte does not take part in the chemical transformations, and therefore its specific gravity does not change with the state of charge of the cell. As long as the plates are covered, the electrical capacity of the battery is independent of the amount of electrolyte present.

In general, silver-zinc batteries require maintenance which is similar in many respects to that which is required of the lead-acid batteries. The state of charge is determined by testing the open circuit voltage of the battery. A silver-zinc battery tester or a voltmeter that reads accurately to 0.1 volt should be used for this test. If the reading is below 25.6 volts, remove the battery cover and inspect the top of the battery for corrosion or damaged cells. If any damage is evident, remove and replace the battery.

Charging

The silver-zinc battery is usually shipped in a dry condition. Only the special electrolyte furnished in the filling kit provided with each new battery should be used. Some battery types may use electrolytes containing special additives; other electrolytes, if used, may degrade the battery. The electrolyte must be kept in a closed alkali-resistant container, or it will absorb carbon dioxide from the air and deteriorate. The filling kit contains detailed instructions for filling and should be followed in detail. (*Note:* Batteries that will not be used within 30 days should be stored in the dry state.)

Silver-zinc batteries are sensitive to excessive voltage during charging and may be damaged if the voltage exceeds 2.05 volts per cell. Precautions must therefore be taken to insure that the charging equipment is adjusted accurately to cut off the current at 28.7 volts.

Where charging is not monitored automatically or periodically, a voltage cut-off system must be used to interrupt the charging current when the voltage rises to 28.7 volts.

If possible, charging should be performed at an ambient temperature of 60° to 90°F., and the battery temperature during charging should not exceed 150°F., as measured at the intercell connections.

While silver-zinc batteries do not generate any harmful gases during normal charge and discharge operations, they do generate both oxygen and hydrogen gases during excessive overcharging. All vent caps and sponge-rubber plugs must be removed from the vent holes during charging operations. If electrolyte is forced from the vent holes or if excessive gassing is evident, it is an indication of overheating, and the charging should be interrupted for eight hours to allow the battery to cool. After charging, the batteries should be allowed to stand idle at least eight hours.

The level of the electrolyte of each cell of the battery should be checked after charging, and the level adjusted either by removing any excess electrolyte or by adding distilled water if it is low.

The safety precautions relating to silver-zinc batteries are the same as those for the nickel-cadmium batteries.

SILVER-CADMIUM BATTERY

One of the most recent developments in storage batteries is the *silver-cadmium battery.* Generally, the most important requirements for evaluating and designing a battery are for high-energy density, good voltage regulation, long shelf life, repeatable number of cycles, and long service life expectancy. The silver-cadmium battery is designed to offer the overall maximum performance in each of these areas.

The silver-cadmium battery has more than twice the wet shelf life of the silver-zinc battery. The long shelf life plus the good voltage regulation make the silver-cadmium battery a highly desirable addition to the family of electric storage batteries. Its limitations include lower cell voltage than other rechargeable batteries and high initial cost.

Chapter 15

Transformers

A *transformer* is a device that has no moving parts and that transfers energy from one circuit to another by electromagnetic induction. The energy is always transferred without a change in frequency, but usually with changes in voltage and current. A step-up transformer receives electrical energy at one voltage and delivers it at a higher voltage. Conversely, a stepdown transformer receives energy at one voltage and delivers it at a lower voltage.

Transformers require little care and maintenance because of their simple, rugged, and durable construction. The efficiency of transformers is high. Because of this, transformers are responsible for the use of alternating current being more extensive than the use of direct current. The conventional constant-potential transformer is designed to operate with the primary connected across a constant-potential source and to provide a secondary voltage that is substantially constant from no load to full load.

Various types of small single-phase transformers are used in electrical equipment. In many installations, transformers are used on switchboards to step down the voltage for indicating lights. Low-voltage transformers are included in some motor control panels to supply control circuits or to operate overload relays.

Instrument transformers include potential, or voltage, transformers and current transformers. Instrument transformers are commonly used with a-c instruments when high voltages or large currents are to be measured.

Electronic circuits and devices employ many types of transformers to provide necessary voltages for proper electron-tube operation, interstage coupling, signal amplification, and so forth. The physical construction of these transformers differs widely.

The *power-supply transformer* used in electronic circuits is a single-phase constant-potential transformer with one or more secondary windings, or a single secondary with several tap connections. These transformers have a low volt-ampere capacity and are less efficient than large constant-potential power transformers. Most power-supply transformers for electronic equipment are designed to operate at a frequency of 50 to 60 Hz. Aircraft power-supply transformers are designed for a frequency of 400 Hz. The higher frequencies permit a saving in size and weight of transformers and associated equipment.

CONSTRUCTION

The typical transformer has two windings insulated electrically from each other. These windings are wound on a common magnetic core made of laminated sheet steel. The principal parts are: (1) The core, which provides a circuit of low reluctance for the magnetic flux; (2) the primary winding, which receives the energy from the a-c source; (3) the secondary winding, which receives the energy by mutual induction from the primary and delivers it to the load, and (4) the enclosure.

When a transformer is used to step up the voltage, the low-voltage winding is the primary. Conversely, when a transformer is used to step down the voltage, the high-voltage winding is the primary. The primary is always connected to the source of the power; the secondary is always connected to the load. It is common practice to refer to the windings as the primary and secondary rather than the high-voltage and low-voltage windings.

The principal types of transformer construction are the core and the shell, as illustrated respectively in Fig. 1. The cores are built of thin stampings of silicon steel. Eddy currents, generated in the core by the alternating flux as it cuts through the iron, are

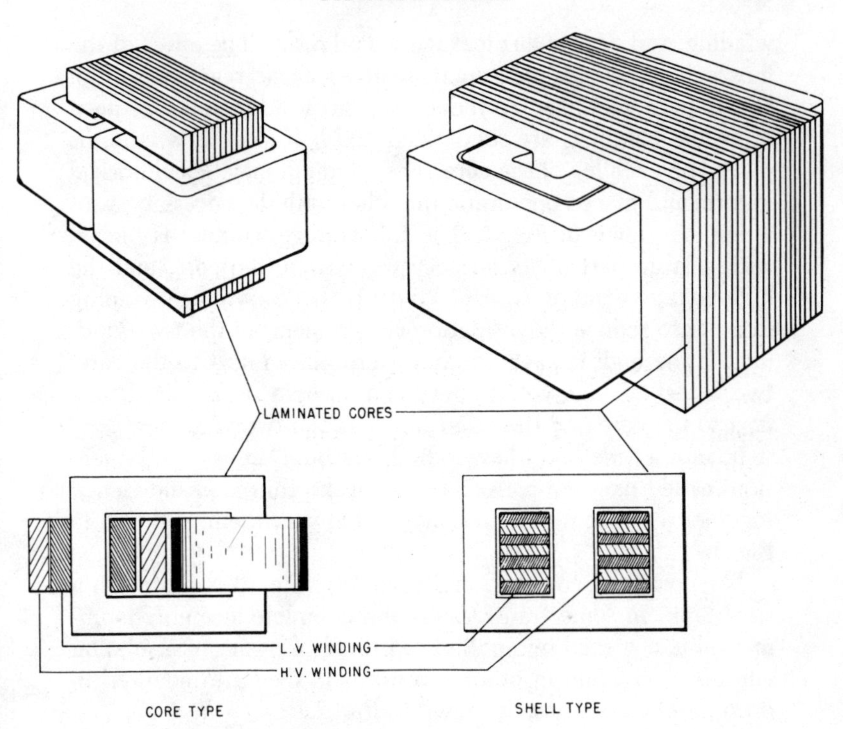

Fig. 1. Types of transformer construction.

minimized by using thin laminations and by coating adjacent laminations with insulating varnish. Hysteresis losses, caused by the friction developed between magnetic particles as they are rotated through each cycle of magnetization, are minimized by using a special grade of heat-treated, grain-oriented, silicon-steel laminations.

In the core type of transformer, the copper windings surround the laminated iron core. In the shell type of transformer, the iron core surrounds the copper windings. Distribution transformers are generally of the core type, whereas some of the largest power transformers are of the shell type.

If the windings of a core-type transformer were placed on separate legs of the core, a relatively large amount of the flux produced by the primary winding would fail to link the secondary winding, and a large flux leakage would result. The effect of the flux leakage would be to increase the leakage reactance drop, IX_L, in both windings. To reduce the flux leakage and reactance drop, the windings are subdivided and half of each winding is placed on each leg of the core. The windings may be cylindrical in form and placed one inside the other with the necessary insulation, as shown in Fig. 1. The low-voltage winding is placed with a large part of its surface area next to the core, and the high-voltage winding is placed outside the low-voltage winding in order to reduce the insulation requirements of the two windings. If the high-voltage winding were placed next to the core, two layers of high-voltage insulation would be required, one next to the core and the other between the two windings.

In another method, the windings are built up in thin flat sections called pancake coils. These pancake coils are sandwiched together with the required insulation between them, as shown in Fig. 1.

The complete core and coil assembly (Fig. 2) is placed in a steel tank. In some transformers the complete assembly is immersed in a special mineral oil to provide a means of insulation and cooling, while in other transformers they are mounted in drop-proof enclosures, as shown in Fig. 2.

Transformers are built in both single-phase and polyphase units. A three-phase transformer consists of separate insulated windings for the different phases, wound on a three-legged core capable of establishing three magnetic fluxes displaced 120 degrees in time phase.

VOLTAGE AND CURRENT RELATIONS

The operation of the transformer is based on the principle that electrical energy can be transferred efficiently by mutual induction from one winding to another. When the primary winding is energized from an a-c source, an alternating magnetic flux is established in the transformer core. This flux links the turns of both primary and secondary, thereby inducing voltages in them. Because the same flux cuts both windings, the same voltage is induced in each turn of both windings. Hence, the

Fig. 2. Single-phase transformer. A, Coil and core assembly, B, enclosure.

total induced voltage in each winding is proportional to the number of turns in that winding; that is,

$$\frac{E_1}{E_2} = \frac{N_1}{N_2}$$

where E_1 and E_2 are the induced voltages in the primary and secondary windings, respectively, and N_1 and N_2 are the number of turns in the primary and secondary windings, respectively. In ordinary transformers the induced primary voltage is almost equal to the applied primary voltage; hence, the applied primary voltage and the secondary induced voltage are approximately proportional to the respective number of turns in the two windings.

A constant-potential single-phase transformer is represented by the schematic diagram in Fig. 3, A. For simplicity, the primary winding is shown as being on one leg of the core and the secondary winding on the other leg. The equation for the voltage induced in one winding of the transformer is

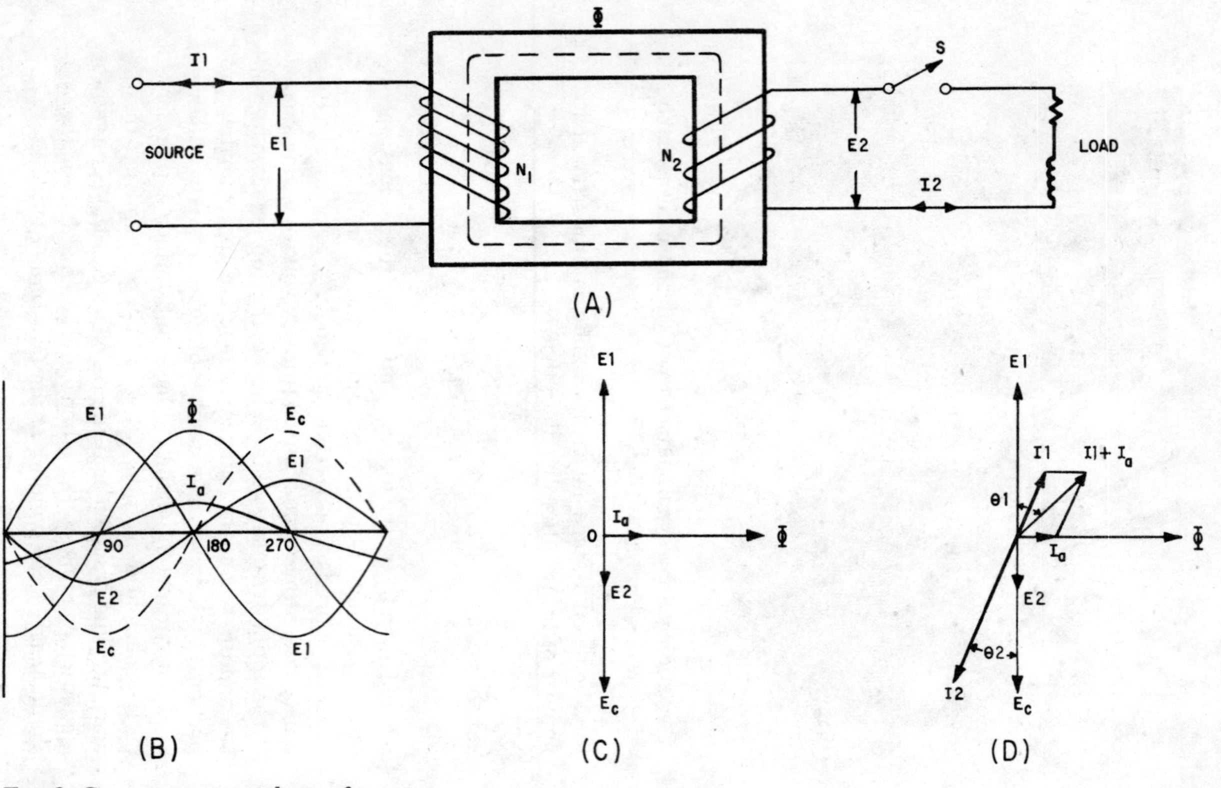

Fig. 3. Constant-potential transformer.

$$E = \frac{4.44\ BSfN}{10^8}$$

where E is the rms voltage, B the maximum value of the magnetic flux density in lines per square inch in the core, S the cross-sectional area of the core in square inches, f the frequency in hertz, and N the number of complete turns in the winding.

For example, if the maximum flux density is 90,000 lines per square inch, the cross-sectional area of the core is 4.18 square inches, the frequency is 60 Hz, and the number of turns in the high-voltage winding is 1,200, the voltage rating of this winding is

$$E_1 = \frac{4.44 \times 90{,}000 \times 4.18 \times 60 \times 1{,}200}{10^8}$$
$$= 1{,}200 \text{ volts}$$

If the primary-to-secondary turns ratio of this transformer is 10 to 1, the number of turns in the low-voltage winding will be

$$\frac{1{,}200}{10} = 120 \text{ turns}$$

and the voltage induced in the secondary will be

$$E_2 = \frac{1{,}200}{10} = 120 \text{ volts}$$

The waveforms of the ideal transformer with no load are shown in Fig. 3, B. When E_1 is applied to the primary winding, N_1, with the switch, S, open, the resulting current, I_a, is small and lags E_1 by almost 90 degrees because the circuit is highly inductive. This no-load current is called the exciting, or magnetizing, current because it supplies the magnetomotive force that produces the transformer core flux Φ. The flux produced by I_a cuts the primary winding, N_1, and induces a counter voltage, E_c, 180° out of phase with E_1, in this winding. The voltage, E_2, induced in the secondary winding is in phase with the induced (counter) voltage, E_c, in the primary winding, and both lag the exciting current and flux, whose variations produce them, by an angle of 90 degrees. These relations are shown in vector form in Fig. 3, C. The values are only approximate and are not drawn exactly to scale.

When a load is connected to the secondary by closing switch S (Fig. 3, A), the secondary current, I_2, depends upon the magnitude of the secondary voltage, E_2, and the load impedance, Z. For example, if E_2 is equal to 120 volts and the load impedance is 20 ohms, the secondary current will be

$$I_2 = \frac{E_2}{Z_2} = \frac{120}{20} = 6 \text{ amperes}$$

If the secondary power factor is 86.6 percent, the phase angle, θ_2, between secondary current and voltage will be the angle whose cosine is 0.866, or 30 degrees.

The secondary load current flowing through the secondary turns comprises a load component of magnetomotive force, which according to Lenz's law is in such a direction as to oppose the flux which is producing it. This opposition tends to reduce the transformer flux by a slight amount. The reduction in flux is accompanied by a reduction in the counter voltage induced in the primary winding of the transformer. Because the internal impedance of the primary winding is low and the primary current is limited principally by the counter emf in the winding, the transformer primary current increases when the counter emf in the primary is reduced.

The increase in primary current continues until the primary ampere-turns are equal to the secondary ampere-turns, neglecting losses. For example, in the transformer being considered, the magnetizing current, I_a, is assumed to be negligible in comparison with the total primary current, $I_1 + I_a$, under load conditions because I_a is small in relation to I_1 and lags it by an angle of 60 degrees. Hence, the primary and secondary ampere-turns are equal and opposite; that is,

$$N_1 I_1 = E_2 I_2$$

In this example,

$$I_1 = \frac{N_2}{N_1} I_2 = \frac{120}{1200} \times 6 = 0.6 \text{ ampere}$$

Neglecting any losses, the power delivered to be primary is equal to the power supplied by the secondary to the load. If the load power is $P_2 = E_2I_2 \cos \theta_2$, or 120×6 ($\cos 30° = 0.866$) = 624 watts, the power supplied to the primary is approximately $P_1 = E_1I_1 \cos \theta$, or 1200×0.6 ($\cos 30° = 0.866$) = 624 watts.

The load component of primary current, I_1, increases with secondary load and maintains the transformer core flux at nearly its initial value. This action enables the transformer primary to take power from the source in proportion to the load demand, and to maintain the terminal voltage approximately constant. The lagging-power-factor load vectors are shown in Fig. 3, D. Note that the load power factor is transferred through the transformer to the primary and that θ_2 is approximately equal to θ_1, the only difference being that θ_1 is slightly larger than θ_2 because of the exciting current, which flows in the primary winding but not in the secondary.

The copper loss of a transformer varies as the square of the load current, whereas the core loss depends on the terminal voltage applied to the primary, and on the frequency of operation. The core loss of a constant-potential transformer is constant from no load to full load because the frequency is constant and the effective values of the applied voltage, exciting current, and flux density are constant.

If the load supplied by a transformer has unity power factor, the kilowatt (true power) output is the same as the kilovolt-ampere (apparent power) output. If the load has a lagging power factor, the kilowatt output is proportionally less than the kilovolt-ampere output. For example, a transformer having a full-load rating of 100 kva can supply a 100-kw load at unity power factor, but only an 80-kw load at a lagging power factor of 80 percent.

Many transformers are rated in terms of the kva load that they can safely carry continuously without exceeding a temperature rise of 80°C when maintaining rated secondary voltage at rated frequency and when operating with an ambient (surrounding atmosphere) temperature of 40°C. The actual temperature rise of any part of the transformer is the difference between the total temperature of that part and the temperature of the surrounding air.

It is possible to operate transformers on a higher frequency than that for which they are designed, but it is not permissible to operate them at more than 10 percent below their rated frequency, because of the resulting overheating. The exciting current in the primary varies directly with the applied voltage and, like any impedance containing inductive reactance, the exciting current varies inversely with the frequency. Thus, at reduced frequency, the exciting current becomes excessively large and the accompanying heating may damage the insulation and the windings.

EFFICIENCY

The *efficiency* of a transformer is the ratio of the output power at the secondary terminals to the input power at the primary terminals. It is also equal to the ratio of the output to the output plus losses. That is,

$$\text{efficiency} = \frac{\text{output}}{\text{input}}$$

$$= \frac{\text{output}}{\text{output} + \text{copper loss} + \text{core loss}}$$

The ordinary power transformer has an efficiency of 97 to 99 percent. The losses are due to the copper losses in both windings and the hysteresis and eddy-current losses in the iron core.

The copper losses vary as the square of the current in the windings and as the winding resistance. In the transformer being considered, if the primary has 1,200 turns of number 23 copper wire, having a length of 1,320 feet, the resistance of the primary winding is 26.9 ohms. If the load current in the primary is 0.5 ampere, the primary copper loss is $(0.5)^2 \times 26.9 = 6.725$ watts.

Similarly, if the secondary winding contains 120 turns of number 13 copper wire, having a length of approximately 132 feet, the secondary resistance will be 0.269 ohm. The secondary copper loss is $I_2^2R_2$, or $(5)^2 \times 0.269 = 6.725$ watts, and the total copper loss is $6.725 \times 2 = 13.45$ watts.

The core losses, consisting of the hysteresis and eddy-current losses, caused by the alternating magnetic flux in the core, are approximately constant from no load to full load, with rated voltage applied to the primary.

In the transformer of Fig. 3, A, if the core loss is 10.6 watts and the copper loss is 13.4 watts, the efficiency is

$$\frac{\text{output}}{\text{output} + \text{copper loss} + \text{core loss}} =$$

$$\frac{624}{624 + 13.4 + 10.6} = \frac{624}{648} = 0.963$$

or 96.3 percent. The rating of the transformer is

$$\frac{E_1I_1}{1{,}000} = \frac{1{,}200 \times 0.5}{1{,}000} = 0.60 \text{ kva}$$

The efficiency of this transformer is relatively low because it is a small transformer and the losses are disproportionately large.

SINGLE-PHASE CONNECTIONS

Single-phase distribution transformers usually have their windings divided into two or more sections, as shown in Fig. 4. When the two secondary windings are connected in series (Fig. 4, A), their voltages add. In Fig. 4, B, the two secondary windings are connected in parallel, and their currents add. For example, if each secondary winding is rated at 120 volts and 100 amperes, the series-connection output rating will be 240 volts at 100 amperes, or 24 kva; the parallel-connection output rating will be 120 volts at 200 amperes, or 24 kva.

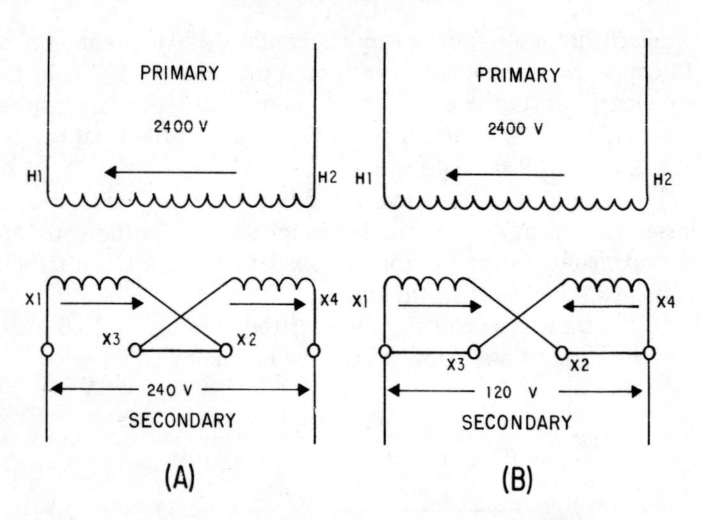

Fig. 4. Single-phase transformer secondary connections.

In the series connection, care must be taken to connect the coils so that their voltages add. The proper arrangement is indicated in Fig. 4, A. A trace made through the secondary circuits from X_1 to X_4 is in the same direction as that of the arrows representing the maximum positive voltages.

In the parallel connection, care must be taken to connect the coils so that their voltages are in opposition. The correct connection is indicated in Fig. 4, B. The direction of a trace made through the secondary windings from X_1 to X_2 to X_4 to X_3 and returning to X_1 is the same as that of the arrow in the right-hand winding. This condition indicates that the secondary voltages have their positive maximum values in directions opposite to each other in the closed circuit, which is formed by paralleling the two secondary windings. Thus, no circulating current will flow in these windings on no load. If either winding were reversed, a short-circuit current would flow in the secondary, and this would cause the primary to draw a short-circuit current from the source. This action would, of course, damage the transformer as well as the source.

THREE-PHASE CONNECTIONS

Power may be supplied through three-phase circuits containing transformers in which the primaries and secondaries are connected in various wye and delta combinations. For example, three single-phase transformers may supply three-phase power with four possible combinations of their primaries and secondaries. These connections are: (1) Primaries in delta and secondaries in delta, (2) primaries in wye and secondaries in wye, (3) primaries in wye and secondaries in delta, (4) primaries in delta and secondaries in wye.

If the primaries of three single-phase transformers are properly connected (either in wye or delta) to a three-phase source, the secondaries may be connected in delta, as shown in Fig. 5. A topographic vector diagram of the three-phase secondary voltages is shown in Fig. 5, A. The vector sum of these three voltages is zero. This may be seen by combining any two vectors—for example, E_A and E_B—and noting that their sum is equal and opposite to the third vector, E_C. A voltmeter inserted within the delta will indicate zero voltage, as shown in Fig. 5, B, when the windings are connected properly.

Assuming all three transformers have the same polarity, the delta connection consists of connecting the X_2 lead of winding A to the X_1 lead of B, the X_2 lead of C to X_1 of A. If any one of the three windings is reversed with respect to the other two windings, the total voltage within the delta will equal twice the value of one phase; and if the delta is closed on itself, the resulting current will be of short-circuit magnitude, with resulting damage to the transformer windings and cores. The delta should never be closed until a test is made to determine that the voltage within the delta is zero or nearly zero. This may be accomplished by using a voltmeter, fuse wire, or test lamp. In the illustration, when the voltmeter is inserted between the X_2 lead of A and the X_1 lead of B, the delta circuit is completed through the voltmeter, and the indication should be approximately zero. Then the delta is completed by connecting the X_2 lead of A to the X_1 lead of B.

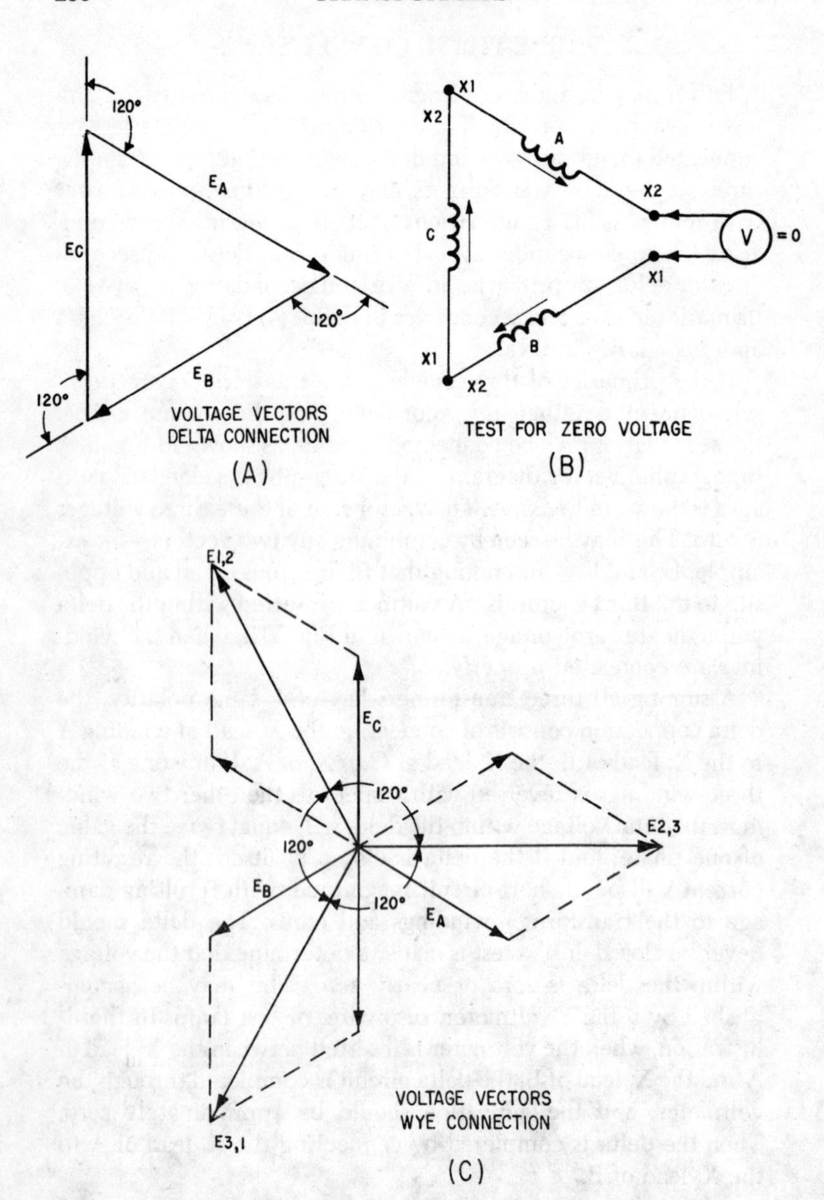

Fig. 5. Delta-connected transformer secondaries.

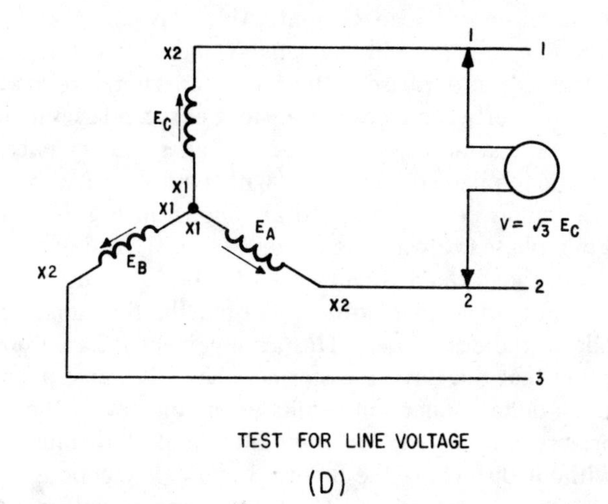

Fig. 5. (continued)

If the three secondaries of an energized transformer bank are properly connected in delta and are supplying a balanced three-phase load, the line current will be equal to 1.73 times the phase current. If the rated current of a phase (winding) is 100 amperes, the rated line current will be 173 amperes. If the rated voltage of a phase is 120 volts, the voltage between any two line wires will be 120 volts.

The three secondaries of the transformer bank may be reconnected in wye in order to increase the output voltage. The voltage vectors are shown in Fig. 5, C. If the phase voltage is 120 volts, the line voltage will be $1.73 \times 120 = 208$ volts. The line voltages are represented by vectors $E_{1,2}$, $E_{2,3}$, and $E_{3,1}$. A voltmeter test for the line voltage is represented in Fig. 5, D. If the three transformers have the same polarity, the proper connections for a wye-connected secondary bank are indicated in the illustration. The X_1 leads are connected to form a common or neutral connection, and the X_2 leads of the three secondaries are brought out to the line leads. If the connections of any one winding are reversed, the voltages between the three line wires will become unbalanced, and the loads will not receive their proper magnitude of load current. Also, the phase angle between the line currents will be changed, and they will no longer be 120 degrees out of phase with each other. Therefore, it is important to properly connect the transformer secondaries in order to preserve the symmetry of the line voltages and currents.

Three single-phase transformers with both primary and secondary windings delta connected are shown in Fig. 6. The H_1 lead of one phase is always connected to the H_2 lead of an adjacent phase, the X_1 lead is connected to the X_2 terminal of the corresponding adjacent phase, and so on; the line connections are made at these junctions. This arrangement is based on the assumption that the three transformers have the same polarity.

An open-delta connection results when any one of the three transformers is removed from the delta-connected transformer bank without disturbing the 3-wire 3-phase connections to the remaining two transformers. These transformers will maintain the correct voltage and phase relations on the secondary to supply a balanced 3-phase load. An open-delta connection is shown in Fig. 7. The 3-phase source supplies the primaries of the two transformers, and the secondaries supply a 3-phase voltage to the load. The line current is equal to the transformer phase current in the open-delta connection. In the closed-delta connection, the transformer phase current, $I_{phase} = \frac{I_{line}}{\sqrt{3}}$. Thus, when one transformer is removed from a delta-connected bank of three transformers, the remaining two transformers will carry a current equal to $\sqrt{3}\, I_{phase}$. This value amounts to an overload current on each transformer of 1.73 times the rated current, or an overload of 73.2 percent.

Therefore, in an open-delta connection, the line current must be reduced so as not to exceed the rated current of the individual transformers if they are not to be overloaded. The open-delta connection therefore results in a reduction in system capacity. The full-load capacity in a delta connection at unity power factor is

$$P_\Delta = 3I_{phase}E_{phase} = \sqrt{3}E_{line}I_{line}$$

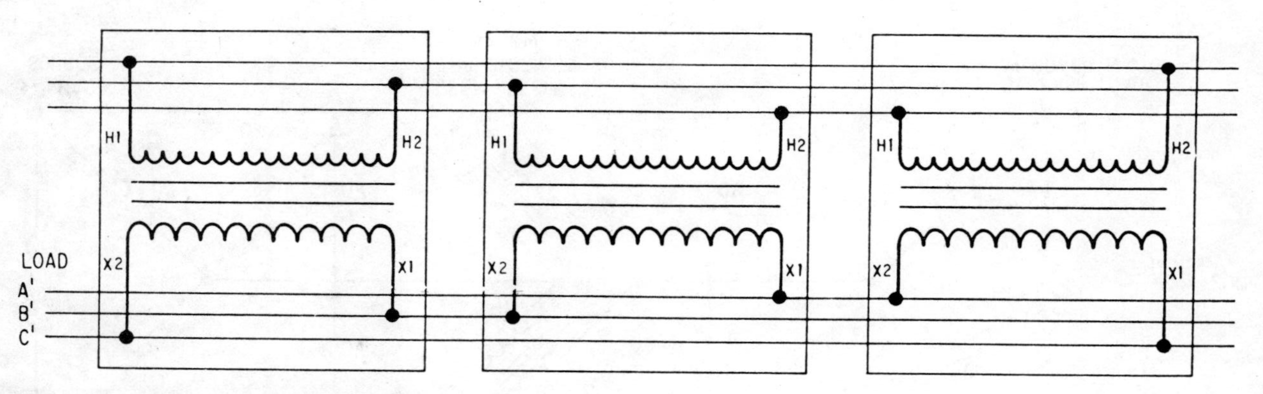

Fig. 6. Delta-delta transformer connections.

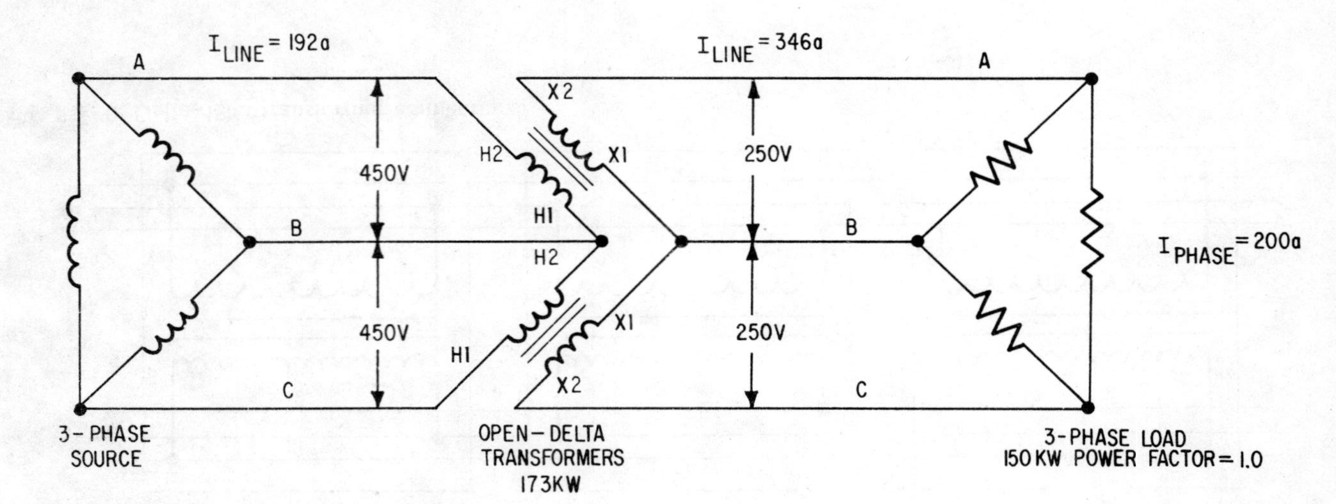

Fig. 7. Open-delta transformer connections.

In an open-delta connection, the line current is limited to the rated phase current of $\frac{I_{line}}{\sqrt{3}}$, and the full-load capacity of the open-delta, or V-connected, system is

$$P\gamma = \sqrt{3}E_{line}\frac{I_{line}}{\sqrt{3}} = E_{line}I_{line}$$

The ratio of the load that can be carried by two transformers connected in open delta to the load that can be carried by three transformers in closed delta is

$$\frac{P\gamma}{P\Delta} = \frac{E_{line}I_{line}}{\sqrt{3}E_{line}I_{line}} = \frac{1}{\sqrt{3}}$$
$$= 0.577, \text{ or } 57.7 \text{ percent}$$

of the closed-delta rating.

For example, a 150-kw 3-phase balanced load operating at unity power factor is supplied at 250 volts. The rating of each of three transformers in closed delta is 150/3 = 50 kw, and the phase current is 50,000/250 = 200 amperes. The line current is 200 $\sqrt{3}$ = 346 amperes. If one transformer is removed from the bank, the remaining two transformers would be overloaded 346 − 200 = 146 amperes, or 146/200 × 100 = 73 percent. To prevent overload on the remaining two transformers, the line current must be reduced from 346 amperes to 200 amperes and the total load reduced to

$$\frac{\sqrt{3} \times 250 \times 200}{1{,}000} = 86.6 \text{ kw}$$

or

$$\frac{86.6}{150} \times 100 = 57.7 \text{ percent}$$

of the original load.

The rating of each transformer in open delta necessary to supply the original 150-kw load is $\frac{E_{phase}I_{phase}}{1{,}000}$, or $250 \times \frac{346}{1{,}000} = 86.6$ kw, and two transformers require a total rating of $2 \times 86.6 = 173.2$ kw, compared with 150 kw for three transformers in closed delta. The required increase in transformer capacity is $173.2 - 150 = 23.2$ kw, or $23.2/150 \times 100 = 15.5$ percent, when two transformers are used in open delta to supply the same load as three 50-kw transformers in closed delta.

Three single-phase transformers with both primary and secondary windings wye connected are shown in Fig. 8. Only 57.7 percent of the line voltage $\left(\frac{E_{line}}{\sqrt{3}}\right)$ is impressed across each winding, but full-line current flows in each transformer winding.

Three single-phase transformers delta connected to the primary circuit and wye connected to the secondary circuit are shown in Fig. 9. This connection provides 4-wire, 3-phase service with 208 volts between line wires A′B′C′, and $\frac{208}{\sqrt{3}}$, or 120 volts, between each line wire and neutral N. The wye-connected secondary is desirable in installations when a large number of single-phase loads are to be supplied from a 3-phase transformer bank. The neutral, or grounded, wire is brought out from the midpoint of the wye connection, permitting the single-phase loads to be distributed evenly across the three phases. At the same time, 3-phase loads can be connected directly across the line wires. The single-phase loads have a voltage rating of 120 volts, and the 3-phase loads are rated at 208 volts. This connection is often used in high-voltage plate-supply transformers. The phase voltage is $\frac{1}{1.73}$, or 0.577 of the line voltage.

Three single-phase transformers with wye-connected primaries and delta-connected secondaries are shown in Fig. 10. This arrangement is used for stepping down the voltage from approximately 4,000 volts between line wires on the primary side to either 120 volts or 240 volts, depending upon whether the sec-

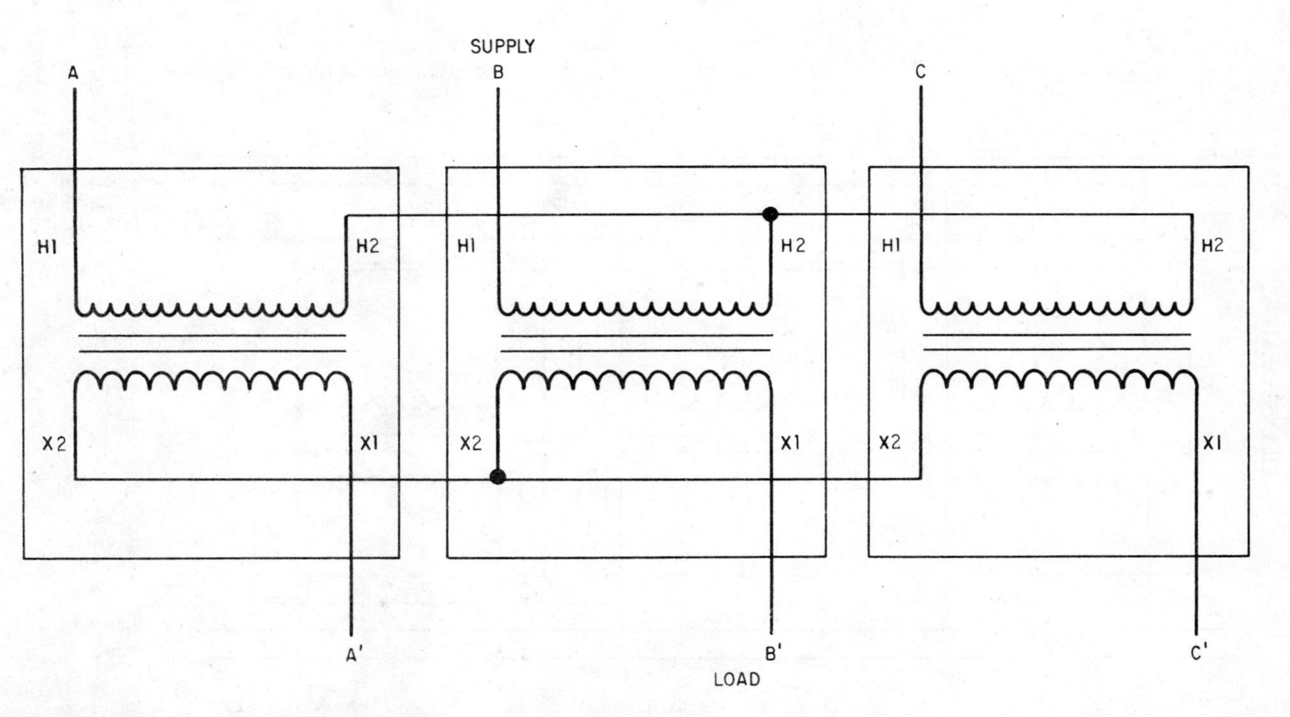

Fig. 8. Wye-wye transformer connections.

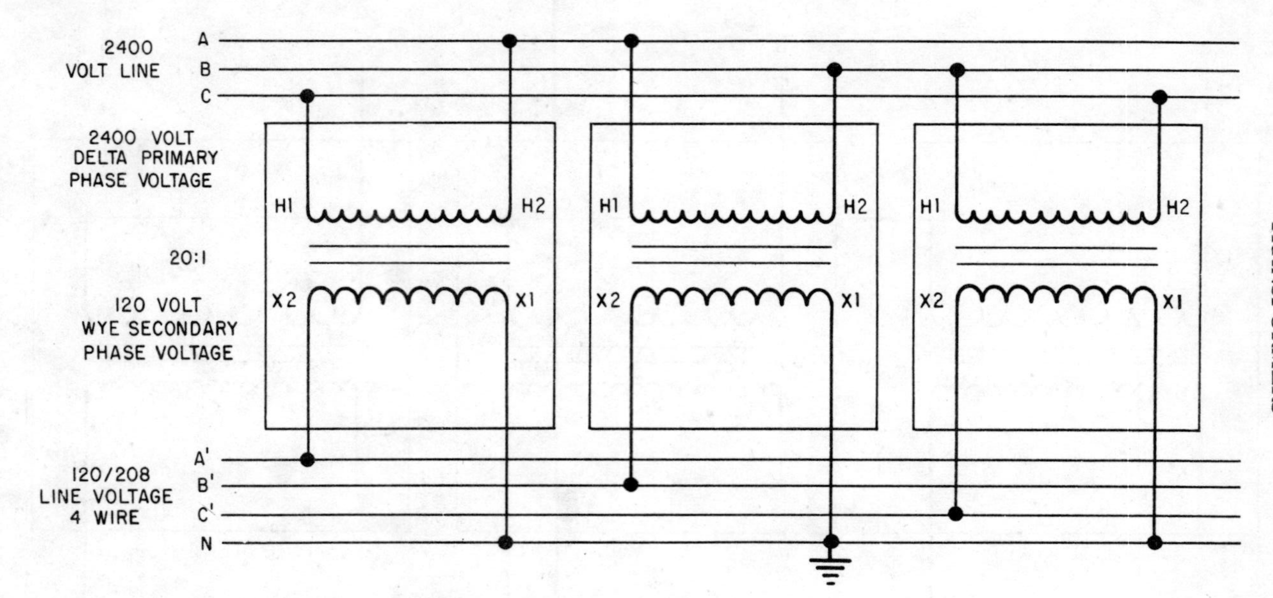

Fig. 9. Delta-wye transformer connections.

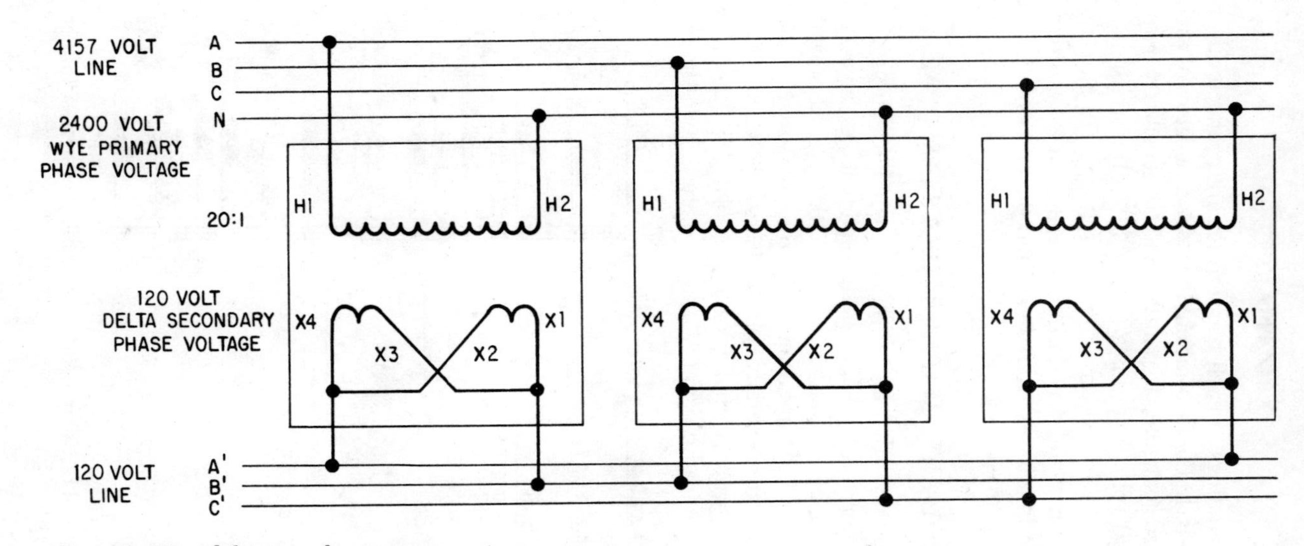

Fig. 10. Wye-delta transformer connections.

ondary windings of each transformer are connected in parallel or in series. In the illustration, the two secondaries of each transformer are connected in parallel, and the secondary output voltage is 120 volts. There is an economy in transmission with the primaries in wye because the line voltage is 73 percent higher than the phase voltage, and the line current is accordingly less. Thus, the line losses are reduced and the efficiency of transmission is improved.

POLARITY MARKING OF TRANSFORMERS

It is essential that all transformer windings be properly connected; therefore, it behooves the technician to have a basic understanding of the coding and the marking of transformer leads.

Small power transformers, of the size used in electronic equipment, are generally color coded, as shown in Fig. 11. In an untapped primary, both leads are black. If the primary is tapped, one lead is common and is colored black, the tap lead is

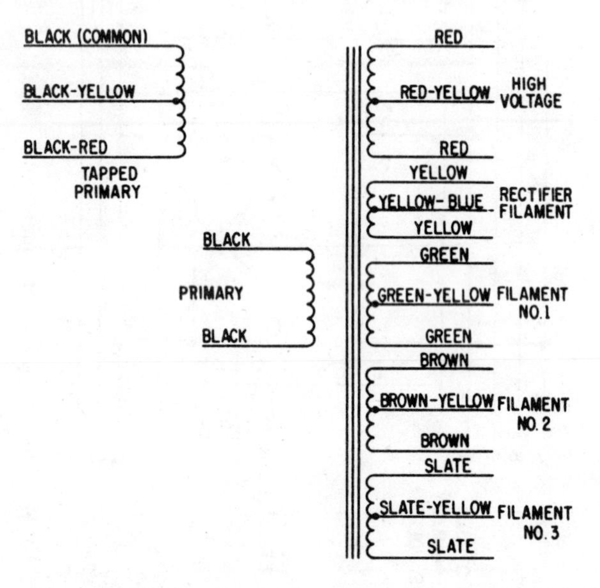

Fig. 11. Color coding of small power transformer leads.

black and yellow, and the other lead is black and red.

On the transformer secondary, the high-voltage winding has two red leads if untapped, or two red leads and a yellow tap lead if tapped. On the rectifier filament windings, yellow leads are used across the whole winding, and the tap lead is yellow and blue. If there are other filament windings, they will be either green, brown, or slate. The tapped wire will be yellow in combination with one of the colors just named, that is, green and yellow, brown and yellow, or slate and yellow.

The leads of large power transformers, such as those used for lighting and public utilities, are marked with numbers, letters, or a combination of both. This type of marking is shown in Fig. 12. Terminals for the high-voltage windings are marked H_1, H_2, H_3, and so forth. The increasing numerical subscript designates an increasing voltage. Thus, the voltage between H_1 and H_3 is higher than the voltage between H_1 and H_2.

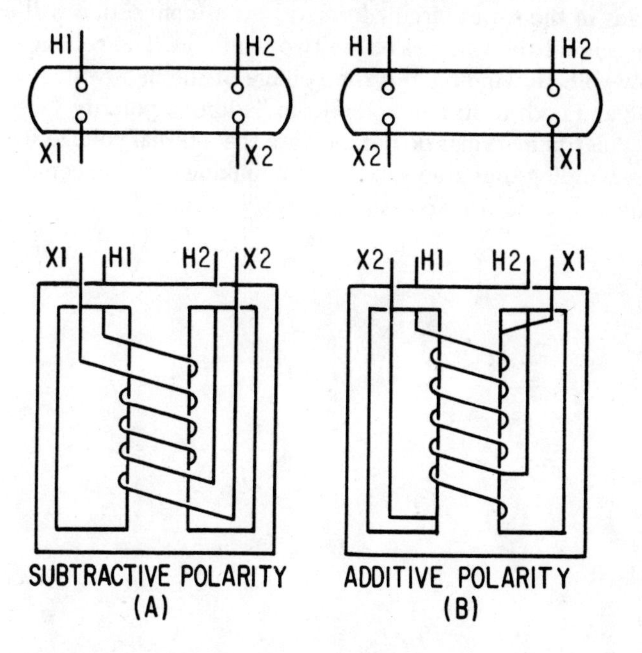

Fig. 12. Polarity markings for large transformers.

The secondary terminals are marked X_1, X_2, X_3, and so forth. There are two types of markings that may be employed on the secondaries. When the H_1 and X_1 leads are brought out on the same side of the transformer (Fig. 12, A), the polarity is called subtractive. The reason this arrangement is called subtractive is as follows: If the H_1 and X_1 leads are connected and a reduced voltage is applied across the H_1 and H_2 leads, the resultant voltage which appears across the H_2 and X_2 leads in the series circuit formed by this connection will equal the difference in the voltages of the two windings. The voltage of the low-voltage winding opposes that of the high-voltage winding and subtracts from it; hence the term "subtractive polarity."

When the H_1 and X_1 leads are brought out on opposite corners of the transformer (Fig. 12, B), the polarity is additive. If the H_1 and X_2 leads are connected and reduced voltage is applied across the H_1 and H_2 leads, the resultant voltage across the H_2 and X_1 leads in the series circuit formed by this connection will equal the sum of the voltages of the two windings. The voltage of the low-voltage winding aids the voltage of the high-voltage winding and adds to it; hence the term "additive polarity."

Polarity markings do not indicate the internal voltage stress in the windings but are useful only in making external connections between transformers, as previously mentioned.

Section V

Safety Rules and Procedures

Primary rules and procedures of safe practice for electricians are divided into four main groupings:

Work area safety
Electrical installation safe practice
Installation operating safety
Causes of electrical shock

Chapter 16

Safe Practices for Electricians

This chapter emphasizes those safety rules and principles in electrical wiring and repair work that, when strictly adhered to, will minimize or eliminate the hazards and dangers in electrical wiring. If the following procedures are strictly enforced and observed by all, both the cost of installing, operating, and maintaining the systems and the number of accidents will be reduced to a minimum. Safety should not be taken for granted. It must be kept in mind and practiced at all times.

WORK AREA SAFETY

Keeping the Job Area Clean

Grease, oil, water, or similar types of material on the floor can create slipping hazards which can cause serious injury to you or others in the area. In addition, loose fittings, wire or conduit ends, or other debris stored or thrown at random on the floor areas in or around the job site create tripping hazards which also can cause personal injury. Thus, development of good habits by workers in and around the work area has an additional advantage of establishing good work habits that are carried over into the installation of wiring.

Storage

All electrical materials and tools used by electricians should be stored in or on shelves, racks, or cabinets. The proper storage of materials will help to minimize the number of grounds or short circuits in any electrical circuit, because the possibility of accidental damage to conductor insulation is reduced. Good storage facilities can reduce the time required for wiring installation, since damaged conduit or fittings, or crushed cable and supports, can create time-consuming problems on the job. Adequate storage areas save money by preventing exposure of materials to damaging elements.

Tools and Equipment

All tools and equipment used by a worker should be maintained in good operating condition and should be replaced when broken. Improperly operating equipment should never be used in wiring practice from a standpoint of both safety and good workmanship. Tool handles should be inspected periodically for correct positioning with reference to their working surface and should also be checked to make sure they are tight. All handles should be insulated against shock hazard whenever possible. Drills, chisels, saws, and all similar cutting tools and equipment should be kept sharp. Workbench surfaces and surrounding floor areas should be covered with rubber insulation.

ELECTRICAL INSTALLATION

Personal Safety

Wire stripping. One of the most common operations of electrical workers is wire stripping. This can be done either with a tool designed for this purpose or with an electrician's knife. When knives are used, care should be taken that the force of cutting is directed away from the body or other people to prevent accidental injury.

Handtools. The general safety principles for electrical tools outlined previously in the section on tools and equipment are also applicable to personal safety in electrical installation. The insulated handles recommended for all tools commonly used by workers, however, should not be relied on for full shock immunity but should be used as an added safety precaution.

Personal clothing. Workers should be careful about the following:

1. Shirts, trousers, overalls, or other apparel with zippers or other metal fasteners should not be worn unless the fasteners are covered.

2. Rings and metal wrist bands or watch chains should not be worn while working.

3. Thin-soled shoes and shoes having heel plates or hobnails should not be worn. High-cut shoes with sewed soles are recommended.

4. Shirt sleeves must not be rolled down and loose; flapping clothing or such flammable articles as celluloid cap visors must not be worn.

5. Gloves or loose clothing must not be worn around moving machinery.

6. Suitable gloves and clothing and approved goggles or head shields must be worn by persons operating welding equipment.

Raising Equipment

Ladders. All ladders used by electricians to reach the work area should be of safe construction, kept in good repair, and never painted. For safety, a ladder should be anchored to the building with a rope or chain or placed against a projection on the supporting surface. If the ladder cannot be secured in either of these ways, the electrician should use an assistant at its base. The distance between the base of the ladder and the vertical surface against which it is supported should be approximately one-quarter the length of the ladder. *Caution* should be used with aluminum or other metal ladders so that they do not come into contact with live circuits or exposed wires.

Scaffolds. When the wiring location is inaccessible from normal work areas and the installation is large or complex, scaffolds rather than ladders should be used as overhead or elevated work platforms. All scaffolds should be sturdy, well-built, and securely anchored and cross-braced. The scaffold planking should be free of knots, checks, and cross-grained sections. Planks used for scaffolding should be load-tested periodically with up to three times their work load. They should not be used if failure or weakness shows up in the test.

Circuits

Deenergized circuits. Whenever possible, all electrical installation work should be done on a deenergized (dead) circuit. The method preferred by the majority of electricians is to trip the main service-entrance switch, and remove the fuses. In addition, the switch should be tagged. The tags should give the date, and the time the switch was turned off. If existing conditions do not permit this procedure, the circuit being repaired or modified should be deenergized. This is done either by removing the branch circuit fuses or by tripping to OFF the circuit breaker connected to the branch circuits.

Hot circuits. The practice of working on "hot" circuits is not recommended except in emergency situations. When the procedure is required, only experienced electricians should work on hot or energized circuits, and they should be aware of the dangers involved. Alternating-current voltages as low as 67 volts and direct-current voltages as low as 110 volts have caused death by electrocution. When an electrician has to work on a hot line, the following procedures should be closely followed.

1. He should insulate himself to prevent his body from becoming a conductor path for the current flow.
2. He should work on only one side of the circuit at a time.
3. He should always use insulating gloves and stand on nonconducting materials.
4. All equipment, tools, and wearing apparel should be dry. Dry materials offer more resistance to the flow of electricity than moistened or wet items.

Circuit Testing

Although an electrician may have deenergized a circuit by pulling the main switch, removing a circuit fuse, or tripping a local switch, it is imperative that he recheck the circuit with an indicating instrument such as a test lamp or an indicating voltmeter before starting work. The voltage rating of the test lamp should be twice the estimated voltage of the circuit under test, since the electrician cannot be certain of the voltage when spot checking. A series combination of test lamps, each rated at the estimated voltage, will also serve this purpose. This is done by placing the test prods of the instrument on the hot and neutral circuit wires either at the fuse, circuit breaker panel, device, or outlet box where the work is to be done. Never attempt to modify or repair an existing circuit until this test is made. Under no circumstances should circuits be tested with the fingers or with tools.

Wire Markers

On equipment or in wiring systems where wires are numerous and circuit identification is difficult, wire markers are a distinct aid to installation. However, they should be used merely as a guide, and the electrician should carefully check the condition of each conductor with a test lamp or voltmeter before working on the circuit. This safety rule also applies to wires that are color coded to conform with the prescribed circuit or equipment-ground requirements.

OPERATIONAL SAFETY (CIRCUIT PROTECTION)

Circuit Protection

Fuses or circuit breakers are installed as overload and short circuit protection for circuit components and connected loads. Their selection with regard to ampere rating must be limited to the maximum value allowable for the smallest conductor or equipment used in the circuit. For example, if an electrical device has a 15-ampere fusible rating and it is connected to a circuit wired with No. 12 gage conductors rated at 20 amperes, the fusing of the circuit should be 15 amperes. Tables 18 and 19 in Appendix 2 list the current requirements for single-phase and 3-phase ac motors at different voltage levels.

Overloading Circuits

Frequently blown fuses or tripped circuit breakers indicate short circuits, grounding, or circuit overloads. The problem cannot be solved merely by installing larger fuses than those recommended for use in the circuit. This action exposes both the building and the people to an electrical fire. First, the cause of trouble must be determined. Overloaded circuits should be rewired, increasing the wire size of the run or dividing the connected load into several circuits.

Bypassing Protection

Under no circumstances should the fuse or circuit breaker protection be bypassed by jumpers or removed from the circuit, except during testing for short circuits. These primary safety devices are of vital importance in a circuit installation. If these protective devices are bypassed, the electrical installation loses its circuit-current limitation and short-circuit protection. If the devices are removed, the troubles that caused the power interruption or circuit failure will not be isolated to the local circuit but will cause the rest of the fuses to blow or the circuit breakers to operate, cutting off the entire system from power.

Wiring Safety Practices

The practices described in this book conform to the rules of the National Electrical Code. Deviations from these practices because of emergency expedients must be corrected as soon as possible after the emergency ceases to exist.

ELECTRICAL SHOCK

Electrical shock can be caused by the following:

1. Equipment failure.
2. Human failure.
3. A combination of equipment failure and human failure.
4. A combination of events so unlikely and unusual that it would be almost impossible to prepare for them in advance. Luckily, accidents of this kind are rare. If they were not, working with electrical equipment would be extremely hazardous.

Equipment Failure

Seldom is electrical shock caused solely by equipment failure, but it can happen. In this category, there could be no human error: the equipment would have to be properly installed, tested for safety after installation, and used according to suitable safety precautions. The equipment failure would have to be completely unpredictable.

Human Failure

Human failures that can lead to electrical shock are as follows:

Failure to observe the necessary safety precautions when using or working on equipment that would be perfectly safe if handled properly. A careless electrician might fail to test equipment to make sure that it is deenergized before working on it. He might not use proper care to avoid contact with equipment or conductors that were known to be energized.

Making unauthorized modifications to equipment that has already caused nonfatal shocks.

Failure to make adequate repairs on equipment that has already caused nonfatal shocks.

Failure to test the insulation resistance and the completion of the ground connection after the equipment has been repaired.

Human failure perhaps is prevalent because people so often have a rather casual attitude toward the deadly potential of electric circuits and equipment. No way has been found to make electrical equipment that will not shock its users when it is improperly used.

Conditions for Shock

Two conditions must be satisfied for current to flow through a person.

1. The person must form part of a closed circuit in which current can flow.
2. Somewhere in the closed circuit there must be an electromotive force or a difference in potential to cause current flow.

Appendix I:

Glossary of Terms

Absorption. When light strikes an object, some of the light is reflected and some is absorbed. The darker the object, the more light it absorbs. Dark colors absorb large amounts of light, much as a blotter soaks up ink.

Absorption chiller. A refrigeration machine using heat as the power input to generate chilled water.

Absorption coefficient. The fraction of the total radiant energy incident on a surface that is absorbed by the surface.

Absorptivity. The physical characteristic of a substance describing its ability to absorb radiation.

Accessible (wiring methods). Not permanently closed in by the structure or finish of the building. Capable of being removed without disturbing the building structure or finish.

Activated carbon. A form of carbon capable of absorbing odors and vapors.

Active materials (storage battery). Materials of plates reacting chemically to produce electrical energy during the discharge.

Air changes. Expression of ventilation rate in terms of room or building volume. Usually air changes/hour.

Algebraic. Pertaining to that branch of mathematics which uses letters and symbols.

Algebraic sum. The addition of letters or numbers where some of them may represent negative quantities.

Alternating current (AC). A current of electrons that move in one direction and then another, alternating 60 times a second.

Alternation. One-half cycle of alternating current.

Alternator. An alternating current generator.

Ambient. Surrounding (i.e., ambient temperature is the temperature in the surrounding space).

Ammeter. Current meter with a scale calibrated in amperes.

Ampacity. Current-carrying capacity expressed in amperes.

Ampere. Unit of electric current equal to a coulomb per second.

Ampere hour. Unit of electrical energy used in rating storage batteries; the product of amperes and hours.

Ampere-hour capacity. The number of watt-hours which can be delivered by a cell or battery under specified conditions as to temperature, rate of discharge, and final voltage.

Ampere-hour-efficiency (electrochemical efficiency). The ratio of the ampere-hours output to the ampere-hours of the recharge.

Ampere turn. Unit of magnetizing force; the product of amperes and turns.

Amplidyne. A rotary magnetic or dynamo-electric amplifier used in servomechanism and control applications.

Amplification. The process of increasing the strength (current, power, or voltage) of a signal.

Amplifier. A device used to increase the signal voltage, current, or power, generally composed of a vacuum tube and associated circuit called a stage. It may contain several stages in order to obtain a desired gain.

Amplitude. In connection with alternating current or any other periodic phenomena, the maximum value of the displacement from the zero position.

Anode. The electrode in a cell (voltaic or electrolytic) that attracts the negative ions and repels the positive; the positive pole.

Antenna. A device used for sending or receiving radio waves.

Apparent power. Product of volts and amperes in ac circuits where the circuit and voltage are out of phase.

Appliance. Current-consuming equipment, fixed or portable.

Appliance, fixed. An appliance which (because of its size, its function and its location) is fixed in position in the home.

Appliance, portable. An appliance capable of being readily moved where established practice or the conditions of use make it necessary or convenient for it to be detached from its source of current by means of flexible cord and attachment plug.

Appliance, stationary. An appliance which is not easily moved but can be moved if necessary.

Approved. Acceptable to the authority enforcing the National Electrical Code.

Arc. A flash caused by an electric current ionizing a gas or vapor.

Armature. The rotating part of an electric motor or generator; also the iron part which completes the magnetic circuit in certain apparatus.

Atom. One of the minute particles of which the universe is composed; a natural group of electrons and protons.

Attenuator. A network of resistors used to reduce voltage, current, or power delivered to a load.

Auto-transformer. A transformer in which the primary and secondary are connected together in one winding.

Average voltage. The average value of the voltage during the period of charge or discharge. It is conveniently obtained from the time integral of the voltage curve.

Ballast. A device used in starting circuit for fluorescent and other types of lamps.

Bare lamp. Incandescent-filament or fluorescent lamp with no shielding.

Barrels (bbls) (energy measurement). 1 barrel equals 42 gallons.

Battery. A device for converting chemical energy into electrical; two or more cells.

Beam spread. In any plane, the angle between the two directions in which the candlepower is equal to a stated percentage (usually 10 percent) of the maximum candlepower in the beam.

Blow down. The discharge of water from a boiler or cooling tower sump that contains a high proportion of total dissolved solids.

Bobêche. A saucer-shaped element at the base of or below the candle socket in a candle-type fixture. In some modern fixtures of traditional styling the bobêche is often used to conceal light sources.

Boost charge (storage battery). A partial charge, usually at a high rate for a short period.

Bowl. Diffusing glass or plastic used to shield light sources from view.

Branch circuit. The portion of the wiring system extending beyond the final overcurrent device protecting the circuit.

Branch circuit light panel. Takes the power from the main switch and feeds it to the light and receptacle circuits through separate circuit breakers or fuses.

Breaker points. Metal contacts that open and close a circuit at timed intervals.

Bridge circuit. The electrical bridge circuit is a term referring to any one of a variety of electric circuit networks, one branch of which, the "bridge" proper, connects two points of equal potential and hence carries no current when the circuit is properly adjusted or balanced.

Brightness. The degree of apparent lightness of any surface emitting or reflecting light. Everything that is visible has some brightness.

British thermal unit (Btu). A standard measurement of energy content no matter what its source. The energy required to increase the temperature of one pound of water by 1° Fahrenheit.

Energy units translated into Btu's:

1 bbl crude oil—5,800,000 Btu's.
1 cubic foot of natural gas—1,024 Btu's.
1 gallon of gasoline—125,000 Btu's.
1 gallon of No. 2 fuel oil—139,000 Btu's.
1 kilowatt-hour—3,413 Btu's.

1 Mcf natural gas—1,024,000 Btu's.
1 therm of gas (or other fuel)—100,000 Btu's.
1 ton of bituminous coal—23,730,000 Btu's.

Brush. The conducting material, usually a block of carbon, bearing against the commutator or sliprings through which the current flows in or out.

Building envelope. All external surfaces which are subject to climatic impact; for example, walls, windows, roof, floor, and the like.

Building load. *See* Cooling load and Heating load.

Bulb. Glass enclosure of incandescent-filament lamp.

Bus bar. A primary power distribution point connected to the main power source.

BX cable. Trade name for armored cable.

Cable. A stranded conductor (single-conductor cable); or a combination of conductors insulated from one another (multiple-conductor cable).

Candela. A unit of luminous intensity, equal to 1/60 of the luminous intensity of a square centimeter of a black body heated to the temperature of the solidification of platinum (1773.5°C).

Candlelight. Fluorescent lamp color similar to deluxe warm white.

Candlepower. Luminous intensity.

Candlepower distribution curve. A curve showing the variation of luminous intensity of a lamp or luminaire.

Canopy. Shield (usually metal) covering joint of lighting fixture to ceiling.

Capacitance. Property of a circuit that opposes any change of voltage.

Capacitive reactance. The effect of capacitance in opposing the flow of alternating or pulsating current.

Capacitor. Two electrodes or sets of electrodes in the form of plates, separated from each other by an insulating material called the dielectric.

Capacity. Output capability over a period of time, expressed in ampere-hours.

Cathode. A fluorescent lamp part which serves to conduct the electricity from the wires into the gas.

Cavity ratio. Number indicating room cavity proportions, which is calculated using length, width, and height.

Cell. A combination of electrodes and electrolyte which converts chemical energy into electrical energy.

Centrifugal chiller. A refrigeration machine using mechanical energy input to drive a centrifugal compressor to generate chilled water.

Centrifugal fan. Device for propelling air by centrifugal action. Forward curved fans have blades which are sloped forward relative to direction of rotation. Backward curved fans have blades which are sloped backward relative to direction of rotation. Backward curved fans are generally more efficient at high pressures than forward curved fans.

Certified. Term applied to portable lamps and fixtures made to meet certain specifications for quantity of light, quality of lighting, sturdy construction, and safety. Fluorescent ballasts and starters meeting certain performance requirements are also *certified.* Such products usually carry a certification tag or label.

Channel. A common term for the metal enclosure containing the ballast, starter, lampholders, and wiring for a fluorescent lamp.

Charge. The process of sending an electric current through a storage battery to renew its action.

Charge, state of. Condition of a cell in terms of the capacity remaining in the cell.

Charging. Process of supplying electrical energy for conversion to stored chemical energy.

Charging rate. The current expressed in amperes at which a battery is charged.

Chemical energy. Energy stored in molecules, such as in fossil fuels. (*See* Fossil fuels.)

Choke coil. A coil of low ohmic resistance and high impedance to alternating current.

Chroma. The attribute of perceived color used to describe its departure from gray of the same lightness.

Circuit. A closed path or mesh of closed paths usually including a source of electromotive force (emf).

Circuit, branch. That portion of a wiring system extending beyond the final overcurrent device protecting the circuit.

Circuit breaker. A device designed to open under abnormal conditions a current-carrying circuit without injury to itself; the automatic type is designed to trip on a predetermined overload of current.

Circular mil. An area equal to that of a circle with a diameter of 0.001 inch. It is used for measuring the cross section of wires.

Closed-circuit voltage. The voltage at the terminals of a cell or battery when current is flowing.

Coaxial cable. A transmission line consisting of two conductors concentric with and insulated from each other.

Coefficient of utilization. Ratio of lumens on a work plane to lumens emitted by the lamps.

Coffer. Recessed panel in ceiling or dome.

Cold deck. A cold-air chamber forming part of a ventilating unit.

Collector rings. Means of conveying current to or from rotating parts of alternate current machinery.

Color rendering. General expression for the effect of a light source on the color appearance of objects in conscious or subconscious comparison with their color appearance under a reference light source.

Commutation. The process of converting alternating current which flows in the armature of direct current generators to direct current.

Commutator. The part of direct current rotating machinery which makes electrical contact with the brushes and connects the armature conductors to the external circuit.

Commutator ripple. The small pulsations which take place in the voltage and current of direct current generators.

Concealed circuit. A circuit that is hidden by the structure or finish of the building.

Concentric-lay cable. A single-conductor cable composed of a central core surrounded by one or more layers of helically laid groups of wires.

Concentric strand. A strand composed of a control core surrounded by one or more layers of helically laid wires or groups of wires.

Condensate. Water obtained by changing the state of water vapor (i.e., steam or moisture in air) from a gas to a liquid, usually by cooling.

Condenser. A heat exchanger which removes latent heat from a vapor, changing it to its liquid state; in refrigeration chillers, the component which rejects heat; a device for inserting the property of capacitance in a circuit; two or more conductors separated by a dielectric.

Conductance. The ability of a material to conduct or carry an electric current. It is the reciprocal of the resistance of the material, and is expressed in units of conductance ohms.

Conductance, thermal. A measure of the thermal conducting properties of a single material expressed in units of Btu inch thickness per (square foot) (hour) (degree F temperature difference).

Conductivity. The ease with which a substance transmits electricity.

Conductor. A wire or combination of wires not insulated from one another, suitable for carrying a single electric current.

Conductor—bare. A conductor with no covering or insulation whatsoever.

Conductor—covered. A conductor with one or more layers of nonconducting material which is not recognized as insulation.

Conductor—insulated. A conductor covered with a recognized insulation.

Conduit. A metal pipe used to enclose electrical conductors.

Constant-current charge. A charge in which the current is maintained at constant value.

Constant-voltage charge. A charge in which the voltage at the terminals of the battery is held at a constant value.

Contactor. A device for closing and opening electrical circuits remotely; a magnetically operated switch.

Continuous load. A load where the maximum current is expected to continue for three or more hours.

Control-center, branch. The term applied to an assembly of circuit breakers for the protection of branch circuits feeding from the control-center.

Control-center, main. The term applied to an assembly of circuit breakers for the protection of feeders and branch circuits feeding from the main control-center.

Control panel. An upright panel, open or closed, where switches, rheostats, and meters are installed for controlling and protecting electrical devices.

Convenience outlet. Electric receptable (often along baseboard) used for connecting portable lamps and appliances.

Converter, rotary. An electrical machine having a commutator at one end and sliprings at the other end of the armature, used for the conversion of alternating to direct current.

Cooling load. The rate of heat gain to the building at a steady state condition when indoor and outdoor temperatures are at their selected design levels, solar gain is at its maximum for the building configuration and orientation, and heat gains due to infiltration, ventilation, lights, and people are present.

Cooling tower. Device that cools water directly by evaporation.

Cool white. Fluorescent lamp *whiter* than filament lamp. Blends well with daylight; red colors may not appear as red under it.

Cord. A small cable, flexible and substantially insulated to withstand wear.

Core. A magnetic material that affords an easy path for magnetic flux lines in a coil.

Cornice. The crown molding used both to cover and to embellish the intersection between side walls and ceiling. In modern lighting practice, it is a horizontal member of wood or plaster attached to the ceiling approximately 6 to 8 inches from the wall, with lighting incorporated between it and the wall.

Cornice lighting. Comprises light sources shielded by a panel parallel to the wall and attached to the ceiling, distributing light downward over the vertical surface.

Coulomb. The unit of static electricity; the quantity of electricity transferred by one ampere in one second.

Counter electromotive force. An emf induced in a coil or armature that opposes the applied voltage.

Couple (storage battery). The element of a cell containing two plates, one positive and one negative. This term is also applied to a positive and a negative plate connected together as one unit for installation in adjacent cells.

Coupling. Term used to represent the means by which energy is transferred from one circuit to another.

Cove. In modern lighting practice, an element of design mounted on an upper wall to conceal light sources and provide indirect lighting.

Cove lighting. Comprises light sources shielded by a ledge and distributing light upward over the ceiling.

Crude oil or "crude." Petroleum in its natural state.

Cubic foot (cf). 1 cubic foot = the volume of a cube whose length, width, and breadth each measure 1 foot.

Current. Gradual drift of free electrons along a conductor; electrical volume.

Current limiter. A protective device similar to a fuse, usually used in high amperage circuits.

Cut-off voltage. Voltage at the end of useful discharge. (*See* End-point voltage.)

Cycle. In periodic phenomena, one complete set of recurring events; one complete positive and one complete negative alternation of a current or voltage.

Damper. A device used to vary the volume of air passing through an air outlet, inlet, or duct.

D'Arsonal galvanometer. A galvanometer in which a moving coil swings between the poles of a permanent horseshoe magnet.

Deep discharge. Withdrawal of all electrical energy to the endpoint voltage before the cell or battery is recharged.

Degree day. The difference between the median temperature of any day and 65°F. when the median temperature is less than 65°F.

Degree hour. The difference between the median temperature for any hour and selected datum.

Deluxe. Trade term applied to fluorescent lamps with good color rendition of all colors. There are cool white and warm white deluxe lamps.

Demand factor. The ratio of the maximum demand of a system, or part of a system, to the total connected load of a system or part of the system under consideration.

Density. Concentration of anything; quantity per unit volume or area.

Desiccant. A substance possessing the ability to absorb moisture.

Dielectric. An insulator; a term that refers to the insulating material between the plates of a capacitor; material which will not conduct an electric current.

Dielectric constant. Ratio of the capacitance of a condenser with a dielectric between the plates to the capacitance of the same condenser with a vacuum between the plates.

Diffuse reflection (diffusion). That process by which incident flux is redirected over a range of angles.

Diffuser. A device to scatter the light from a source primarily by the process of diffuse transmission.

Dimmer. A device for providing variable light output from lamps.

Diode. Vacuum tube—a two-element tube that contains a cathode and plate; semi-conductor—a material of either germanium or silicon that is manufactured to allow current to flow in only one direction. Diodes are used as rectifiers and detectors.

Direct current (dc). Current which is constant in magnitude and direction.

Direct expansion. Generic term used to describe refrigeration systems where the cooling effect is obtained directly from the refrigerant (e.g., refrigerant is evaporated directly in a cooling coil in the air stream).

Direct glare. Glare resulting from areas of excessive luminance and insufficiently shielded light sources in the field of view.

Direct lighting. A lighting system in which all or nearly all of the light is distributed downward by the luminaire. Fixtures that send somewhat more light upward but are still predominantly of the direct type are called semi-direct.

Disability glare. Spurious light from any source, which impairs a viewer's ability to discern a given object.

Discharge. Withdrawal of electrical energy from a cell or battery, usually to operate connected equipment.

Diversity factor. The ratio of the sum of the individual maximum demands of the various subdivisions of a system, or a part of a system, to the maximum demand of the whole system, or part under consideration.

Double bundle condenser. Condenser (usually in refrigeration machine) that contains two separate tube bundles allowing the option of either rejecting heat to the cooling tower or to another building system requiring heat input.

Double pole (D.P.). Having two contacts; e.g., a double pole switch.

Downlight. Luminaire which directs all the luminous flux down. Usually recessed, though it may be surface mounted.

Drain. Withdrawal of current from a cell.

Dry battery. A battery in which the electrolyte in a cell is immobilized, being either in the form of a paste or gel or absorbed in the separator material.

Dry bulb temperature. The measure of the sensible temperature of air.

Duplex cable. Two insulated single-conductor cables twisted together.

Duplex receptacle. A double outlet receptacle used in house wiring.

Dynamo. A machine for converting mechanical energy into electrical energy or vice versa.

Economizer cycle. A method of operating a ventilation system to reduce refrigeration load. Whenever the outdoor air conditions are more favorable (lower heat content) than return air conditions, outdoor air quantity is increased.

Eddy current. Induced circulating currents in a conducting material that are caused by a varying magnetic field.

Edison socket. A socket that accepts a light bulb with a screw-in base.

Efficacy of fixtures. Ratio of usable light to energy input for a lighting fixture or system (lumens/watt).

Efficiency. The ratio of the output of a cell or battery to the input required to restore the initial state of charge under specified conditions of temperature, current rate, and final voltage.

Electric circuit. A completed conducting pathway, consisting not only of the conductor, but including the path through the voltage source.

Electric power. Rate of doing electrical work.

Electricity. One of the fundamental quantities in nature consisting of elementary particles, electrons, and protons, which is manifested as a force of attraction or repulsion, and also in work that can be performed when electrons are caused to move; a material agency which exhibits magnetic, chemical, and thermal effects when in motion, and when at rest is accompanied by an interplay of forces between associated localities in which it is present.

Electrodes. The solid conductors of a cell or battery which are placed in contact with the liquid conductor that makes electrical contact with a liquid or gas.

Electrolyte. A solution of a substance which is capable of conducting electricity. An electrolyte may be in the form of either liquid or paste.

Electromagnet. Temporary magnet which is constructed by winding a number of turns of insulated wire about an iron core.

Electromotive force (emf). Difference of electrical potential or pressure measured in volts.

Electron. One of the ultimate subdivisions of matter having about 1/1845 of the mass of a hydrogen atom (carrying a negative charge of electricity); one of the negative particles of an atom.

Electronics. That brand of science and technology which relates to the conduction of electricity through gases or in vacuum.

Electro-strip wall receptacle. An insulated strip where several receptacles are installed on the wall for convenience.

Element (storage battery). The positive and negative groups with separators assembled for a cell.

End cells. The cells of a battery which may be cut in or out of the circuit for the purpose of adjusting the battery voltage.

End-point voltage. Cell voltage below which operation is not recommended.

Energy. The capacity to perform work.

Energy density. Ratio of cell energy to weight or volume (watt-hours per pound or watt-hours per cubic inch).

Energy requirement. The total yearly energy used by a building to maintain the selected inside design conditions under the dynamic impact of a typical year's climate. It includes raw fossil fuel consumed in the building and all electricity used for lighting and power. Efficiencies of utilization are applied, and all energy is expressed in the common unit of Btu's.

Enthalpy. For the purpose of air conditioning enthalpy is the total heat content of air above a datum usually in units of Btu/lb. It is the sum of sensible and latent heat and ignores internal energy changes due to pressure change.

Equalizing charge. An extended charge given to a battery to ensure the complete restoration of the active materials in all the plates of all the cells.

Equipment, electrical. The term (as used in this handbook) applies to electrical appliances and lighting fixtures—fixed and portable.

Equivalent sphere illumination. That illumination which would fall upon a task covered by an imaginary transparent hemisphere which passes light of the same intensity through each unit area.

Evaporator. A heat exchanger which adds latent heat to a liquid, changing it to a gaseous state. (In a refrigeration system it is the component which absorbs heat.)

Excitor. Small generator for supplying direct current to the alternator's field windings.

Farad. Unit of capacitance equal to the amount of capacitance present when one volt can store one coulomb of electricity.

Feedback. A transfer of energy from the output circuit of a device back to its input.

Feeders. Any conductors of a wiring system between the service equipment and the branch circuit overcurrent device.

Field. The space containing electric or magnetic lines of force.

Field magnet. The magnet used to produce a magnetic field (usually in motors or generators).

Field of force. Region in space filled with force which spreads out in all directions and will act through a vacuum.

Field winding. The coil used to provide for magnetizing force in motors and generators.

Filter. A device which changes, by transmission, the magnitude and/or the spectral composition of the flux incident upon it.

Final voltage. The prescribed voltage upon reaching which the discharge is considered complete. The final voltage is usually chosen so that the useful capacity of the cell is realized. Final voltages vary with the type of battery, the rate of the discharge, temperature, and the service in which the battery is used.

Finishing rate. The rate of charge expressed in amperes to which the charging current for some types of lead batteries is reduced near the end of charge to prevent excessive gassing and temperature rise.

Fixture. *See* Luminaire.

Float charging. Method of recharging in which a secondary cell is continuously connected to a constant-voltage supply that maintains the cell in fully charged condition.

Floating. A method of operation in which a constant voltage is applied to the battery terminals sufficient to maintain an approximately constant state of charge.

Flood lamp (R or PAR). Incandescent filament lamp providing a relatively wide beam pattern.

Floor duct. Large square conduit buried under a concrete floor. It has receptacle outlets flush with the floor.

Fluorescence. A property of phosphorus which, under controlled conditions, radiates light when absorbing electrons.

Fluorescent lamp (tube). A low-pressure mercury electric-discharge lamp in which a fluorescing coating (phosphor) transforms ultraviolet energy into visible light.

Flux. Magnetic field which is established in a magnetic circuit.

Foot-candle. Energy of light at a distance of one foot from a standard (sperm oil) candle.

Footlambert (fL). A quantitative unit for measuring luminance. The footcandles striking a diffuse reflecting surface, times the reflectance of that surface, equals the luminance in footlamberts.

Force. That which tends to change the state of rest or motion of matter.

Fossil fuels. Fuels derived from the remains of carbonaceous fossils, including petroleum, natural gas, coal, oil shale (a fine-grained laminated sedimentary rock that contains an oil-yeilding material called kerogen), and tar sands.

Free electrons. Electrons which are loosely held and consequently tend to move at random among the atoms of the material.

Frequency. In periodic phenomena the number of complete reoccurrences in unit time; in alternating current the number of cycles per second.

Full-wave rectifier circuit. A circuit which utilizes both the positive and the negative alternations of an alternating current to produce a direct current.

Fuse. A circuit-protecting device which makes use of a substance that has a low melting point.

Gain. The ratio of the output power, voltage, or current to the input power, voltage, or current, respectively.

Galvanometer. An instrument used to measure small dc currents.

Gassing (storage battery). The evolution of oxygen or hydrogen, or both.

General lighting. The lighting designed to provide a substantially uniform level of illumination throughout an area, exclusive of any provision for special local requirements.

Generator. A device for converting mechanical energy into electrical energy.

Geothermal energy. Energy extracted from the heat of the earth's interior.

Glare. Any brightness or brightness relationship that annoys, distracts, or reduces visibility.

Glitter. Small areas of high brightness, sometimes desirable to provide sensory stimulation.

Grid. A metal wire mesh placed between the cathode and plate.

Grid battery. The battery used to supply the desired potential to the grid.

Grid leak. A very high resistance placed in parallel with the grid condenser.

Ground. A metallic connection with the earth to establish ground potential. Also, a common return to a point of zero potential.

Group (storage battery). Assembly of a set of plates of the same polarity for one cell.

Harp. A rigid wire device which fits around the socket and bulb and supports the shade on some styles of floor and table lamps.

Heat gain. As applied to heating, ventilating, and air conditioning (HVAC) calculations, it is that amount of heat gained by a space from all sources, including people, lights, machines, sunshine, and so on. The total heat gain represents the amount of heat that must be removed from a space to maintain indoor comfort conditions.

Heating load. The rate of heat loss from the building at steady state conditions when the indoor and outdoor temperatures are at their selected design levels (design criteria). The heating load always includes infiltration and may include ventilation loss and heat gain credits for lights and people.

Heat, latent. The quantity of heat required to effect a change in state.

Heat loss. The sum cooling effect of the building structure when the outdoor temperature is lower than the desired indoor temperature. It represents the amount of heat that must be provided to a space to maintain indoor comfort conditions.

Heat pump. A refrigeration machine capable of reversing the flow so that its output can be either heating or cooling. When used for heating, it extracts heat from a low temperature source to the point where it can be used.

Heat, specific. Ratio of the amount of heat required to raise a unit mass of material one degree to that required to raise a unit mass of water one degree.

Heat transmission coefficient. Any one of a number of coefficients used in the calculation of heat transmission by conduction, convection, and radiation, through various materials and structures.

Henry. Unit of inductance; the inductance present which will cause one volt to be induced if the current changes at the rate of one ampere per second.

Hertz. A unit of frequency equal to one cycle per second.

Hickey. A tool used to bend metal pipe or conduit (commonly called a pipe bender).

High-rate discharge. Withdrawal of large currents for short intervals of time, usually at a rate that would completely discharge a cell or battery in less than one hour.

Horsepower. The English unit of power, equal to work done at the rate of 550 foot-pounds per second. Equal to 746 watts of electrical power.

Horsepower-hour. Energy expended if work is done for one hour at the rate of one horsepower (hp).

Hot deck. A hot-air chamber forming part of a ventilating unit.

Hue. The attribute of perceived color which determines whether it is red, yellow, green, blue, or the like.

Humidifier. A device designed to increase the humidity within a room or a house by means of the discharge of water vapor. It may consist of individual room-size units or larger units attached to the heating plant to condition the entire house.

Humidity, relative. A measurement indicating moisture content of air.

Hydrometer. Device for measuring the specific gravity of liquids.

Hydropower energy. Energy possessed by matter in motion.

Hysteresis. A lagging or retardation of effect when the forces acting upon a body are changed; encountered both in magnetic and dielectric phenomena.

Illumination. The density of luminous flux on a surface.

Impedance. The total opposition that a battery offers to the flow of alternating current or any other varying current at a particular frequency. Impedance is a combination of resistance and reactance.

Incandescent filament lamp (bulb). A lamp in which light is produced by a filament heated to incandescence by an electric current.

Indirect lighting. Lighting provided by luminaires which distribute all light upward. Luminaires sending some luminous flux downward but still predominantly of the indirect type are called semi-direct.

Inductance. The property of a circuit which tends to oppose a change in the existing current.

Induction. The act or process of producing voltage by the relative motion of a magnetic field across a conductor.

Induction coil. Two coils so arranged that an interrupted current in the first produces a voltage in the second.

Inductive reactance. The opposition to the flow of alternating or pulsating current caused by the inductance of a circuit. It is measured in ohms.

Infiltration. The process by which outdoor air leaks into a building by natural forces through cracks around doors and windows, and the like.

Initial drain. Current that a cell or battery supplies at nominal voltage.

Initial voltage. The voltage of a cell or battery at the beginning of a charge or discharge. It is usually taken after the current has been flowing for a sufficient period of time for the rate of change of voltage to become practically constant.

Inphase. Applied to the condition that exists when two waves of the same frequency pass through their maximum and minimum values of like polarity at the same instant.

Insolation. The amount of solar radiation on a given plane. Expressed in Langleys or Btu/ft.^{2}.

Instantaneous value. The value at any particular instant of a quantity that is continually varying.

Insulation board, rigid. A structural building board made of coarse wood or cane fiber in 1/2- and 25/32-inch thicknesses.

Insulation, reflective. Sheet material with one or both surfaces of comparatively low heat emissivity, such as aluminum foil.

Insulation, thermal. Any material high in resistance to heat transmission that, when placed in the walls, ceiling, or floors of a structure, will not conduct electricity.

Insulator. A substance containing very few free electrons (a nonconductor); e.g., plastic, rubber.

Intercom. An audio communication system.

Internal resistance. Opposition to current flow within a cell (ohms).

Interreflectance. The portion of the lumens reaching the work plane that has been reflected one or more times in the space.

Interrupter. A device for the automatic making and breaking of an electrical circuit.

Inversely. Inverted or reversed in position or relationship.

Ion. An electrically charged atom.

Isogonic line. An imaginary line drawn through points on the earth's surface where the magnetic variation is equal.

Jar (storage battery). The container for the element and electrolyte of a cell. Specifically a jar for lead-acid cells is usually of hard-rubber composition or glass; but for nickel-iron-alkaline cells it is a nickel-plated steel container frequently referred to as a "can."

Joule. A unit of energy or work. A joule of energy is liberated by one ampere flowing for one second through a resistance of one ohm.

Junction box. A box in which connections of several wires are made.

Kilo. A prefix meaning 1,000.

Kilowatt (kW). A common unit of measurement of electrical energy, equal to 1,000 watts. One kilowatt is the equivalent of about 1½ horsepower.

Kilowatt-hour (kWh). 1,000 watt-hours. A unit of electrical energy equal to the energy delivered by the flow of 1 kilowatt of electrical power for 1 hour. (A 100-watt bulb burning for 10 hours will consume 1 kilowatt-hour of energy.) One barrel of oil equals 500 kWh. (Rounded from 492.7 kWh.)

Lag. The amount one wave is behind another in time; expressed in electrical degrees.

Laminated core. A core built up from thin sheets of metal and used in transformers and relays.

Laminations. The thin sheets or discs making up an iron core.

Lamps. A generic term for a man-made source of light. By extension, the term is also used to denote sources that radiate in regions of the spectrum adjacent to the visible. (A lighting unit consisting of a lamp with shade, reflector, enclosing globe housing, or other accessories is also called a *lamp.* To distinguish between the assembled unit and the light source within it, the latter is often called a *bulb* or a *tube,* if it is electrically powered.)

Langley. Measurement of radiation intensity. One Langley = 3.68 $Btu/ft.^2$.

Lead. The opposite of Lag. Also, a wire or connection.

Leakage. Term used to express current loss through imperfect insulators.

Lens. Device for optical control of luminous flux by the process of refraction.

Life cycle cost. The cost of the equipment over its entire life including operating and maintenance costs.

Lighting outlet. Means by which branch circuits are made available for connection to lampholders, to surface-mounted fixtures, to flush or recessed fixtures, or for extension to mounting devices for light sources in valances, cornices, or coves.

Line of force. A path through space along which a field of force acts, and shown by a line on a sketch.

Load. The power that is being delivered by any power-producing device. The equipment that uses the power from the power-producing device.

Load leveling. Deferment of certain loads to limit electrical power demand to a predetermined level.

Load profile. Time distribution of building heating, cooling, and electrical load.

Local action or self-discharge. The internal loss of charge which goes on continuously within a cell regardless of connections to an external circuit.

Local lighting. The lighting that illuminates a relatively small area or confined space.

Louver. A series of baffles used to shield a source from view at certain angles or to absorb unwanted light.

Low-rate discharge. Withdrawal of small currents for long periods of time, usually longer than one hour.

Lumen. The unit of luminous flux.

Lumiline. A tubular incandescent lamp with a filament extending the length of the tube and connected at each end to a disc base.

Luminaire. A complete light unit consisting of a lamp, or lamps, together with parts designed to distribute the light, to position and protect the lamps and to connect the lamps to the power supply.

Luminaire efficiency. The ratio of the luminous flux leaving a luminaire to that emitted by the lamp, or lamps, used therein.

Luminance (photometric brightness). The luminous intensity of any surface in a given direction per unit area of that surface as viewed from that direction.

Luminous ceiling. A ceiling area lighting system comprising a continuous surface of transmitting material of a diffusing or light-controlling character with light sources mounted above it.

Luminous flux. The descriptive term for the time rate of flow of light.

Lux (lx). A quantitative unit for measuring illumination; the illumination on a surface of one meter square on which there is a uniformly distributed flux of one lumen.

Magnetic amplifier. A saturable reactor-type device that is used in a circuit to amplify or control.

Magnetic circuit. The complete path of magnetic lines of force.

Magnetic field. The space in which a magnetic force exists.

Magnetic flux. The total number of lines of force issuing from a pole of a magnet.

Magnetic pole. Region where the majority of magnetic lines of force leave or enter the magnet.

Magnetism. The property of the molecules of certain substances, such as iron, by virtue of which they may store energy in the form of a field of force; it is caused by the motion of the electrons in the atoms of the substance; a manifestation of energy due to the motion of a dielectric field of force.

Magnetize. To convert a material into a magnet by causing the molecules to rearrange.

Magneto. A generator which produces alternating current and has a permanent magnet as its field.

Magnetomotive force. The force necessary to establish flux in a magnetic circuit or to magnetize an unmagnetized object.

Main. A main is a supply circuit to which other energy-consuming circuits are connected through automatic cutouts, such as fuses or breakers.

Main disconnect switch. Cuts off the power to the entire building. Its fuses will blow if a major short occurs, thus protecting the wiring.

Make-up. Water supplied to a system to replace that lost by blow down, leakage, evaporation, and the like.

Matter. Anything which has weight and occupies space.

Maximum value. The greatest instantaneous value of an alternating voltage or current.

Mcf. 1,000 cubic feet (of natural gas).

Measuring energy. Specific forms of energy are measured in different ways—barrels of oil, therms or cubic feet of gas, kilowatts or kilowatt-hours of electricity, and tons of coal, for example.

Megawatt (mW). 1 million watts, or 1,000 kilowatts.

Megger. A test instrument used to measure insulation resistance and other high resistances. It is a portable hand-operated dc generator used as an ohmmeter.

Megohm. A large unit of resistance; equal to one million ohms.

Micro. A prefix meaning one-millionth.

Microfarad. Practical unit of capacitance; one-millionth of a farad.

Mil. A prefix meaning one-thousandth of an inch.

Milliammeter. An ammeter that measures current in thousandths of an ampere.

Milliampere. Small unit of electric current; equal to one-thousandth of an ampere.

Millivoltmeter. A voltmeter reading thousandths of a volt.

Modular. A system arrangement whereby the demand for energy (heating, cooling) is met by a series of units sized to meet a portion of the load.

Molecule. A small natural particle of matter, usually composed of two or more atoms.

Motor. A device for converting electrical energy into mechanical energy.

Motor-generator. A motor and a generator with a common shaft used to convert line voltages to other voltages or frequencies.

Motor starter. Device for protecting electric motors from excessive current while they are reaching full speed.

Multimeter. A combination volt, ampere, and ohm meter.

Multiple or **parallel circuits.** Those circuits in which the components are so arranged that the current divides between them.

Munsell color system. A system of object color specification based on perceptually uniform color scaled for the three variables—hue, value, and chroma.

Mutual inductance. Inductance associated with more than one circuit.

Mutual induction. The inducting of an electromotive force (emf) in a circuit by the field of a nearby circuit.

Negative charge. The electrical charge carried by a body which has an excess of electrons.

Negative plate (storage battery). The grid and active material from which the current flows to the external circuit when the battery is discharging.

Neon-glow lamp testers. Device for determining if a circuit is live, for determining polarity of dc circuits, and for determining if a circuit is alternating or direct current.

Neutron. A particle having the weight of a proton but carrying no electric charge. It is located in the nucleus of an atom.

Norminal voltage. Voltage of a fully charged cell when delivering rated capacity.

Nuclear energy. Energy, largely in the form of heat, produced during nuclear chain reaction. This thermal energy can be transformed into electrical energy. (*See* Power.)

Nucleus. The central part of an atom that is mainly comprised of protons and neutrons. It is the part of the atom that has the most mass.

Null. Zero.

Office lights. Usually fluorescent fixtures hung end to end so that the power feeds into the end fixtures and the wires are run from fixture to fixture in troughs or housings. This is also called trough wiring.

Ohm. Fundamental unit of resistance.

Ohmmeter. Device for measuring resistance by merely placing test prods across the resistor to be measured and reading the indication on a calibrated scale.

Ohm's law. The relationship which exists among current, pressure, and resistance.

Open-circuit voltage. The voltage of a cell or battery at its terminals when no current is flowing.

Orifice plate. Device inserted in a pipe or duct which causes a pressure drop across it. Depending on orifice size it can be used to restrict flow or form part of a measuring device.

Orsat apparatus. A device for measuring the combustion components of boiler or furnace flue gases.

Outlet. Point on wiring system at which current is taken to supply fixtures, heaters, motors, and current-consuming equipment generally.

Outline lighting. An arrangement of incandescent lamps or gaseous tubes used to outline or call attention to certain features such as the shape of a building or the decoration of a window.

Overload. A greater load applied to a circuit than it was designed to carry.

Parabolic lamp (PAR). Parabolic aluminized reflector lamp, spot or flood distribution, made of hard glass for indoor or outdoor use.

Parallel circuit. Two or more paths for electrons to follow.

Peak value. *See* as Maximum value.

Perceived object color. The color perceived to belong to an object, resulting from characteristics of the object, of the incident light, and of the surround, the viewing direction, and observer adaptation.

Period. The time required for the completion of one cycle.

Permalloy. An alloy of nickel and iron having an abnormally high magnetic permeability.

Permanent magnet. Piece of steel or alloy which has its molecules lined up in such a way that a magnetic field exists without the application of a magnetizing force.

Permeability. Recriprocal of reluctance; a measure of the ease with which flux can be established in the magnetic circuit; a ratio of the flux produced by a current-carrying coil with a core to one without a core.

Phase. The portion of a whole period which has elapsed since the thing in question passed through its zero position in a positive direction.

Phase difference. The time in electrical degree by which one wave leads or lags another.

Physical. Of or pertaining to matter and material things involving no chemical changes.

Piggyback operation. Arrangement of chilled-water generation equipment whereby exhaust steam from a steam turbine-driven centrifugal chiller is used as the best source for an absorption chiller.

Pigtail. A flexible wire extending from a component for ease of connection.

Pilot cell. A selected cell whose temperature, voltage, and specific gravity of electrolyte are assumed to indicate the condition of the entire battery.

Polarity. An electrical condition determining the direction in which current tends to flow.

Pole. The section of a magnet where the flux lines are concentrated; also, where they enter and leave the magnet. An electrode of a battery.

Polyphase. A circuit that utilizes more than one phase of alternating current.

Portable batteries. Batteries designed to be transported during service.

Portable lighting. Table or floor lamp, or unit, which is not permanently affixed to the electrical power supply.

Positive charge. The electrical charge carried by a body which has become deficient in electrons.

Positive plate (storage battery). The grid and active material to which the current flows from the external circuit when the battery is discharging.

Potential. A characteristic of a point in an electric field or circuit indicated by the work necessary to bring a unit positive charge to it from infinity; the degree of electrification as referred to some standard as that of the earth.

Potential difference. The arithmetical difference between two electrical potentials; same as electromotive force, electrical pressure, or voltage.

Potential energy. Energy that is stored in matter because of its position or because of the arrangements of its parts.

Potentiometer. A variable voltage divider; a resistor which has a variable contact arm so that any portion of the potential applied between its ends may be selected.

Power. The rate at which work is done. Power commonly is measured in units such as horsepower or kilowatts. Most bulk electric power is generated in this country by converting chemical energy to thermal, then mechanical, then electrical energy in steam, gas turbine, or large diesel powerplants, all requiring coal or petroleum resources. A lesser amount of the nation's electricity is generated by nuclear power.

Power distribution panel. A panel that takes the power from the main switch and feeds it to the utilization equipment through separate circuit breakers or fuses.

Power factor. Ratio of true power to apparent power; equal to the cosine of the phase angle between the voltage and current.

Pressure. That which causes the current to flow.

Primary. Cell or battery which cannot be recharged efficiently or safely after any amount of discharge.

Primary line of sight. The line connecting the point of observation and the fixation point.

Prime mover. The source of mechanical power used to drive the rotor of a generator.

Protective disconnect (modern). *See* Circuit breaker.

Proton. The positive particles of an atom.

Pulsating direct current. Current which varies in magnitude but not in direction.

Quad. One quadrillion Btu's.

Quality of lighting. The distribution of brightness and color rendition in a visual environment. The term is used in a positive sense and implies that these attributes contribute favorably to visual performance, visual comfort, ease of seeing, safety, and aesthetics for the specific visual tasks involved.

Raceway. Any channel for holding wires, cables, or bus bars, designed expressly for, and used solely for, this purpose.

Radiant heating. A method of heating, usually consisting of a forced hot-water system with pipes placed in the floor, wall, or ceiling, or with electrically heated panels.

Range outlet. An electrical cook stove connection.

Ratio. The value obtained by dividing one number by another, indicating their relative proportions.

Raw source energy. The quantity of energy input at a generating station required to produce electrical energy, including all thermal and power conversion losses.

Reactance. The opposition offered to the flow of an alternating current by the inductance, capacitance, or both, in any circuit.

Readily accessible. Capable of being reached quickly for operation, renewal, or inspection, without requiring those to whom access is requisite to climb over or remove obstacles or to resort to portable ladders, chairs, and the like.

Receptacle. A contact device installed at an outlet for the connection of plug and flexible cord.

Recessed or flush unit. A luminaire mounted above the ceiling (or behind a wall or other surface) with the opening of the luminaire level with the surface.

Rechargeable. Capable of being recharged; refers to secondary cells or batteries.

Rectifier. Device for changing alternating current to pulsating direct current.

Reflectance. The ratio of the flux reflected by a surface or medium to the incident flux. This general term may be restricted by the use of one or more of the following adjectives: specular (regular), diffuse, and spectral.

Reflected glare. A glare resulting from specular reflections of high luminance in polished or glossy surfaces in the field of view.

Reflection. The process by which flux leaves a surface or medium from the incident side.

Reflector. A device used to redirect the luminous flux from a source by the process of reflection.

Reflector lamp (R). Incandescent filament lamp with reflector of silver or aluminum on inner surface.

Refraction. The process by which the direction of a ray of light changes as it passes obliquely from one medium to another.

Regressed unit. A luminaire designed with the control medium above the ceiling line.

Relay. Device for controlling electrical circuits from remote position; a magnetic switch.

Reluctance. The opposition to magnetic flux.

Resistance. The opposition to the flow of current caused by the nature and physical dimensions of a conductor.

Resistor. A circuit element whose chief characteristic is resistance; used to oppose the flow of current.

Retentivity. The measure of the ability of a material to hold its magnetism.

Reversal. Change in normal polarity of a storage cell.

Rheostat. A variable resistance for limiting the current in a circuit.

Roof spray. A system that reduces heat gain through a roof by cooling the outside surface with a water spray.

Rope-lay cable. A single-conductor cable composed of a central core surrounded by one or more layers of helically laid groups of wires.

Rotor. The rotating part of an ac induction motor.

Run windings. Those windings of large wire in an electric motor which are energized throughout the motor.

R-value. The resistance to heat flow expressed in units of square foot hour degree F/Btu.

Saturable reactor. A control device that uses a small dc current to control a large ac current by controlling core flux density.

Saturation. The condition existing in any circuit when an increase in the driving signal produces no further change in the resultant effect.

Schematic. A diagram of electronic circuits showing connections.

Seasonal efficiency. Ratio of useful output to energy input for a piece of equipment over an entire heating and cooling season. It can be derived by integrating part load efficiencies against time.

Secondary. Cell or battery which can be recharged; refers to secondary cells or batteries.

Self-inductance. Inductance associated with only one circuit.

Self-induction. The process by which a circuit induces an emf into itself by its own magnetic field.

Sensitivity. The degree of responsiveness measured inversely; in connection with current meters it is the current required for full-scale deflection; in connection with voltmeters it is the ohms per volt of scale on the meter.

Separator (storage battery). A device for preventing metallic contact between the plates of opposite polarity within the cell.

Series circuit. A circuit that contains only one possible path for electrons to follow.

Series connection. An arrangement of cells, generators, condensers, or conductors so that each carries the entire current of a circuit.

Series-wound. A motor or generator in which the armature is wired in series with the field winding.

Service. The conductors and equipment necessary for delivering electrical energy from the electrical source to the wiring system of the premises served.

Service cable. Service conductors (wires) made into cables.

Service conductors. That portion of the supply conductors which extends from the street main or duct or from transformers to the service equipment of the premises supplied. For overhead conductors this includes the conductors from the last line pole to the service equipment.

Service drop. That portion of overhead service conductors between the pole and the first point of attachment to the building.

Service drop line. The line that carries power from the transformer to the building.

Service entrance. The equipment needed to bring power into the building.

Service entrance conductors. That portion of service conductors between the terminals of service equipment and a point outside the building, clear of building walls, where joined by tap or splice to the service drop or to street mains or other source of supply.

Service equipment. The necessary equipment, usually consisting of circuit breaker or switch and fuses, and their accessories, located near point of entrance of supply conductors to a building and intended to constitute the main control and means of cutoff for the supply to that building.

Service feeders. Lines that carry power from the service drop to the meters.

Servo. A device used to convert a small movement into a greater movement or force.

Servomechanism. A closed-loop system that produces a force to position an object in accordance with the information that originates at the input.

Shelf life. For a dry cell, the period of time (measured from date of manufacture), at a storage temperature of 70°F., after which the cell retains a specified percentage (usually 90 percent) of its original energy content.

Shielding angle (of a luminaire). The angle between a horizontal line through the light center and the line of sight at which the bare source first becomes visible.

Short circuit. A direct connection across the source of current.

Silver bowl lamp. Incandescent filament lamp with silver reflector on the lower half of bowl. Provides indirect distribution.

Single phase. Standard alternating current which phases from positive to negative in a sine curve at the rate of 60 times per second.

Single-pole switch. A switch which has open and closed positions only.

Sol-air temperature. The theoretical air temperature that would give a heat flow rate through a building surface equal in magnitude to that obtained by the addition of conduction and radiation effects.

Solar energy. Energy radiated directly from the sun.

Solenoid. A tubular coil for the production of a magnetic field; electromagnet with a core which is free to move in and out.

Space charge. The cloud of electrons existing in the space between the cathode and plate in a vacuum tube, formed by the electrons emitted from the cathode in excess of those immediately attracted to the plate.

Special-purpose outlet. A point of connection to the wiring system for particular equipment.

Specific gravity. The ratio of the mass of a body to the mass of an equal volume of water at four degrees centigrade.

Specular angle. That angle between the perpendicular to a surface and the reflected ray. It is numerically equal to the angle of incidence.

Specular reflection. That process by which incident flux is redirected at the specular angle.

Specular surface. Shiny or glossy surfaces (including mirror and polished metals) that reflect incident flux of a specular angle.

Speed. Time rate of motion measured by the distance moved in unit time; in rotating equipment it is the revolution per minute, or rpm.

Spot lamp (R or Par). Incandescent filament lamp providing a relatively narrow beam pattern.

Starter. A device used in conjunction with a ballast for the purpose of starting an electric discharge (fluorescent) lamp.

Start windings. Those windings of smaller wire in an electric motor which are energized only to start the motor and which are turned off after the motor has achieved a set number of revolutions per minute.

Stationary batteries. Batteries designed for service in a permanent location.

Stator. The part of an ac generator or motor that has the stationary winding on it.

Step-down transformer. A transformer with fewer turns in the secondary than in the primary.

Step-up transformer. A transformer with more turns in the secondary than in the primary.

Storage battery. A device which may be used repeatedly for storing energy at one time in the form of chemical energy, for use at another time in the form of electrical energy.

Storage cell. Fundamental unit of any storage battery.

Stranded conductor. A conductor composed of a group of wires or any combination of groups of wires. (These wires are usually twisted or braided together.)

Stranded wire. A group of small wires, used as a single wire.

Subjective brightness. The subjective attribute of any light sensation giving rise to the percept of luminous intensity including the whole scale of qualities of being bright, light, brilliant, dim, or dark. (The term brightness occasionally is used when referring to measurable *photo-metric brightness.*)

Surface-mounted unit. A luminaire mounted directly on the ceiling.

Suspended unit. A luminaire hung from the ceiling by supports.

Switch. A device for opening or closing an electric circuit.

Synchronous. Having the same period and phase; happening at the same time.

Synchroscope. An instrument used to indicate a difference in frequency between two ac sources.

Synchro system. An electrical system that gives remote indications or control by means of self-synchronizing motors.

System ground wire. A wire that grounds the entire conduit system so that a shorted or grounded hot wire will blow a fuse or breaker rather than electrifying the entire conduit system.

Tachometer. An instrument for indicating revolutions per minute.

Task plane. *See* Work plane.

Temperature. The condition of a body which determines the transfer of heat to or from other bodies; condition as to heat or cold; degree of heat or cold.

Terminal posts. The points of the cell or battery to which the external circuit is connected.

Tertiary winding. A third winding on a transformer or magnetic amplifier that is used as a second control winding.

Test lamp. A weatherproof rubber-insulated socket into which is screwed an incandescent lamp of the highest voltage rating of the circuits involved. Used for rough tests on interior-lighting and motor-wiring systems.

Therm. A unit of heat equal to 100,000 Btu's.

Thermal energy. A form of energy whose effect (heat) is produced by accelerated vibration of molecules.

Thermistor. A resistor that is used to compensate for temperature variations in a circuit.

Thermocouple. A device for directly converting heat energy into electrical energy.

Three-phase. Alternating current with staggered phases of 120 degrees in a sine curve. Each operates similarly to a single phase.

Three-way switch. A three-screw switch in which the current may take either of two paths.

Time rate. The rate in amperes at which a battery will be fully discharged in a specified time, under specified conditions of temperature and final voltage.

Ton of refrigeration. A means of expressing cooling capacity—1 ton = 12,000 Btu/hour cooling.

Torchere. An indirect floor lamp which directs all, or nearly all, of the luminous flux upward.

Torque. The effectiveness of a force to produce rotation about a center.

Transformer. An apparatus for converting electrical power in an ac system at one voltage or current into electrical power at some other voltage or current without the use of rotating parts; a device for raising or lowering ac voltage.

Transmission. The characteristic of many materials such as glass, plastics, and textiles. The process by which incident flux leaves a surface or medium on a side other than the incident side.

Transmission lines. Any conductor or system of conductors used to carry electrical energy from its source to a load.

Transmittance. The ratio of the flux transmitted by a medium to the incident flux.

Tray (storage battery). A support or container for one or more cells.

Trickle charging. Method of recharging in which a secondary cell is either continuously or intermittently connected to a constant-current supply that maintains the cell in fully charged condition.

Triplex cable. Three insulated single-conductor cables twisted together.

Troffer. A long recessed lighting unit, usually installed with the opening flush with the ceiling.

True power. The actual power consumed by an ac circuit, equal I^2R; expression used to distinguish from apparent power.

Tube. *See* Fluorescent lamp.

Tungsten Halogen lamp. Compact incandescent filament lamp with initial efficacy essentially maintained over life of the lamp.

Twin cable. Two insulated single-conductor cables laid parallel, having a common covering.

Twin wire. Two small insulated conductors laid parallel, with a common covering.

Two-phase current. Two different alternating currents out of phase 90 degrees with each other.

Unidirectional. As applied to a current of electricity, a current that flows in one direction only.

Usable wall space. Where this term is used as the basis of requirements for spacing of outlets, it is defined as all portions of the wall except that which is not usable when a door is in its normal open position.

"U" value. A coefficient expressing the thermal conductance of a composite structure in Btu per square foot hour degree F temperature difference.

Vacuum tube. A tube from which the air has been pumped out. The tube contains an element that emits electrons when properly excited and an electrode to attract the electrons and set up a current in an external circuit.

Valance. A longitudinal shielding member mounted across the top of a window. Usually parallel to the window, it conceals the light sources and usually gives both upward and downward distributions. The same device applied to a wall is a wall bracket.

Value. The attribute of perceived color by which it seems to transmit, or reflect, a greater, or lesser, fraction of the incident light.

Vapor barrier. A moisture-impervious layer designed to prevent moisture migration.

Vars. Abbreviation for volt-ampere, reactive.

Vector. A line used to represent both direction and magnitude.

Veiling reflection. Reflection of light from a task, or work surface, into the viewer's eyes.

Visual acuity. The ability to distinguish fine details.

Visual angle. The angle which an object or detail subtends at the point of observation.

Visual field. The focus of objects or points in space which can be perceived when the head and eyes are kept fixed.

Visual surround. All portions of the visual field except the visual task.

Visual task. Those details and objects which must be seen for the performance of a given activity.

Volt. Unit of potential, potential difference, electromotive force, or electrical pressure.

Volt efficiency (storage battery). The ratio of the average voltage during the discharge to the average voltage during the recharge.

Voltage. Electromotive forcing electrical pressure.

Voltage regulator. Device used in connection with generators to keep voltage constant as load or speed is changed.

Voltages. When described as 115 or 230 volts, voltages should be understood to be nominal voltages and to include, respectively, voltage lf 110 to 125 and 220 to 250. In any location in which the electrical service is furnished at 120/208 volts from a three-phase, four-wire system, the local utility should be consulted regarding necessary changes in the requirements.

Voltmeter. An instrument for measuring potential difference or electrical pressure.

Wall switch. A switch on the wall, not a part of any fixture for the control of one or more outlets.

Watt. A unit of measure of electrical power; equal to a joule per second.

Watt-hour capacity. The number of watt-hours which can be delivered by a cell or battery under specified conditions as to temperature, rate of discharge, and final voltage.

Watt-hour efficiency (energy efficiency). The ratio of the watt-hours output to the watt-hours of the recharge.

Wattmeter. An instrument for measuring electrical power in watts.

Weight. The force with which a body is attracted toward the center of the earth by the gravitational field of force.

Wet. Indication that the liquid electrolyte in a cell is free-flowing.

Wet bulb temperature. The lowest temperature attainable by evaporating water in the air without the addition or subtraction of energy.

Wind energy. Energy derived from the wind.

Wire. A slender rod or filament of drawn metal.

Wire sizes. The size of wire is usually expressed according to some wire gage.

Work. The result of a force acting against opposition to produce motion; it is measured in terms of the product of the force and the distance it acts.

Work centers, kitchen. Refrigerator and preparation center; sink and dishwashing center; range and serving center.

Work plane. The plane at which work is done and at which illumination is specified; this is assumed to be a horizontal plane at the level of the task.

Appendix II:

Electrical Data Tables 7 Through 27

TABLE 7

CHARACTERISTICS OF ELECTRICAL SYSTEMS

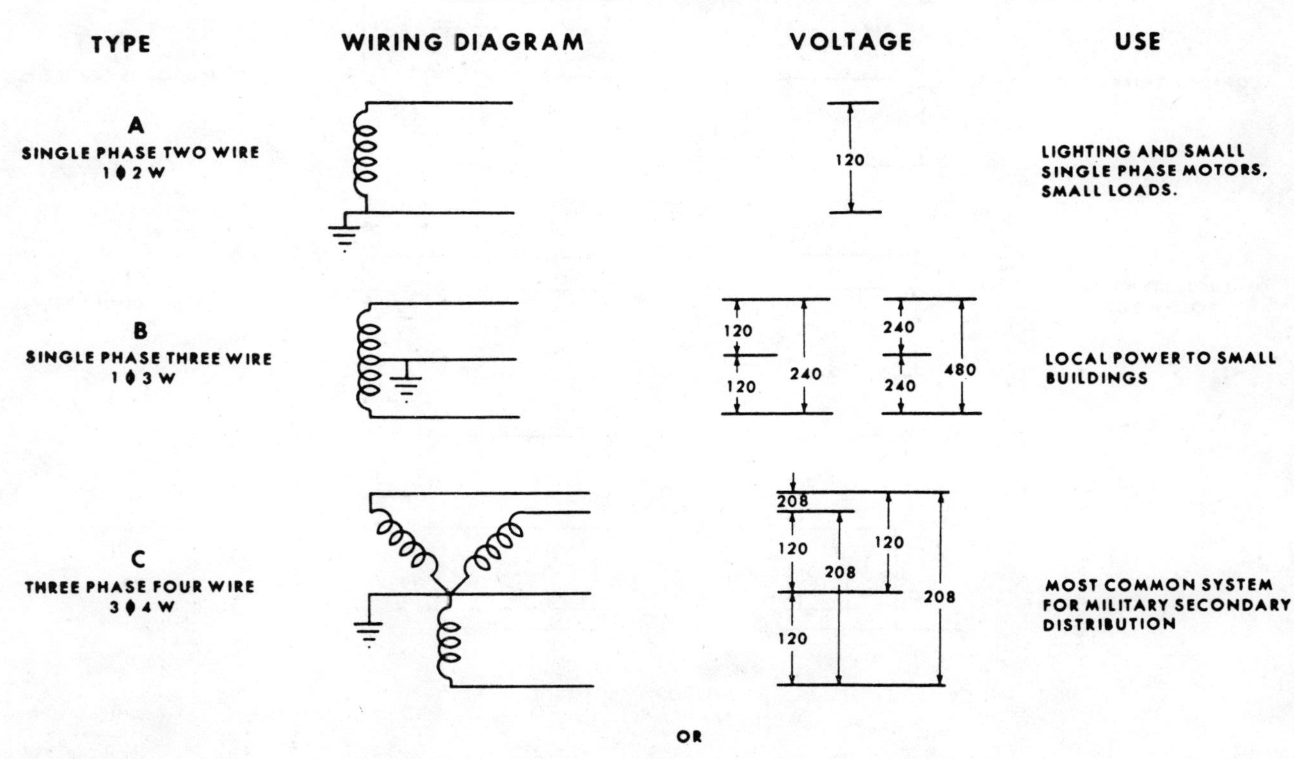

TABLE 7 (*continued*)

System	Connection	Voltages	Use
	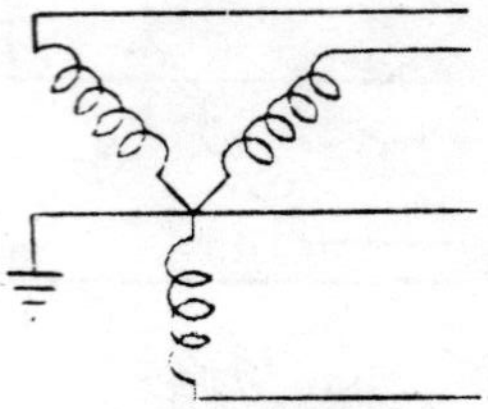	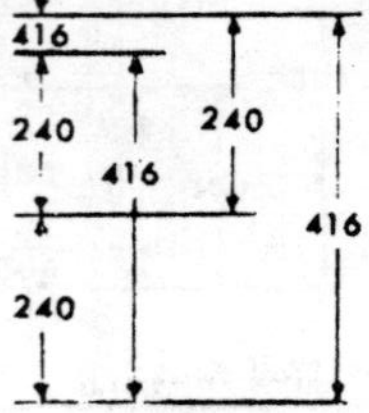 416, 240, 416, 240, 240, 416	
D **THREE PHASE THREE WIRE** **3φ 3 W**	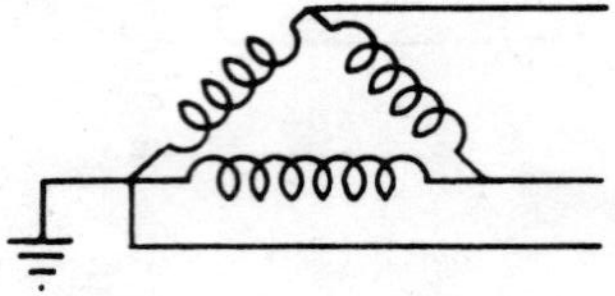	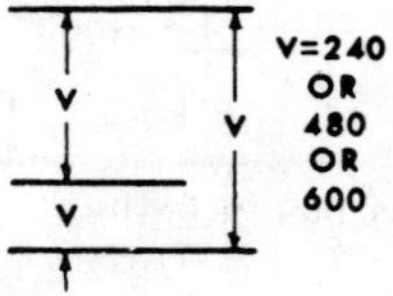 V, V, V V=240 OR 480 OR 600	FOR LARGE MOTOR LOADS, SMALL LIGHTING LOADS.
E **THREE PHASE FOUR WIRE** **3φ 4 W**	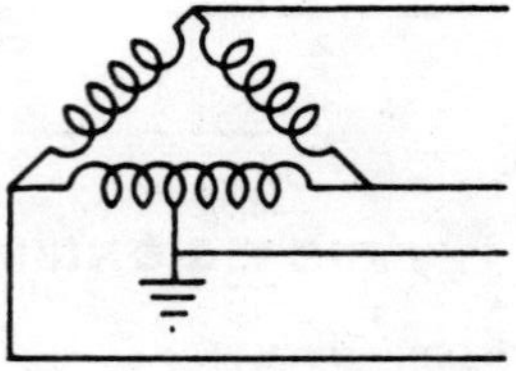	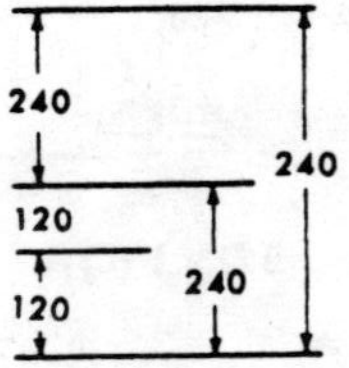 240, 120, 120, 240, 240	MOTOR AND LIGHTING LOADS

TABLE 8
CONDUCTOR INSULATION

Trade Name	Type Letter	Temp. rating	Application Provisions
Rubber-Covered Fixture Wire	*RF-1	60°C 140°F	Fixture wiring. Limited to 300 V.
Solid or 7-Strand	*RF-2	60°C 140°F	Fixture wiring.
Rubber-Covered Fixture Wire	*FF-1	60°C 140°F	Fixture wiring. Limited to 300 V.
Flexible Stranding	*FF-2	60°C 140°F	Fixture wiring.
Heat-Resistant Rubber-Covered Fixture Wire	*RFH-1	75°C 167°F	Fixture wiring. Limited to 300 V.
Solid or 7-Strand	*RFH-2	75°C 167°F	Fixture wiring.
Heat-Resistant Rubber-Covered Fixture Wire	*FFH-1	75°C 167°F	Fixture wiring. Limited to 300 V.
Flexible Stranding	*FFH-2	75°C 167°F	Fixture wiring.
Thermoplastic-Covered Fixture Wire—Solid or Stranded	*TF	60°C 140°F	Fixture wiring.

TABLE 8 (*continued*)

Trade Name	Type Letter	Temp. rating	Application Provisions
Thermoplastic-Covered Fixture Wire—Flexible Stranding	*TFF	60°C 140°F	Fixture wiring.
Cotton-Covered, Heat-Resistant, Fixture Wire	*CF	90°C 194°F	Fixture wiring. Limited to 300 V.
Asbestos-Covered Heat-Resistant, Fixture Wire	*AF	150°C 302°F	Fixture wiring. Limited to 300 V. and Indoor Dry Location.
Silicone Rubber Insulated Fixture Wire	*SF-1	200°C 392°F	Fixture wiring. Limited to 300 V.
Solid or 7 Strand	*SF-2	200°C 392°F	Fixture wiring
Silicone Rubber Insulated Fixture Wire	*SFF-1	150°C 302°F	Fixture wiring. Limited to 300 V.
Flexible Stranding	*SFF-2	150°C 302°F	Fixture wiring.
Code Rubber	R	60°C 140°F	Dry locations.
Heat-Resistant Rubber	RH	75°C 167°F	Dry locations.

*Fixture wires are not intended for installation as branch circuit conductors nor for the connection of portable or stationary appliances.

TABLE 8 (*continued*)

Heat Resistant Rubber	RHH	90°C 194°F	Dry locations.
Moisture-Resistant Rubber	RW	60°C 140°F	Dry and wet locations. For over 2000 volts, insulation shall be ozone-resistant.
Moisture and Heat Resistant Rubber	RH-RW	60°C 140°F	Dry and wet locations. For over 2000 volts, insulation shall be ozone-resistant.
		75°C	Dry locations.
Thermoplastic and Fibrous Outer Braid	TBS	90°C 194°F	Switchboard wiring only.
Synthetic Heat-Resistant	SIS	90°C 194°F	Switchboard wiring only.
Mineral Insulation (Metal Sheathed)	MI	85°C 185°F	Dry and wet locations with Type O termination fittings. Max. operating temperature for special applications 250°C.
Silicone-Asbestos	SA	90°C	Dry locations—max. operating temperature for special application 125°C.

TABLE 8 (*continued*)

Trade Name	Type Letter	Temp. rating	Application Provisions
Fluorinated Ethylene Propylene	FEP or FEPB	90°C 194°F 200°C 392°F	Dry locations. Dry locations—special applications.
		167°F	For over 2000 volts, insulation shall be ozone-resistant.
Moisture and Heat Resistant Rubber	RHW	75°C 167°F	Dry and wet locations. For over 2000 volts, insulation shall be ozone-resistant.
Latex Rubber	RU	60°C 140°F	Dry locations.
Heat Resistant Latex Rubber	RUH	75°C	Dry locations.
Moisture Resistant Latex Rubber	RUW	60°C 140°F	Dry and wet locations.
Thermoplastic	T	60°C 140°F	Dry locations.
Moisture-Resistant Thermoplastic	TW	60°C 140°F	Dry and wet locations.

TABLE 8 (*continued*)

Heat-Resistant Thermoplastic	THHN	90°C 194°F	Dry locations.
Moisture and Heat-Resistant Thermoplastic	THW	75°C 167°F	Dry and wet locations.
Moisture and Heat-Resistant Thermoplastic	THWN	75°C 167°F	Dry and wet locations.
Thermoplastic and Asbestos	TA	90°C 194°F	Switchboard wiring only.
Varnished Cambric	V	85°C 185°F	Dry locations only. Smaller than No. 6 by special permission.
Asbestos and Varnished Cambric	AVA	110°C 230°F	Dry locations only.
Asbestos and Varnished Cambric	AVL	110°C 230°F	Dry and wet locations.
Asbestos and Varnished Cambric	AVB	90°C 194°F	Dry locations only.
		75°C	Dry locations.
Thermoplastic and Fibrous Outer Braid	TBS	90°C 194°F	Switchboard wiring only.

TABLE 8 (*continued*)

Synthetic Heat-Resistant	SIS	90°C 194°F	Switchboard wiring only.
Mineral Insulation (Metal Sheathed)	MI	85°C 185°F	Dry and wet locations with Type O termination fittings. Max. operating temperature for special applications 250°C.
Silicone-Asbestos	SA	90°C	Dry locations—max. operating temperature for special application 125°C.
Fluorinated Ethylene Propylene	FEP or FEPB	90°C 194°F 200°C 392°F	Dry locations. Dry locations—special applications.

TABLE 9

ALLOWABLE CURRENT-CARRYING CAPACITY OF COPPER CONDUCTORS

(NOT MORE THAN THREE CONDUCTORS IN RACEWAY OR CABLE)

Size	Temperature Rating of Conductor *					
AWG MCM	60°C (140°F)	75°C (167°F)	85°–90°C (185°F)	110°C (230°F)	125°C (257°F)	200°C (392°F)
14	15	15	25†	30	30	30
12	20	20	30†	35	40	40
10	30	30	40†	45	50	55
8	40	45	50	60	65	70
6	55	65	70	80	85	95
4	70	85	90	105	115	120
3	80	100	105	120	130	145
2	95	115	120	135	145	165
1	110	130	140	160	170	190
0	125	150	155	190	200	225
00	145	175	185	215	230	250
000	165	200	210	245	265	285
0000	195	230	235	275	310	340
250	215	255	270	315	335	---
300	240	285	300	345	380	---
350	260	310	325	390	420	---
400	280	335	360	420	450	---
500	320	380	405	470	500	---

TABLE 9 (*continued*)

600	355	420	455	525	545	---
700	385	460	490	560	600	---
750	400	475	500	580	620	---
800	410	490	515	600	640	---
900	435	520	555	---	---	---
1000	455	545	585	680	730	---
1250	495	590	645	---	---	---
1500	520	625	700	785	---	---
1750	545	650	735	---	---	---
2000	560	665	775	840	---	---

CORRECTION FACTORS, ROOM TEMPS. OVER 30°C. 86°F.

C.	F.					
40	104	.82	.88	.90	.94	.95	---
45	113	.71	.82	.85	.90	.92	---
50	122	.58	.75	.80	.87	.89	---
55	131	.41	.67	.74	.83	.86	---
60	140	---	.58	.67	.79	.83	.91
70	158	---	.35	.52	.71	.76	.87
75	167	---	---	.43	.66	.72	.86
80	176	---	---	.30	.61	.69	.84

TABLE 9 (*continued*)

90	194	---	---	---	.50	.61	.80
100	212	---	---	---	---	.51	.77
120	248	---	---	---	---	---	.69
140	284	---	---	---	---	---	.50

° See Table 8 for temperature rating of conductor.

These current capacities relate only to conductors described in Table 8.

† The current capacities for types FEP, FEPB, FHH, and THHN conductors for sizes AWG 14, 12, and 10 shall be the same as designated for 75° C. conductors in this table.

TABLE 10

ALLOWABLE CURRENT-CARRYING CAPACITY OF COPPER CONDUCTORS IN FREE AIR

Size	Temperature Rating of Conductor *						
AWG MCM	60°C (140°F)	75°C (167°F)	85°–90°C (185°F)	110°C (230°F)	125°C (257°F)	200°C (392°F)	Bare & Covered Conductor
14	20	20	30†	40	40	45	30
12	25	25	40†	50	50	55	40
10	40	40	55†	65	70	75	55
8	55	65	70	85	90	100	70
6	80	95	100	120	125	135	100
4	105	125	135	160	170	180	130
3	120	145	155	180	195	210	150
2	140	170	180	210	225	240	175
1	165	195	210	245	265	280	205
0	195	230	245	285	305	325	235
00	225	265	285	330	355	370	275
000	260	310	330	385	410	430	320
0000	300	360	385	445	475	510	370
250	340	405	425	495	530	....	410
300	375	445	480	555	590	....	460
350	420	505	530	610	655	....	510
400	455	545	575	665	710	....	555
500	515	620	660	765	815	....	630

TABLE 10 (*continued*)

600	575	690	740	855	910	----	710
700	630	755	815	940	1005	----	780
750	655	785	845	980	1045	----	810
800	680	815	880	1020	1085	----	845
900	730	870	940	----	----	----	905
1000	780	935	1000	1165	1240	----	965
1250	890	1065	1130	----	----	----	----
1500	980	1175	1260	1450	----	----	1215
1750	1070	1280	1370	----	----	----	----
2000	1155	1385	1470	1715	----	----	1405

CORRECTION FACTORS, ROOM TEMPS. OVER 30°C. 86°F.

C. F.							
40 104	.82	.88	.90	.94	.95	----	----
45 113	.71	.82	.85	.90	.92	----	----
50 122	.58	.75	.80	.87	.89	----	----
55 131	.41	.67	.74	.83	.86	----	----
60 140	----	.58	.67	.79	.83	.91	----
70 158	----	.35	.52	.71	.76	.87	----
75 167	----	----	.43	.66	.72	.86	----
80 176	----	----	.80	.61	.69	.84	----

TABLE 10 (*continued*)

90	**194**	----	----	----	**.50**	**.61**	**.80**	----
100	**212**	----	----	----	----	**.51**	**.77**	----
120	**248**	----	----	----	----	----	**.69**	----
140	**284**	----	----	----	----	----	**.50**	----

These current capacities relate only to conductors described in Table 8.

† The current capacities for types FEP, FEPB, RHH, and THHN conductors for sizes AWG 14, 12, and 10 shall be the same as designated for 75° C. conductors in this table.

* See Table 8 for temperature rating of conductor.

TABLE 11

ALLOWABLE CURRENT-CARRYING CAPACITY OF ALUMINUM CONDUCTORS

(NOT MORE THAN THREE CONDUCTORS IN RACEWAY OR CABLE)

Size	Temperature Rating of Conductor **					
AWG MCM	60°C (140°F)	75°C (167°F)	85°–90°C (185°F)	110°C (230°F)	125°C (257°F)	200°C (392°F)
12	15	15	25†	25	30	30
10	25	25	30†	35	40	45
8	30	40	40†	45	50	55
6	40	50	55	60	65	75
4	55	65	70	80	90	95
3	65	75	80	95	100	115
*2	75	90	95	105	115	130
*1	85	100	110	125	135	150
*0	100	120	125	150	160	180
*00	115	135	145	170	180	200
*000	130	155	165	195	210	225
*0000	155	180	185	215	245	270
250	170	205	215	250	270	---
300	190	230	240	275	305	---
350	210	250	260	310	335	---
400	225	270	290	335	360	---
500	260	310	330	380	405	---

TABLE 11 (*continued*)

600	285	340	370	425	440	---
700	310	375	395	455	485	---
750	320	385	405	470	500	---
800	330	395	415	485	520	---
900	355	425	455	---	---	---
1000	375	445	480	560	600	---
1250	405	485	530	---	---	---
1500	435	520	580	650	---	---
1750	455	545	615	---	---	---
2000	470	560	650	705	---	---

CORRECTION FACTORS, ROOM TEMPS. OVER 30°C. 86°F.

C. F.						
40 104	.82	.88	.90	.94	.95	---
45 113	.71	.82	.85	.90	.92	---
50 122	.58	.75	.80	.87	.89	---
55 131	.41	.67	.74	.83	.86	---
60 140	---	.58	.67	.79	.83	.91
70 158	---	.35	.52	.71	.76	.87
75 167	---	---	.43	.66	.72	.86
80 176	---	---	.30	.61	.69	.84

TABLE 11 (*continued*)

90	194	---	---	---	.50	.61	.80
100	212	---	---	---	---	.51	.77
120	248	---	---	---	---	---	.69
140	284	---	---	---	---	---	.50

These current capacities relate only to conductors described in Table 8.

° For three wire, single phase service and sub-service circuits, the allowable current capacity of RH, RH-RW, RHH, RHW, and THW aluminum conductors shall be for sizes #2-100 Amp., #1/0-150 amp., #3/0-170 Amp., and #4/0-200 Amp.

† The current capacities for types RHH and THHN conductors for sizes AWG 12, 10, and 8 shall be the same as designated for 75° C. conductors in this table.

°° See Table 8 for temperature rating of conductor.

TABLE 12

ALLOWABLE CURRENT-CARRYING CAPACITY OF ALUMINUM CONDUCTORS IN FREE AIR

Size	Temperature Rating of Conductor *						
AWG MCM	60°C (140°F)	75°C (167°F)	85°–90°C (185°F)	110°C (230°F)	125°C (257°F)	200°C (392°F)	Bare & Covered Conductor
12	20	20	30†	40	40	45	30
10	30	30	45†	50	55	60	45
8	45	55	55†	65	70	80	55
6	60	75	80	95	100	105	80
4	80	100	105	125	135	140	100
3	95	115	120	140	150	165	115
2	110	135	140	165	175	185	135
1	130	155	165	190	205	220	160
0	150	180	190	220	240	255	185
00	175	210	220	255	275	290	215
000	200	240	255	300	320	335	250
0000	230	280	300	345	370	400	290
250	265	315	330	385	415	....	320
300	290	350	375	435	460	....	360
350	330	395	415	475	510	....	400
400	355	425	450	520	555	....	435
500	405	485	515	595	635	....	490

TABLE 12 (*continued*)

600	455	545	585	675	720	----	560
700	500	595	645	745	795	----	615
750	515	620	670	775	825	----	640
800	535	645	695	805	855	----	670
900	580	700	750	----	----	----	725
1000	625	750	800	930	990	----	770
1250	710	855	905	----	----	----	----
1500	795	950	1020	1175	----	----	985
1750	875	1050	1125	----	----	----	----
2000	960	1150	1220	1425	----	----	1165

CORRECTION FACTORS, ROOM TEMPS. OVER 30°C. 86°F.

C. F.							
40 104	.82	.88	.90	.94	.95	----	----
45 113	.71	.82	.85	.90	.92	----	----
50 122	.58	.75	.80	.87	.89	----	----
55 131	.41	.67	.74	.83	.86	----	----
60 140	---	.58	.67	.79	.83	.91	----
70 158	---	.35	.52	.71	.76	.87	----
75 167	---	----	.48	.66	.72	.86	----
80 176	---	----	.30	.61	.69	.84	----

TABLE 12 (*continued*)

90	194	---	----	----	.50	.61	.80	----
100	212	---	----	----	----	.51	.77	----
130	248	---	----	----	----	----	.69	----
140	284	---	----	----	----	----	.50	----

These current capacities relate only to conductors described in Table 8.

† The current capacities for types RHH and THHN conductors for sizes AWG 12, 10, and 8 shall be the same as designated for 75° C. conductors in this table.

* See Table 8 for temperature rating of conductor.

TABLE 13

REDUCTION OF CURRENT-CARRYING CAPACITY FOR MORE THAN THREE CONDUCTORS TO RACEWAY OR CABLE

Number of conductors	Percent reduction of Tables 8 and 11
4 to 6	80
7 to 24	70
25 to 42	60
43 and above	50

TABLE 14

FLEXIBLE CORDS

Trade Name	Type Letter	Size AWG	No. of Conductors	Insulation	Braid on Each Conductor	Outer Covering	Use		
Parallel Tinsel Cord	TP	27	2	Rubber	None	Rubber	Attached to an Appliance	Damp Places	Not Hard Usage
	TPT	27	2	Thermo-plastic	None	Thermo-plastic	Attached to an Appliance	Damp Places	Not Hard Usage
Jacketed Tinsel Cord	TS	27	2 or 3	Rubber	None	Rubber	Attached to an Appliance	Damp Places	Not Hard Usage
	TST	27	2 or 3	Thermo-plastic	None	Thermo-plastic	Attached to an Appliance	Damp Places	Not Hard Usage
Asbestos-Covered Heat-Resistant Cord	AFC	18–10	2 or 3	Impreg-nated Asbestos	Cotton or Rayon	None	Pendant	Dry Places	Not Hard Usage
	AFPO		2			Cotton, Rayon or Saturated Asbestos			
	AFPD		2 or 3		None				
Cotton-Covered Heat-Resistant Cord	CFC	18–10	2 or 3	Impreg-nated Cotton	Cotton or Rayon	None	Pendant	Dry Places	Not Hard Usage
	CFPO		2			Cotton or Rayon			
	CFPD		2 or 3		None				
Parallel Cord	PO-1	18	2	Rubber	Cotton	Cotton or Rayon	Pendant or Portable	Dry Places	Not Hard Usage
	PO-2	18–16							
	PO	18–10							
All Rubber Parallel Cord	SP-1	18	2	Rubber	None	Rubber	Pendant or Portable	Damp Places	Not Hard Usage
	SP-2	18–16							
	SP-3	18–12		Rubber	None	Rubber	Refriger-ators or Room Air Condi-tioners	Damp Places	Not Hard Usage

TABLE 14 (*continued*)

All Plastic Parallel Cord	SPT–1	18	2	Thermo-plastic	None	Thermoplastic	Pendant or Portable	Damp Places	Not Hard Usage
	SPT–2	18–16							
All Plastic Parallel Cord	SPT–3	18–10	2	Thermo-plastic	None	Thermoplastic	Refriger-ators or Room Air Condi-tioners	Damp Places	Not Hard Usage
Lamp Cord	C	18–10	2 or more	Rubber	Cotton	None	Pendant or Port.	Dry Places	Not Hard Usage
Twisted Port-able Cord	PD	18–10	2 or more	Rubber	Cotton	Cotton or Rayon	Pendant or Port.	Dry Places	Not Hard Usage
Reinforced Cord	P–1	18	2 or more	Rubber	Cotton	Cotton over Rubber Filler	Pendant or Portable	Dry Places	Not Hard Usage
	P–2	18–16							Hard Usage
	P	18–10							
Braided Heavy Duty Cord	K	18–10	2 or more	Rubber	Cotton	Two Cotton, Mois-ture-Resistant Finish	Pendant or Portable	Damp Places	Hard Usage
Vacuum Cleaner Cord	SV, SVO	18	2	Rubber	None	Rubber	Pendant or Portable	Damp Places	Not Hard Usage
	SVT, SVTO			Thermopl'		Thermoplastic			
Junior Hard Service Cord	SJ	18–16	2, 3, or 4	Rubber	None	Rubber	Pendant or Portable	Damp Places	Hard Usage
	SJO					Oil Resistant Compound			
	SJT SJTO			Thermopl' or Rubber		Thermoplastic			
Hard Service Cord	S	18–2	2 or more	Rubber	None	Rubber	Pendant or Portable	Damp Places	Extra Hard Usage
	SO					Oil Resist. Compound			
	ST			Thermopl' or Rubber		Thermoplastic			
	STO					Oil Resistant Thermoplastic			
Rubber-Jack-eted Heat-Re-sistant Cord	AFSJ	18–16	2 or 3	Impreg-nated Asbestos	None	Rubber	Portable	Damp Places	Portable Heaters
	AFS	18–16–14							

TABLE 14 (*continued*)

Trade Name	Type Letter	Size AWG	No. of Conductors	Insulation	Braid on Each Conductor	Outer Covering	Use		
Heater Cord	HC	18–12	2, 3 or 4	Rubber & Asbestos	Cotton	None	Portable	Dry Places	Portable Heaters
	HPD	18–12	2, 3 or 4	Rubber with Asbestos or All Neoprene	None	Cotton or Rayon			
Rubber Jacketed Heater Cord	HSJ	18–16	2, 3 or 4	Rubber with Asbestos or All Neoprene	None	Cotton and Rubber	Portable	Damp Places	Portable Heaters
Jacketed Heater Cord	HSJO	18–16	2, 3 or 4	Rubber with Asbestos or All Neoprene	None	Cotton and Oil Resistant Compound	Portable	Damp Places	Portable Heaters
	HS	14–12				Cotton and Rubber or Neoprene			
	HSO	14–12				Cotton and Oil Resistant Compound			
Parallel Heater Cord	HPN	18–16	2	Thermosetting	None	Thermosetting	Portable	Damp Places	Portable Heaters
Heat & Moisture-Resistant Cord	AVPO	18–10	2	Asbestos & Var. Cam.	None	Asbestos, Flame-Ret. Moisture Resistant	Pendant or Portable	Damp Places	Not Hard Usage
	AVPD		2 or 3						
Range, Dryer Cable	SRD	10–4	3 or 4	Rubber	None	Rubber or Neoprene	Portable	Damp Places	Ranges, Dryers
	SRDT	10–4	3 or 4	Thermoplastic	None	Thermoplastic	Portable	Damp Places	Ranges, Dryers

TABLE 15

DEEP BOXES

Box dimension, inches	Maximum number of conductors			
	No. 14	No. 12	No. 10	No. 8
1½ x 3¼ octagonal	5	5	4	0
1½ x 4 octagonal	8	7	6	5
1½ x 4 square	11	9	7	5
1½ x 4 11/16 square	16	12	10	8
2⅛ x 4 11/16 square	20	16	12	10
1¾ x 2¾ x 2	5	4	4	4
1¾ x 2¾ x 2½	6	6	5	0
1¾ x 2¾ x 3	7	7	6	0

TABLE 16

SHALLOW BOXES

(LESS THAN 1½ INCHES DEEP)

Box dimensions, inches trade size	Maximum number of conductors		
	No. 14	No. 12	No. 10
3¼	4	4	3
4	6	6	4
1¼ x 4 square	9	7	6
4 11/16	8	6	6

TABLE 17

SUPPORT FOR NONMETALLIC CONDUIT RUNS

Conduit size inches	*Maximum spacing between supports feet*
½	4
¾	4
1	5
1¼	5
1½	5
2	5
2½	6
3	6
3½	7
4	7
5	7
6	8

TABLE 18

FULL LOAD CURRENT—SINGLE PHASE AC MOTORS

HP	*115V*	*230V*	*440V*
1/6	4.4	2.2	--
¼	5.8	2.9	--
1/3	7.2	3.6	--
½	9.8	4.9	--
¾	13.8	6.9	--
1	16	8	--
1½	20	10	--
2	24	12	--
3	34	17	--
5	56	28	--
7½	80	40	21
10	100	50	26

TABLE 19

FULL LOAD CURRENT—THREE PHASE AC MOTORS

HP	Induction Type Squirrel-Cage and Wound Rotor Amperes					Synchronous Type Unity Power Factor Amperes			
	110V	220V	440V	550V	2300V	220V	440V	550V	2300V
½	4	2	1	.8					
¾	5.6	2.8	1.4	1.1					
1	7	3.5	1.8	1.4					
1½	10	5	2.5	2.0					
2	13	6.5	3.3	2.6					
3		9	4.5	4					
5		15	7.5	6					
7½		22	11	9					
10		27	14	11					
15		40	20	16					
20		52	26	21					
25		64	32	26	7	54	27	22	5.4
30		78	39	31	8.5	65	33	26	6.5
40		104	52	41	10.5	86	43	35	8
50		125	63	50	13	108	54	44	10
60		150	75	60	16	128	64	51	12
75		185	93	74	19	161	81	65	15
100		246	123	98	25	211	106	85	20
125		310	155	124	31	264	132	106	25
150		360	180	144	37		158	127	30
200		480	240	192	48		210	168	40

TABLE 20

WATTAGE CONSUMPTION OF ELECTRICAL APPLIANCES

Appliance	*Average wattage*
Clock	3
Coffeemaker	1000
Fan, 8-inch	30
Fan, 10-inch	35
Fan, 12-inch	50
Heater (radiant)	1300
Hotplate	1250
Humidifier	500
Iron	1000
Mixer	200
Phonograph	40
Refrigerator	500
Radio	100
Soldering Iron	200
Television	300
Toaster	1200
Washing Machine	1300
Water heater	4500

TABLE 21

REQUIREMENTS FOR BRANCH CIRCUITS

(Type FEP, FEPB, R, RW, RU, RUW, RH-RW, SA, T, TW, RH, RUH, RHW RHH, THHN, THW, and THWN conductors in raceway or cable.)

Circuit rating	15 Amp.	20 Amp.	30 Amp.	40 Amp.	50 Amp.
Conductors: (Min. size)					
Circuit wires*	**14**	**12**	**10**	**8**	**6**
Taps	**14**	**14**	**14**	**12**	**12**
Overcurrent protection	**15 Amp.**	**20 Amp.**	**30 Amp.**	**40 Amp.**	**50 Amp.**
Outlet devices:					
Lampholders permitted	**Any Type**	**Any Type**	**Heavy Duty**	**Heavy Duty**	**Heavy Duty**
Receptacle rating	**15 Max. Amp.**	**15 or 20 Amp.**	**30 Amp.**	**40 and 50 Amp.**	**50 Amp.**
Maximum load	**15 Amp.**	**20 Amp.**	**30 Amp.**	**40 Amp.**	**50 Amp.**

* These current capacities are for copper conductors with no correction factor for temperature (see Tables 9 through 13).

TABLE 22

VOLTAGE DROP TABLE (BASED ON 3 PERCENT DROP)*
VERIFY SELECTED CONDUCTOR FOR CURRENT-CARRYING CAPACITY
(*See* TABLES 8 THROUGH 13)**

10—Aluminum wire
12—Copper wire

	FOR 110V CIRCUIT DISTANCE TO LOAD IN FEET									
LOAD IN AMPS.	**50**	**75**	**100**	**125**	**150**	**200**	**250**	**300**	**400**	**500**
15	10 12	8 10	8 10	6 8	6 8	4 6	4 6	3 4	2 4	1 3
20	10 12	8 10	6 8	6 8	4 6	4 6	3 4	2 4	1 3	0 2
25	8 10	6 8	6 8	4 6	4 6	3 4	2 4	1 3	0 2	2/0 1
30	6 10	6 8	4 6	4 6	3 4	2 4	1 3	0 2	2/0 1	3/0 0
40	6 8	4 6	4 6	3 4	2 4	1 3	0 2	2/0 1	3/0 0	4/0 2/0
50	4 8	4 6	3 4	2 4	1 3	0 2	2/0 1	3/0 0	4/0 2/0	300 3/0
60	4 6	2 4	2 4	1 3	0 2	2/0 1	3/0 0	4/0 2/0	250 3/0	350 4/0
70	4 6	2 4	1 3	0 2	2/0 2	3/0 0	4/0 2/0	250 2/0	300 4/0	400 250
80	4 6	2 4	1 3	0 2	2/0 1	3/0 0	4/0 2/0	250 3/0	350 4/0	500 250
90	2 4	1 3	0 2	2/0 1	3/0 1	4/0 2/0	250 3/0	300 3/0	400 250	500 300
100	2 4	1 3	0 2	2/0 1	3/0 0	4/0 2/0	300 3/0	350 4/0	500 250	600 350

TABLE 22 (*Continued*)

LOAD IN AMPS.	FOR 220V CIRCUIT DISTANCE TO LOAD IN FEET									
	100	200	300	400	500	600	700	800	900	1000
15	$\frac{12}{12}$	$\frac{8}{10}$	$\frac{6}{8}$	$\frac{4}{6}$	$\frac{4}{6}$	$\frac{3}{4}$	$\frac{2}{4}$	$\frac{2}{4}$	$\frac{1}{3}$	$\frac{1}{3}$
20	$\frac{10}{12}$	$\frac{6}{8}$	$\frac{4}{6}$	$\frac{4}{6}$	$\frac{3}{4}$	$\frac{2}{4}$	$\frac{1}{3}$	$\frac{1}{3}$	$\frac{0}{2}$	$\frac{0}{2}$
25	$\frac{8}{10}$	$\frac{6}{8}$	$\frac{4}{6}$	$\frac{3}{4}$	$\frac{2}{4}$	$\frac{1}{3}$	$\frac{0}{2}$	$\frac{0}{2}$	$\frac{2/0}{1}$	$\frac{2/0}{1}$
30	$\frac{6}{10}$	$\frac{4}{6}$	$\frac{3}{4}$	$\frac{2}{4}$	$\frac{1}{3}$	$\frac{0}{2}$	$\frac{2/0}{2}$	$\frac{2/0}{1}$	$\frac{3/0}{0}$	$\frac{3/0}{0}$
40	$\frac{4}{8}$	$\frac{4}{6}$	$\frac{2}{4}$	$\frac{1}{3}$	$\frac{0}{2}$	$\frac{2/0}{1}$	$\frac{3/0}{0}$	$\frac{3/0}{0}$	$\frac{4/0}{2/0}$	$\frac{4/0}{2/0}$
50	$\frac{4}{8}$	$\frac{3}{4}$	$\frac{1}{3}$	$\frac{0}{2}$	$\frac{2/0}{1}$	$\frac{3/0}{0}$	$\frac{4/0}{2/0}$	$\frac{4/0}{2/0}$	$\frac{250}{3/0}$	$\frac{300}{3/0}$
60	$\frac{4}{6}$	$\frac{2}{4}$	$\frac{0}{2}$	$\frac{2/0}{1}$	$\frac{3/0}{0}$	$\frac{4/0}{2/0}$	$\frac{250}{2/0}$	$\frac{250}{3/0}$	$\frac{300}{4/0}$	$\frac{350}{4/0}$
70	$\frac{4}{6}$	$\frac{1}{3}$	$\frac{2/0}{2}$	$\frac{3/0}{0}$	$\frac{4/0}{2/0}$	$\frac{250}{2/0}$	$\frac{300}{3/0}$	$\frac{300}{4/0}$	$\frac{350}{4/0}$	$\frac{400}{250}$
80	$\frac{4}{6}$	$\frac{1}{3}$	$\frac{2/0}{1}$	$\frac{3/0}{0}$	$\frac{4/0}{2/0}$	$\frac{250}{3/0}$	$\frac{300}{4/0}$	$\frac{350}{4/0}$	$\frac{400}{250}$	$\frac{500}{250}$
90	$\frac{2}{4}$	$\frac{0}{2}$	$\frac{3/0}{0}$	$\frac{4/0}{2/0}$	$\frac{250}{3/0}$	$\frac{300}{4/0}$	$\frac{350}{4/0}$	$\frac{400}{250}$	$\frac{500}{300}$	$\frac{500}{300}$
100	$\frac{2}{4}$	$\frac{0}{2}$	$\frac{3/0}{0}$	$\frac{4/0}{2/0}$	$\frac{300}{3/0}$	$\frac{350}{4/0}$	$\frac{400}{250}$	$\frac{500}{250}$	$\frac{500}{300}$	$\frac{600}{350}$

* For other voltage see Appendix III.

** *Example.* A building using open wiring requires 20 amperes (at 110V) to be supplied to a load located 150 ft. from the circuit breaker. From Tables 8 and 9 (assuming R type copper wire is used), the minimum size wire for this circuit is No. 14. From the above table, a No. 6 copper wire is required to limit the voltage drop to 3%. Therefore, a No. 6 copper wire should be used. If the wire available were aluminum, then the minimum size is No. 12, and for a maximum voltage drop of 3%, a No. 4 must be used.

TABLE 23

SUPPORTS FOR RIGID METAL CONDUIT RUNS

Conduit size (inches)	*Maximum distance between rigid metal conduit supports (feet)*
½	10
¾	10
1	12
1¼	14
1½	14
2	16
2½	16
3	20

TABLE 24

RADIUS OF CONDUIT BENDS

Size of conduit	*Conductors without lead sheath*	*Conductors with lead sheath*
½ in.	4 in.	6 in.
¾ in.	5 in.	8 in.
1 in.	6 in.	11 in.
1¼ in.	8 in.	14 in.
1½ in.	10 in.	16 in.
2 in.	12 in.	21 in.
2½ in.	15 in.	25 in.
3 in.	18 in.	31 in.
3½ in.	21 in.	36 in.
4 in.	24 in.	40 in.
5 in.	30 in.	50 in.
6 in.	36 in.	61 in.

TABLE 25

MINIMUM SIZE (INCHES) OF A CONDUIT OF ELECTRICAL METALLIC TUBING TO CONTAIN A GIVEN NUMBER OF 600–VOLT CONDUCTORS

Wire size gage No.	Number of conductors—types R, RW, RH, RU, RUW, TF, T, and TW								
	1	2	3	4	5	6	7	8	9
18	½	½	½	½	½	½	½	¾	¾
16	½	½	½	½	½	½	¾	¾	¾
14	½	½	½	½	¾	¾	1	1	1
12	½	½	½	¾	¾	1	1	1	1¼
10	½	¾	¾	¾	1	1	1	1¼	1¼
8	½	¾	¾	1	1¼	1¼	1¼	1½	1½
6	½	1	1	1¼	1½	1½	2	2	2
4	½	1¼	1¼	1½	1½	2	2	2	2½
3	¾	1¼	1¼	1½	2	2	2	2½	2½
2	¾	1¼	1¼	2	2	2	2½	2½	2½
1	¾	1½	1½	2	2½	2½	2½	3	3
0	1	1½	2	2	2½	2½	3	3	3
00	1	2	2	2½	2½	3	3	3	3½
000	1	2	2	2½	3	3	3	3½	3½
0000	1¼	2	2½	3	3	3	3½	3½	4

TABLE 26

SIZE CONDUIT OF ELECTRICAL METALLIC TUBING FOR COMBINATIONS OF CONDUCTOR. PERCENTAGE OF CROSS SECTIONAL AREA OF CONDUIT OR TUBING

Conductors	Number of conductors				
	1	2	3	4	over 4
Not lead covered	53	31	40	40	40
Lead covered	55	30	40	38	35
For rewiring existing conduits	60	40	40	50	50

TABLE 27

DIMENSIONS AND PERCENT OF AREA OF CONDUIT AND TUBING FOR COMBINATIONS

Conduit size	Internal diameter in.	Area, square in.									
			Not lead covered				Lead covered				
		Total 100%	1 cond. 53%	2 cond. 31%	3 cond. 43%	4 cond. and over 40%	1 cond. 55%	2 cond. 30%	3 cond. 40%	4 cond. 38%	Over 4 cond. 35%
½	0.622	0.30	0.16	0.09	0.13	0.12	0.17	0.09	0.12	0.11	0.11
¾	0.824	0.53	0.28	0.16	0.23	0.21	0.29	0.16	0.21	0.20	0.19
1............	1.049	0.86	0.46	0.27	0.37	0.34	0.47	0.26	0.34	0.33	0.30
1¼	1.380	1.50	0.80	0.47	0.65	0.60	0.83	0.65	0.60	0.57	0.53
1½	1.610	2.04	1.08	0.63	0.88	0.82	1.112	0.61	0.82	0.78	0.71
2............	2.067	3.36	1.78	1.04	1.44	1.34	1.85	1.01	1.34	1.28	1.18
2½	2.469	4.79	2.54	1.48	2.06	1.92	2.63	1.44	1.92	1.82	1.68
3............	3.068	7.38	3.91	2.29	3.17	2.95	4.06	2.21	2.95	2.80	2.58
3½	3.548	9.90	5.25	3.07	4.26	3.96	5.44	2.97	3.96	3.76	3.47

Appendix III:

Design Procedures for Electrical Wiring

The names of the circuits used in interior wiring are service, feeder, sub-feeder, main, sub-main, and branch. Figure 1 shows the layout of the various circuits.

The *service* is the name given the conductors carrying electric power to a building. It may bring power to a building from the exterior distribution feeder, local transformer, or a generator. The service terminates at the main distribution center or main fuse box of the building.

Feeders are circuits that supply power directly between a distribution center and a sub-distribution center. The only cutout for a feeder is at the main distribution center.

A *sub-feeder* is an extension of a feeder or another sub-feeder through a cutout (the cutout being the sub-distribution center).

A *main* is a circuit to which other energy-consuming circuits are attached through automatic cutouts (fuses or circuit breakers) along its length. A main has the same size wire throughout its length and has no cutouts in series with it.

A *sub-main* is fed through a cutout from a main or another sub-main and has branch circuits or loads connected to it through cutouts. A sub-main has the same size wire throughout its length, but it is usually of smaller size than the main or sub-main feeding it.

A *branch* circuit connects one or more energy-consuming devices (loads) through *one* cutout to a source (distribution center, sub-distribution center, main, or sub-main). Interior lighting circuits are usually branch circuits, since many lights are connected to one circuit which is controlled by one fuse or circuit breaker.

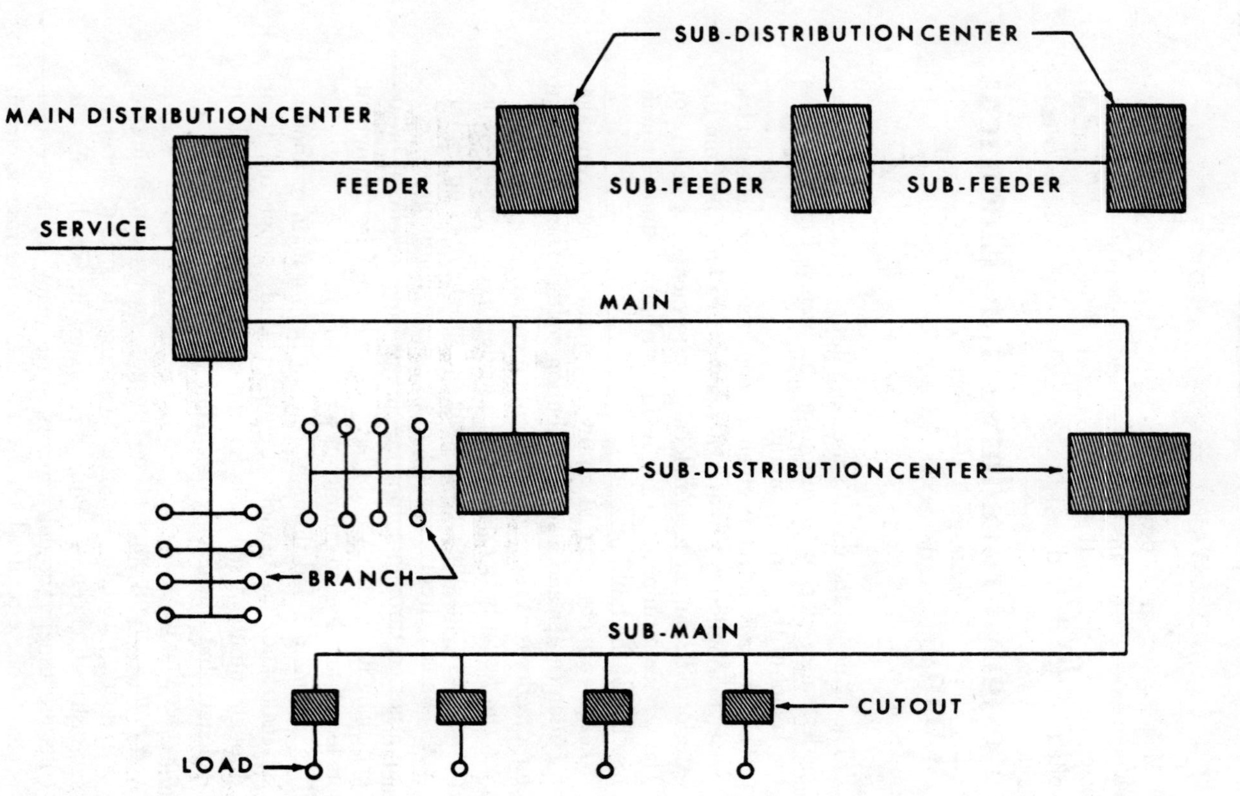
SUB-DISTRIBUTION CENTER
MAIN DISTRIBUTION CENTER
FEEDER
SUB-FEEDER
SUB-FEEDER
SERVICE
MAIN
SUB-DISTRIBUTION CENTER
BRANCH
SUB-MAIN
CUTOUT
LOAD

Figure 1

ALLOWABLE VOLTAGE DROP

Most equipment is designed to allow for a 10 percent *total* voltage drop. This means that the voltage at the equipment must be equal to or greater than 90 percent of the equipment's rated voltage and no more than 110 percent. The National Electrical Code is followed for large installations having central power plants to determine the electrical installation requirements. The Code allows a maximum of 5 percent voltage drop from the generator to the most distant service entrance. Therefore, inside a building another 5 percent drop may be tolerated. This normally is divided into a 3 percent drop for branch circuits and a 2 percent drop for feeders or main. The above requirements are necessary for large, city-like complexes.

CALCULATING VOLTAGE DROPS

The design procedures described in Chapter 5 are straightforward, but they do not permit flexibility in design. As an example, if a load of 20 amperes at 120 volts must be located 150 feet from the cutout, a No. 6 copper wire is required for a voltage drop of 3 percent or less. If No. 6 copper wire is not available, a larger wire could be used. This would eliminate voltage-drop problems but would not be economical. It may be possible to use a smaller wire, but this must be based on an analysis of the total electrical system. Thc following design procedure does not account for inductance, capacitance, or skin effect, which exists in alternating-current circuits. This procedure, therefore, is valid for wire sizes of 4/0 and smaller for lighting circuits and No. 1 and smaller for motor circuits. As the load increases and the power factor decreases, AC design procedures from electrical engineering references must be used.

Table 28 gives the values of resistance for copper and aluminum wire in ohms per thousand feet. Since the first step in designing an interior wiring system is to estimate or calculate the load, the current needed in each wire is thus ascertained. (*See* Chapter 5, Design and Layout of Interior Wiring.) The operating voltage is determined by the available power source or the

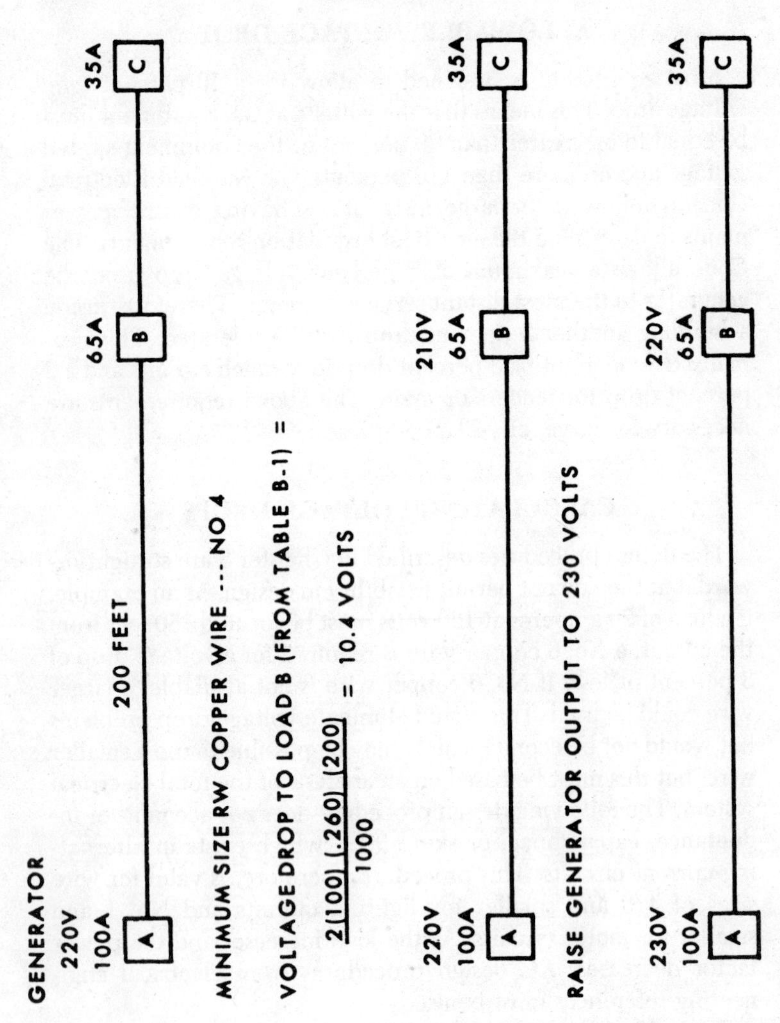
GENERATOR
220V
100A
A
200 FEET
65A
B
35A
C
MINIMUM SIZE RW COPPER WIRE ... NO 4
VOLTAGE DROP TO LOAD B (FROM TABLE B-1) =
2(100) (.260) (200) / 1000 = 10.4 VOLTS
220V
100A
A
210V
65A
B
35A
C
RAISE GENERATOR OUTPUT TO 230 VOLTS
230V
100A
A
220V
65A
B
35A
C

Figure 2

equipment to be operated. Knowing the current and distance, the voltage drop for any size wire may be calculated. Since the operating voltage and allowable voltage drop are known, the system may be designed using minimum-size wires.

TABLE 28

CHARACTERISTICS OF WIRE

Size AWG or MCM	No. wires	Diameter		Resistance Ohms per 1,000 feet	
		in.	mm	Copper	Aluminum
14	solid	.0641	1.63	2.57	4.22
12	solid	.0808	2.05	1.62	2.66
10	solid	.102	2.59	1.02	1.67
8	solid	.129	3.27	.640	1.05
6	7	,184	4.67	.410	.674
4	7	.232	5.89	.260	.424
3	7	.260	6.60	.205	.336
2	7	.292	7.41	.162	.266
1	19	.332	8.42	.129	.211
0	19	.373	9.46	.102	.168
2/0	19	.418	10.6	.0811	.133
3/0	19	.470	11.9	.0642	.105
4/0	19	.528	13.3	.0509	.0836
250	37	.575	14.6	.0431	.0780
300	37	.630	16.0	.0360	.0590
350	37	.681	17.3	.0308	.0505
400	37	.728	18.5	.0270	.0442
500	37	.814	20.6	.0216	.0354
600	61	.893	22.7	.0180	.0295
700	61	.964	24.5	.0154	.0253
150	61	.998	25.4	.0144	.0236

The benefit of this system may be shown by choosing the wire for the example in the first paragraph in this section. Based on the tables mentioned in Chapter 5, the 120-volt, 20-ampere load at a distance of 150 feet required a No. 6 copper wire at the minimum. The voltage drop from the main distribution center to any individual load within the building may be allowed to be 5 percent, or $120 \times .05 = 6$ volts. By trial and error it will be found that the smallest size of copper wire is No. 8. Table 28 shows that No. 8 copper wire has a resistance of 0.64 ohm per thousand feet. Two wires are needed for the circuit.

$$\begin{aligned} \text{voltage drop} &= \text{load in amps} \times \text{resistance of wire} \\ &= \frac{20(2)\ (.64)\ (150)}{1{,}000} \\ &= 3.84 \text{ volts} \end{aligned}$$

Since this is less than a five percent drop it is acceptable. Finally, Tables 7 through 12, Appendix 2, must be checked to determine if the type of wire to be used is capable of carrying the design current.

MINIMIZING VOLTAGE DROPS

One method of reducing low-voltage problems caused by unavailability of proper wire sizes is to increase the power supplied to the system by increasing the voltage output of the generator. In other words, the increase in generator output may be used to compensate for the voltage drop between the generator and the nearest service entrance or building. The voltage at any building should not exceed the rated value. As an example, the building nearest the generator site must be identified. The maximum voltage at this point is to be the circuit's rated voltage. The voltage drop of the wire between the generator and this point is the value that the generator output voltage may be increased (Fig. 2). For building A, the voltage drop that may now be allowed for interior wiring would be 10 percent, since the rated voltage of 220 volts appears at the service entrance. The wire needed in the example problem described previously in the section on calculating voltage drops may, if located in building A, be a No. 14. Thus, if the variety of sizes of wire is limited, this method may permit the supply of power to a location previously unattainable. Buildings further from the generator would have different design limits. The allowable voltage drops would depend on if the voltage drop (line AB) has been compensated for; the voltage at all buildings will be higher. This will permit greater allowable drops for building interiors and, therefore, smaller wire sizes may be used.

DESIGN PROCEDURE

A systematic procedure should be employed when designing electrical circuits. If the exterior distribution system is in existence, then the voltage at the main distribution center of each building must be determined. This voltage will influence the voltage drop calculation of the interior systems. Each type of interior circuit may then be chosen based on required load and location. The wire sizes for the circuits are determined by load, wire capacity, distance, and voltage drop. *Variations from National Electrical Code standards, as discussed in this Appendix, should be made only when the situation dictates their necessity.*

Appendix IV:

Greek Alphabet

Name	Capital	Lower Case	Designates
Alpha	A	α	Angles.
Beta.	B	β	Angles, flux density.
Gamma.	Γ	γ	Conductivity.
Delta	Δ	δ	Variation of a quantity, increment.
Epsilon.	E	ε	Base of natural logarithm (2. 71828).
Zeta.	Z	ζ	Impedance, coefficients, coordinates.
Eta	H	η	Hysteresis coefficient, efficiency, magnetizing force.
Theta	Θ	θ	Phase angle.
Iota	I	ι	
Kappa	K	κ	Dielectric constant, coupling coefficient, susceptibility.
Lambda	Λ	λ	Wavelength.

Mu.	M	μ	Permeability, micro, amplification factor.
Nu	N	ν	Reluctivity.
Xi	Ξ	ξ	
Omicron	O	ο	
Pi	Π	π	3.1416
Rho	P	ρ	Resistivity.
Sigma.	Σ	σ	
Tau	T	τ	Time constant, time-phase displacement.
Upsilon.	Υ	υ	
Phi	Φ	ϕ	Angles, magnetic flux.
Chi	X	χ	
Psi	Ψ	ψ	Dielectric flux, phase difference.
Omega	Ω	ω	Ohms (capital), angular velocity (2 π f).

Appendix V:

Electrical Abbreviations

a-c	alternating-current
a-f	audio-frequency
B	flux density
C	capacitance
cemf	instantaneous counter electromotive force
cm	centimeters
cps	cycles per second
C_T	total capacitance
d	distance between points
d-c	direct-current
de	change in voltage
di	change in current
dq	change in charge
dt	change in time
E	voltage
e	instantaneous voltage
E_c	capacitive voltage
e_c	instantaneous capacitive voltage
E_L	inductive voltage
e_L	instantaneous inductive voltage
E_m	maximum voltage
E_{max}	maximum voltage
emf	electromotive force
E_p	primary voltage
E_s	secondary voltage
f	frequency
f_r	frequency at resonance
H	magnetizing flux

h	henry
I	current
i	instantaneous current
I_c	capacitive current
i_c	instantaneous capacitive current
I_{eff}	effective current
i-f	intermediate-frequency
I_L	inductive current
i_L	instantaneous inductive current
I_m	maximum current
I_{max}	maximum current
I_p	plate current
I_R	current through resistance
i_R	instantaneous current through resistance
I_s	secondary current
I_T	total current
IØ	phase current
K	coefficient of coupling
kc	kilocycle
L	inductance
L-C	inductance-capacitance
L-C-R	inductance-capacitance-resistance
L_T	total inductance
mh	millihenry
N	revolutions per minute
N_p	primary turns
N_s	secondary turns
P	power
p	instantaneous power
P_{ap}	apparent power
P_{av}	average power
P_p	primary power
P_s	secondary power
Q	charge or quality
q	instantaneous charge
r-f	radio-frequency

R_G	grid resistance
R_o	load resistance
rpm	revolutions per minute
sq cm	square centimeters
t	time constant
t	time (seconds)
μf	microfarad
$\mu\mu$f	micromicrofarad
V, v	volt
X_c	capacitive reactance
X_L	inductive reactance
Z	impedance
Z_o	load impedance
Z_p	primary impedance
Z_s	secondary impedance
Z_T	total impedance
$\emptyset_M$	mutual flux
$\emptyset_P$	primary flux
$\emptyset_S$	secondary flux

Appendix VI:

Common Markings and Symbols Used on Electrical Circuit Diagrams

Common markings are used on electrical circuit diagrams to designate the functional use of a device. Table 29 gives a list of markings most frequently used.

TABLE 29
COMMON MARKINGS AND SYMBOLS USED ON ELECTRICAL CIRCUIT DIAGRAMS

Device	Contractor designation	Relay designation	Other equipment designation
Accelerating	A	AR	
Ammeter switch			AS
Autotransformer			AT
Brake	B	BR	
Capacitor			C
Circuit breaker			CB
Closing coil		CCR	CC
Control		CR	
Control switch			CS
Counter EMF		CEMF	
Current limit		CLR	
Current transformer			CT
Down	D		
Dynamic braking	DB	DBR	
Emergency switch			ES
Exciter field	EF	EFR	
Field	F	FR	
Field accelerating	FA	FAR	
Field discharge	FD	FDR	

TABLE 29 (*continued*)

Device	Contractor designation	Relay designation	Other equipment designation
Field economy	FE	FER	
Field loss (failure)	FL	FLR	
Field weakening	FW	FWR	
Float flow switch			FS
Forward	F	FR	
Full field	FF	FFR	
Ground detector			GD
High speed	HS	HSR	
Hoist	H	HR	
Jam, jog	J	JR	
Kickoff	KO	KOR	
Limit switch			LS
Lowering	L	LR	
Low speed	LS	LSR	
Main breaker			MB
Master switch			MS
Motor circuit switch			MCS
Motor field			MF
Overload		OL	
Oversupeed		OSR	
Overspeed switch			OSS
Plugging	P	PR	
Plugging forward		PF	
Plugging reverse		PR	
Potential transformer			PT
Power factor		PFR	
Power factor meter			PF
Pushbutton			PB
Rectifier			REC
Resistor			RES
Reverse, run. raise	R		
Sequence protective		SPR	
Slow down		SR	
Squirrel-cage protective.		SCR	
Start	S		
Switch			SW
Time closing			TC
Time opening			TO

TABLE 29 (*continued*)

Time relay		TR	
Transfer relay		TRR	
Trip coil			TC
Undervoltage	UV	UVR	
Up	U		
Voltage regulator		VRG	
Voltmeter switch			VS

POWER-TERMINAL MARKINGS

Common markings for designating power terminals on electrical circuit diagrams consist of a capital letter followed by a suffix numeral. If a multiplicity of equal devices are used, they are further designated by a numeral followed by a letter. Thus, 1A and 2A are device designations. A1 and A2 are terminal markings. Table 30 gives a partial list of terminal markings.

TABLE 30

POWER-TERMINAL MARKINGS

	Direct current	Alternating current
Brake	B1, B2, B3	B1, B2, B3.
Brush on commutator (armature).	A1, A2	A1, A2, A3.
Brush on slipring (rotor)		M1, M2, M3.
Field (series)	S1, S2	
Field (shunt)	F1, F2	F1, F2.
Line	L1, L2	L1, L2, L3.
Resistance (armature)	R1, R2, R3	R1, R2, R3.
Resistance (shunt field)	V1, V2, V3	
Stator		T1, T2, T3.
Transformer (high voltage).		H1, H2, H3.
Transformer (low voltage).		X1, X2, X3.

ELECTRICAL SYMBOLS OF AMERICAN STANDARDS ASSOCIATION

It is common practice to use symbols to designate various pieces of equipment, and everyone recognizes the equipment represented by the symbols, although there may be no resemblance between the symbol and the physical appearance of the article represented. Following is a list of symbols generally used in electrical circuit diagrams.

NAME	SYMBOL
BATTERY	− + OR
CAPACITOR, FIXED	
CIRCUIT BREAKERS	
AIR CIRCUIT BREAKER	
THREE-POLE POWER CIRCUIT BREAKER (SINGLE THROW) (WITH TERMINALS)	
THERMAL TRIP AIR CIRCUIT BREAKER	
COILS	
NON-MAGNETIC CORE-FIXED	
MAGNETIC CORE-FIXED	

NAME	SYMBOL
MAGNETIC CORE–ADJUSTABLE TAP OR SLIDE WIRE	
OPERATING COIL	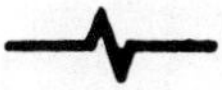
BLOWOUT COIL	
BLOWOUT COIL WITH TERMINALS	
SERIES FIELD	
SHUNT FIELD	
COMMUTATING FIELD	

CONNECTIONS (MECHANICAL)

NAME	SYMBOL
MECHANICAL CONNECTION OF SHIELD	 – – – –
MECHANICAL INTERLOCK	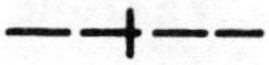
DIRECT CONNECTED UNITS	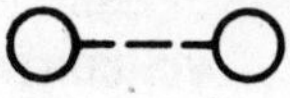

CONNECTIONS (WIRING)

NAME	SYMBOL
ELECTRIC CONDUCTOR–CONTROL	
ELECTRIC CONDUCTOR–POWER	
JUNCTION OF CONDUCTORS	•
WIRING TERMINAL	∘

NAME	SYMBOL
GROUND	
CROSSING OF CONDUCTORS — NOT CONNECTED	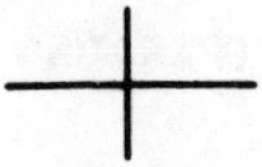
CROSSING OF CONNECTED CONDUCTORS	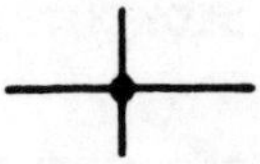
JOINING OF CONDUCTORS — NOT CROSSING	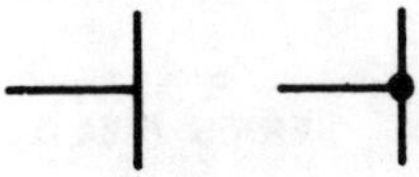

CONTACTS (ELECTRICAL)

NAME	SYMBOL
NORMALLY CLOSED CONTACT (NC)	
NORMALLY OPEN CONTACT (NO)	
NO CONTACT WITH TIME CLOSING (TC) FEATURE	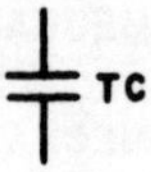 TC
NC CONTACT WITH TIME OPENING (TO) FEATURE	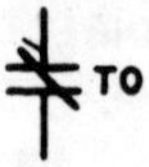 TO

NOTE: NO (NORMALLY OPEN) AND NC (NORMALLY CLOSED) DESIGNATES THE POSITION OF THE CONTACTS WHEN THE MAIN DEVICE IS IN THE DE-ENERGIZED OR NONOPERATED POSITION.

NAME	SYMBOL
CONTACTOR, SINGLE-POLE, ELECTRICALLY OPERATED, WITH BLOWOUT COIL NOTE: FUNDAMENTAL SYMBOLS FOR CONTACTS, COILS, MECHANICAL CONNECTIONS, etc., ARE THE BASIS OF CONTACTOR SYMBOLS	
FUSE	
INDICATING LIGHTS	
INDICATING LAMP WITH LEADS	
INDICATING LAMP WITH TERMINALS	
INSTRUMENTS	
AMMETER, WITH TERMINALS	A OR
VOLTMETER, WITH TERMINALS	V OR
WATTMETER, WITH TERMINALS	OR 
MACHINES (ROTATING)	
MACHINE OR ROTATING ARMATURE	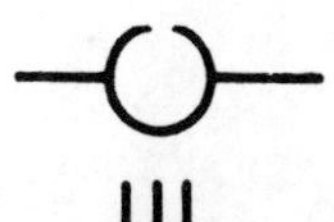
SQUIRREL-CAGE INDUCTION MOTOR	
WOUND-ROTOR INDUCTION MOTOR OR GENERATOR	

NAME	SYMBOL
SYNCHRONOUS MOTOR, GENERATOR OR CONDENSER	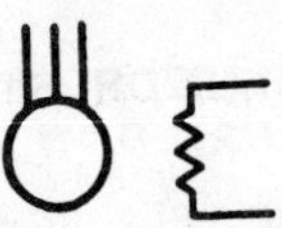
D-C COMPOUND MOTOR OR GENERATOR NOTE: COMMUTATING, SERIES, AND SHUNT FIELDS MAY BE INDICATED BY 1, 2 AND 3 ZIGZAGS RESPECTIVELY. SERIES AND SHUNT COILS MAY BE INDICATED BY HEAVY AND LIGHT LINES OR 1 AND 2 ZIGZAGS RESPECTIVELY.	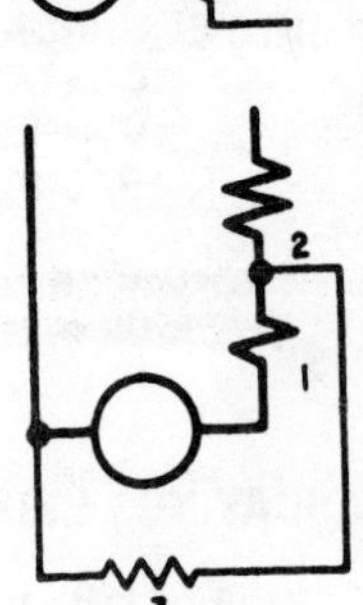

WINDING SYMBOLS

NAME	SYMBOL
THREE PHASE WYE (UNGROUNDED)	
THREE PHASE WYE (GROUNDED)	
THREE PHASE DELTA NOTE: WINDING SYMBOLS MAY BE SHOWN IN CIRCLES FOR ALL MOTOR AND GENERATOR SYMBOLS.	

NAME	SYMBOL
RECTIFIER, DRY OR ELECTROLYTIC, FULL WAVE	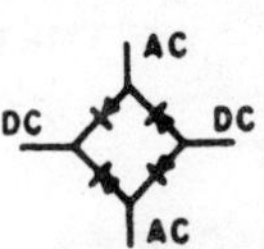

RELAYS

NAME	SYMBOL
OVERCURRENT OR OVERVOLTAGE RELAY WITH 1 *NO* CONTACT	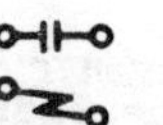 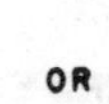 OR 
THERMAL OVERLOAD RELAY HAVING 2 SERIES HEATING ELEMENTS AND 1 *NC* CONTACT	 OR 

RESISTORS

RESISTOR, FIXED, WITH LEADS

RESISTOR, FIXED, WITH TERMINALS

RESISTOR, ADJUSTABLE TAP OR SLIDE WIRE

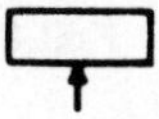

RESISTOR, ADJUSTABLE BY FIXED LEADS

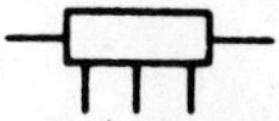

RESISTOR, ADJUSTABLE BY FIXED TERMINALS

INSTRUMENT OR RELAY SHUNT

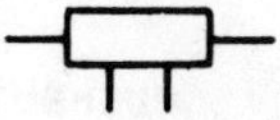

SWITCHES

KNIFE SWITCH, SINGLE-POLE (SP)

KNIFE SWITCH, DOUBLE-POLE SINGLE-THROW (DPST)

KNIFE SWITCH, TRIPLE-POLE SINGLE-THROW (TPST)

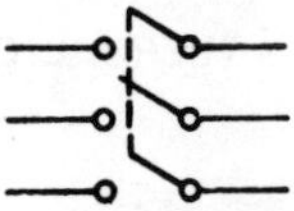

KNIFE SWITCH, SINGLE-POLE DOUBLE-THROW (SPDT)

KNIFE SWITCH, DOUBLE-POLE DOUBLE-THROW (DPDT)

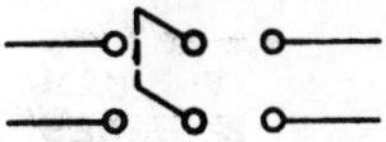

KNIFE SWITCH, TRIPLE-POLE DOUBLE-THROW (TPDT)

FIELD-DISCHARGE SWITCH WITH RESISTOR

PUSHBUTTON NORMALLY OPEN (NO)

PUSHBUTTON NORMALLY CLOSED (NC)

PUSHBUTTON OPEN AND CLOSED (SPRING-RETURN)

NORMALLY CLOSED LIMIT SWITCH CONTACT

LS

NORMALLY OPEN LIMIT SWITCH CONTACT

LS

THERMAL ELEMENT

TRANSFORMERS

I PHASE TWO-WINDING TRANSFORMER

AUTOTRANSFORMER SINGLE-PHASE

Appendix VII:

Current-Carrying Capacities of Conductors

TABLE 31

CURRENT-CARRYING CAPACITIES OF CONDUCTORS
(BASED ON THE NATIONAL ELECTRICAL CODE, 1951)

Size AWG or Thousand Cir. Mils	Current Capacity in Amperes		
	Rubber or Thermo-plastic Insulation	Paper or Varnished Cambric Insulation	Impregnated Asbestos Insulation
14	15	25	30
12	20	30	40
10	30	40	50
8	40	50	65
6	55	70	85
4	70	90	115
3	80	105	130
2	95	120	145
1	110	140	170
0	125	155	200
00	145	185	230
000	165	210	265
0000	195	235	310
250	215	270	335
300	240	300	380
350	260	325	420
400	280	360	450
500	320	405	500
600	355	455	545
700	385	490	600
750	400	500	620
800	410	515	640
900	435	555	...
1,000	455	585	730
1,250	495	645	...
1,500	520	700	...
1,750	545	735	...
2,000	560	775	...

Notes:
1. The ratings are for not more than three conductors in a cable or raceway, with a room temperature of 30° C or 86° F.
2. The above tables are for copper wires. The ratings for aluminum wires are 84 percent of these values.
3. Consult the National Electrical Code for more information. For example, higher ratings are allowed for single conductors in free air.

Appendix VIII:

Formulas

Ohm's Law for D-C Circuits

$$I = \frac{E}{R} = \frac{P}{E} = \sqrt{\frac{P}{R}}$$

$$R = \frac{E}{I} = \frac{P}{I^2} = \frac{E^2}{P}$$

$$E = IR = \frac{P}{I} = \sqrt{PR}$$

$$P = EI = \frac{E^2}{R} = I^2R$$

Resistors in Series

$$R_T = R_1 + R_2 + \ldots$$

Resistors in Parallel

Two resistors

$$R_T = \frac{R_1R_2}{R_1 + R_2}$$

More than two

$$\frac{1}{R_T} = \frac{1}{R_1} + \frac{1}{R_2} + \frac{1}{R_3} + \ldots$$

RL Circuit Time Constant

$$\frac{L \text{ (in henrys)}}{R \text{ (in ohms)}} = t \text{ (in seconds), or}$$

$$\frac{L \text{ (in microhenrys)}}{R \text{ (in ohms)}} = t \text{ (in microseconds)}$$

RC Circuit Time Constant

R (ohms) x C (farads) = t (seconds)

R (megohms) x C (microfarads) = t (seconds)

R (ohms) x C (microfarads) = t (microseconds)

R (megohms) x C (micromicrofarads) = t (microseconds)

Capacitors in Series

Two capacitors

$$C_T = \frac{C_1 C_2}{C_1 + C_2}$$

More than two

$$\frac{1}{C_T} = \frac{1}{C_1} + \frac{1}{C_2} + \frac{1}{C_3} + \ldots$$

Capacitors in Parallel: $C_T = C_1 + C_2 + \ldots$

Capacitive Reactance: $X_C = \frac{1}{2\pi fC}$

Impedance in an RC Circuit (Series)

$$Z = \sqrt{R^2 + (X_C)^2}$$

Inductors in Series

$L_T = L_1 + L_2 + \ldots$ **(No coupling between coils)**

Inductors in Parallel

Two inductors

$L_T = \frac{L_1 L_2}{L_1 + L_2}$ **(No coupling between coils)**

More than two

$$\frac{1}{L_T} = \frac{1}{L_1} + \frac{1}{L_2} + \frac{1}{L_3} + \ldots$$ (No coupling between coils)

Inductive Reactance

$$X_L = 2\pi fL$$

Q of a Coil

$$Q = \frac{X_L}{R}$$

Impedance of an RL Circuit (Series)

$$Z = \sqrt{R^2 + (X_L)^2}$$

Impedance with R, C, and L in Series

$$Z = \sqrt{R^2 + (X_L - X_C)^2}$$

Parallel Circuit Impedance

$$Z = \frac{Z_1 Z_2}{Z_1 + Z_2}$$

Sine-Wave Voltage Relationships

Average value

$$E_{ave} = \frac{2}{\pi} \times E_{max} = 0.637E_{max}$$

Effective or r.m.s. value

$$E_{eff} = \frac{E_{max}}{\sqrt{2}} = \frac{E_{max}}{1.414} = 0.707E_{max}$$

$$= 1.11E_{ave}$$

Maximum value

$$E_{max} = \sqrt{2}(E_{eff}) = 1.414E_{eff}$$
$$= 1.57E_{ave}$$

Voltage in an a-c circuit

$$E = IZ = \frac{P}{I \times P.F.}$$

Current in an a-c circuit

$$I = \frac{E}{Z} = \frac{P}{E \times P.F.}$$

Power in A-C Circuit

Apparent power: $P = EI$

True power: $P = EI \cos \theta = EI \times P.F.$

Power Factor

$$P.F. = \frac{P}{EI} = \cos \theta$$

$$\cos \theta = \frac{\text{true power}}{\text{apparent power}}$$

Transformers

Voltage relationship

$$\frac{E_p}{E_s} = \frac{N_p}{N_s} \text{ or } E_s = E_p \times \frac{N_s}{N_p}$$

Current relationship

$$\frac{I_p}{I_s} = \frac{N_s}{N_p}$$

Induced voltage

$$E_{eff} = 4.44 \times BAfN \times 10^{-8}$$

Turns ratio

$$\frac{N_p}{N_s} = \sqrt{\frac{Z_p}{Z_s}}$$

Secondary current

$$I_s = I_p \times \frac{N_p}{N_s}$$

Secondary voltage

$$E_s = E_p \times \frac{N_s}{N_p}$$

Three-Phase Voltage and Current Relationships

With wye connected windings

$$E_{line} = \sqrt{3}(E_{coil}) = 1.732E_{coil}$$

$$I_{line} = I_{coil}$$

With delta connected windings

$$E_{line} = E_{coil}$$

$$I_{line} = 1.732I_{coil}$$

With wye or delta connected winding

$$P_{coil} = E_{coil}I_{coil}$$

$$P_t = 3P_{coil}$$

$$P_t = 1.732E_{line}I_{line}$$

(To convert to true power multiply by cos θ)

Synchronous Speed of Motor

$$\text{r.p.m.} = \frac{120 \times \text{frequency}}{\text{number of poles}}$$

Comparison of Units in Electric and Magnetic Circuits

	Electric circuit	Magnetic circuit
Force.	Volt, E, or e. m. f.	Gilberts, F, or m. m. f.
Flow	Ampere, I	Flux, Φ, in maxwells
Opposition.	Ohms, R	Reluctance, $\mathcal{R}$
Law.	Ohm's law, $I = \frac{E}{R}$	Rowland's law, $\Phi = \frac{F}{\mathcal{R}}$
Intensity of force	Volts per cm. of length.	$H = \frac{1.257IN}{L}$, gilberts per centimeter of length.
Density.	Current density—for example, amperes per cm^2.	Flux density—for example, lines per $cm.^2$, or gausses.

Appendix IX:

Laws of Exponents

The International Symbols Committee has adopted prefixes for denoting decimal multiples of units. The National Bureau of Standards has followed the recommendations of this committee, and has adopted the following list of prefixes:

Numbers	Powers of ten	Prefixes	Symbols
1, 000, 000, 000, 000	10^{12}	tera	T
1, 000, 000, 000	10^{9}	giga	G
1, 000, 000	10^{6}	mega	M
1, 000	10^{3}	kilo	k
100	10^{2}	hecto	h
10	10	deka	da
.1	10^{-1}	deci	d
.01	10^{-2}	centi	c
.001	10^{-3}	milli	m
.000001	10^{-6}	micro	u
.000000001	10^{-9}	nano	n
.000000000001	10^{-12}	pico	p
.000000000000001	10^{-15}	femto	f
.000000000000000001	10^{-18}	atto	a

To multiply like (with same base) exponential quantities, add the exponents. In the language of alegebra the rule is $a^m \times a^n = a^{m+n}$

$$10^4 \times 10^2 = 10^{4+2} = 10^6$$

$$0.003 \times 825.2 = 3 \times 10^{-3} \times 8.252 \times 10^2$$

$$= 24.756 \times 10^{-1} = 2.4756$$

To divide exponential quantities, subtract the exponents. In the language of algebra the rule is

$$\frac{a^m}{a^n} = a^{m-n} \text{ or}$$

$$10^8 \div 10^2 = 10^6$$

$$3,000 \div 0.015 = (3 \times 10^3) \div (1.5 \times 10^{-2})$$

$$= 2 \times 10^5 = 200,000$$

To raise an exponential quantity to a power, multiply the exponents. In the languague of algebra $(x^m)^n = x^{mn}$.

$$(10^3)^4 = 10^{3\times 4} = 10^{12}$$

$$2,500^2 = (2.5 \times 10^3)^2 = 6.25 \times 10^6 = 6,250,000$$

Any number (except zero) raised to the zero power is one. In the language of algebra $x^0=1$

$$x^3 \div x^3 = 1$$

$$10^4 \div 10^4 = 1$$

Any base with a negative exponent is equal to 1 divided by the base with an equal positive exponent. In the language of algebra $x^{-a}=\frac{1}{x^a}$

$$10^{-2} = \frac{1}{10^2} = \frac{1}{100}$$

$$5a^{-3} = \frac{5}{a^3}$$

$$(6a)^{-1} = \frac{1}{6a}$$

To raise a product to a power, raise each factor of the product to that power.

$$(2 \times 10)^2 = 2^2 \times 10^2$$

$$3,000^3 = (3 \times 10^3)^3 = 27 \times 10^9$$

To find the nth root of an exponential quantity, divide the exponent by the index of the root. Thus, the nth root of $a^m = a^{m/n}$.

$$\sqrt{x^6} = x^{6/2} = x^3$$

$$\sqrt[3]{64 \times 10^3} = 4 \times 10 = 40$$

Appendix X:

Squares and Square Roots

TABLE 32

SQUARES AND SQUARE ROOTS

N	N^2	$\sqrt{N}$	N	N^2	$\sqrt{N}$	N	N^2	$\sqrt{N}$
1	1	1.000	41	1681	6.4031	81	6561	9.0000
2	4	1.414	42	1764	6.4807	82	6724	9.0554
3	9	1.732	43	1849	6.5574	83	6889	9.1104
4	16	2.000	44	1936	6.6332	84	7056	9.1652
5	25	2.236	45	2025	6.7082	85	7225	9.2195
6	36	2.449	46	2116	6.7823	86	7396	9.2736
7	49	2.646	47	2209	6.8557	87	7569	9.3274
8	64	2.828	48	2304	6.9282	88	7744	9.3808
9	81	3.000	49	2401	7.0000	89	7921	9.4340
10	100	3.162	50	2500	7.0711	90	8100	9.4868
11	121	3.3166	51	2601	7.1414	91	8281	9.5394
12	144	3.4641	52	2704	7.2111	92	8464	9.5917
13	169	3.6056	53	2809	7.2801	93	8649	9.6437
14	196	3.7417	54	2916	7.3485	94	8836	9.6954
15	225	3.8730	55	3025	7.4162	95	9025	9.7468
16	256	4.0000	56	3136	7.4833	96	9216	9.7980
17	289	4.1231	57	3249	7.5498	97	9409	9.8489
18	324	4.2426	58	3364	7.6158	98	9604	9.8995
19	361	4.3589	59	3481	7.6811	99	9801	9.9499
20	400	4.4721	60	3600	7.7460	100	10000	10.0000
21	441	4.5826	61	3721	7.8102	101	10201	10.0499
22	484	4.6904	62	3844	7.8740	102	10404	10.0995
23	529	4.7958	63	3969	7.9373	103	10609	10.1489
24	576	4.8990	64	4096	8.0000	104	10816	10.1980
25	625	5.0000	65	4225	8.0623	105	11025	10.2470
26	676	5.0990	66	4356	8.1240	106	11236	10.2956
27	729	5.1962	67	4489	8.1854	107	11449	10.3441
28	784	5.2915	68	4624	8.2462	108	11664	10.3923
29	841	5.3852	69	4761	8.3066	109	11881	10.4403
30	900	5.4772	70	4900	8.3666	110	12100	10.4881
31	961	5.5678	71	5041	8.4261	111	12321	10.5357
32	1024	5.6569	72	5184	8.4853	112	12544	10.5830
33	1089	5.7447	73	5329	8.5440	113	12769	10.6301
34	1156	5.8310	74	5476	8.6023	114	12996	10.6771
35	1225	5.9161	75	5625	8.6603	115	13225	10.7238
36	1296	6.0000	76	5776	8.7178	116	13456	10.7703
37	1369	6.0828	77	5929	8.7750	117	13689	10.8167
38	1444	6.1644	78	6084	8.8318	118	13924	10.8628
39	1521	6.2450	79	6241	8.8882	119	14161	10.9087
40	1600	6.3246	80	6400	8.9443	120	14400	10.9545

For numbers up to 120. For larger numbers divide into factors smaller than 120.

Appendix XI:

Wire and Motor Data

TABLE 33

PROPERTIES OF COPPER CONDUCTORS

Size A W G	Area Cir Mils	Concentric lay stranded conductors		Bare conductors		D-C resistance ohms/M ft at 25° C. 77° F.	
		No. wires	Diameter each wire, inches	Diameter inches	Area,* square inches	Bare conductor	Tinned conductor
18	1624	Solid	0.0403	0.0403	0.0013	6.510	6.77
16	2583	Solid	.0508	.0508	.0020	4.094	4.25
14	4107	Solid	.0641	.0641	.0032	2.575	2.68
12	6530	Solid	.0808	.0808	.0051	1.619	1.69
10	10380	Solid	.1019	.1019	.0081	1.018	1.06
8	16510	Solid	.1285	.1285	.0130	.641	.660
6	26250	7	.0612	.184	.027	.410	.426
4	41740	7	.0772	.232	.042	.259	.269
3	52640	7	.0867	.260	.053	.205	.213
2	66370	7	.0974	.292	.067	.162	.169
1	83690	19	.0664	.332	.087	.129	.134
0	105500	19	.0745	.373	.109	.102	.106
00	133100	19	.0837	.418	.137	.0811	.0844
000	167800	19	.0940	.470	.173	.0642	.0668
0000	211600	19	.1055	.528	.219	.0509	.0524
2500	250000	37	.0822	.575	.260	.0431	.0444
	300000	37	.0900	.630	.312	.0360	.0371
	350000	37	.0973	.681	.364	.0308	.0318
	400000	37	.1040	.728	.416	.0270	.0278
	500000	37	.1162	.814	.520	.0216	.0225
	600000	61	.0992	.893	.626	.0180	.0185
	700000	61	.1071	.964	.730	.0154	.0159
	750000	61	.1109	.998	.782	.0144	.0148
	800000	61	.1145	1.031	.835	.0135	.0139
	900000	61	.1215	1.093	.938	.0120	.0124
	1000000	61	.1280	1.152	1.042	.0108	.0111
	1250000	91	.1172	1.289	1.305	.00864	.00890
	1500000	91	.1284	1.412	1.566	.00719	.00740
	1750000	127	.1174	1.526	1.829	.00617	.00636
	2000000	127	.1255	1.631	2.089	.00539	.00555

* Area given is that of a circle having a diameter equal to the overall diameter of a stranded conductor.

Other values given in the table are those given in Circular 31 of the National Bureau of Standards, except that those shown in the last column are those given in specifications B33 of the American Society for Testing Materials.

The resistance values given in the last two columns are applicable only to direct current. When conductors larger than No. 4/0 are used with alternating current, the multiplying factors in Fig. 2 should be used to compensate for skin effect.

TABLE 34

MULTIPLYING FACTORS FOR CONVERTING DC RESISTANCE TO AC RESISTANCE

Size C M	Multiplying factor	
	25 Hertz	60 Hertz
250000	--------	1.005
300000	--------	1.006
350000	--------	1.009
400000	--------	1.011
500000	--------	1.018
600000	1.005	1.025
700000	1.006	1.034
750000	1.007	1.039
800000	1.008	1.044
900000	1.010	1.055
1000000	1.012	1.067
1250000	1.019	1.102
1500000	1.027	1.142
1750000	1.037	1.185
2000000	1.048	1.233

TABLE 35

CONDUCTOR SIZES AND OVERCURRENT PROTECTION FOR MOTORS

(1)	(2)	(3)	(5)	(6)	(7) With code letters	(8) With code letters	(9) With code letters	(10) With code letters
					Maximum allowable rating or setting of branch circuit protective devices			
Full load current rating of motor in amperes.	Minimum-sized conductor in raceways. For conductors in air, or for other insulations, see tables B-4 and B-5. AWG and MCM		For running protection of motors.[a]		Single-phase and squirrel-cage and synchronous. Full voltage, resistor, or reactor starting. Code letters F to V inc.	Single-phase and squirrel-cage and synchronous. Full voltage, resistor, or reactor starting. Code letters B to E inc.	Squirrel-cage and synchronous. Auto-transformer starting. Code letters B to E inc.	All motors, Code letter A
	Rubber Type R Type T	Rubber Type RH	Maximum rating of non-adjustable protective devices. Amperes	Maximum setting of adjustable protective devices. Amperes	Without code letters	Without code letters	Without code letters	Without code letters
					Same as above.	Squirrel-cage and synchronous, auto-transformer starting. High reactance squirrel-cage.[b] Both not more than 30 amperes.	Squirrel-cage and synchronous, auto-transformer starting. High reactance squirrel-cage.[b] Both more than 30 amperes.	DC and wound-rotor motors.
1	14	14	2	1.25	15	15	15	15
2	14	14	3	2.50	15	15	15	15
3	14	14	4	3.75	15	15	15	15
4	14	14	6	5.00	15	15	15	15
5	14	14	8	6.25	15	15	15	15
6	14	14	8	7.50	20	15	15	15
7	14	14	10	8.75	25	20	15	15
8	14	14	10	10.00	25	20	20	15
9	14	14	12	11.25	30	25	20	15
10	14	14	15	12.50	30	25	20	15
11	14	14	15	13.75	35	30	25	20
12	14	14	15	15.00	40	30	25	20

TABLE 35 (*continued*)

(1)	(2)	(3)	(5)	(6)	Maximum allowable rating or setting of branch circuit protective devices			
					(7) With code letters	(8) With code letters	(9) With code letters	(10) With code letters
13	12	12	20	16.25	40	35	30	25
14	12	12	20	17.50	45	35	30	25
15	12	12	20	18.75	45	40	30	25
16	12	12	20	20.00	50	40	35	25
17	10	10	25	21.35	60	45	35	30
18	10	10	25	22.50	60	45	40	30
19	10	10	25	23.75	60	50	40	30
20	10	10	25	25.00	60	50	40	30
22	10	10	30	27.50	70	60	45	35
24	10	10	30	30.00	80	60	50	40
26	8	8	35	32.50	80	70	60	40
28	8	8	35	35.00	90	70	60	45
30	8	8	40	37.50	90	70	60	45
32	8	8	40	40.00	100	80	70	50
34	6	8	45	42.50	110	90	70	60
36	6	8	45	45.00	110	90	80	60
38	6	6	50	47.50	125	100	80	60
40	6	6	50	50.00	125	100	80	60
42	6	6	50	52.50	125	110	90	70
44	6	6	60	55.00	125	110	90	70
46	4	6	60	57.50	150	125	100	70
48	4	6	60	60.00	150	125	100	80
50	4	6	60	62.50	150	125	100	80
52	4	6	70	65.00	175	150	110	80
54	4	4	70	67.50	175	150	110	90
56	4	4	70	70.00	175	150	120	90
58	3	4	70	72.50	175	150	120	90
60	3	4	80	75.00	200	150	120	90
62	3	4	80	77.50	200	175	125	100
64	3	4	80	80.00	200	175	150	100
66	2	4	80	82.50	200	175	150	100
68	2	4	90	85.00	225	175	150	110
70	2	3	90	87.50	225	175	150	110
72	2	3	90	90.00	225	200	150	110
74	2	3	90	92.50	225	200	150	125
76	2	3	100	95.00	250	200	175	125
78	1	3	100	97.50	250	200	175	125
80	1	3	100	100.00	250	200	175	125
82	1	2	110	102.50	250	225	175	125
84	1	2	110	105.00	250	225	175	150
86	1	2	110	107.50	300	225	175	150
88	1	2	110	110.00	200	225	200	150
90	0	2	110	112.50	300	225	200	150
92	0	2	125	115.00	300	250	200	150
94	0	1	125	117.50	300	250	200	150
96	0	1	125	120.00	300	250	200	150
98	0	1	125	122.50	300	250	200	150
100	0	1	125	125.00	300	250	200	150
105	00	1	150	131.50	350	300	225	175
110	00	0	150	137.50	350	300	225	175
115	00	0	150	144.00	350	300	250	175
120	000	0	150	150.00	400	300	250	200
125	000	00	175	156.50	400	350	250	200

TABLE 35 (*continued*)

(1)	(2)	(3)	(5)	(6)	Maximum allowable rating or setting of branch circuit protective devices (7) With code letters	(8) With code letters	(9) With code letters	(10) With code letters
130	000	00	175	162.50	400	350	300	200
135	0000	00	175	169.00	450	350	300	225
140	0000	00	175	175.00	450	350	300	225
145	0000	000	200	181.50	450	400	300	225
150	0000	000	200	187.50	450	400	300	225
155	0000	000	200	194.00	500	400	350	250
160	250	000	200	200.00	500	400	350	250
165	250	0000	225	206.00	500	450	350	250
170	250	0000	225	213.00	500	450	350	300
175	300	0000	225	219.00	600	450	350	300
180	300	0000	225	225.00	600	450	400	300
185	300	0000	250	231.00	600	500	400	300
190	300	250	250	238.00	600	500	400	300
195	350	250	250	244.00	600	500	400	300
200	350	250	250	250.00	600	500	400	300
210	400	300	250	263.00	...	600	450	350
220	400	300	300	275.00	...	600	450	350
230	500	300	300	288.00	...	600	500	350
240	500	350	300	300.00	...	600	500	400
250	500	350	300	313.00	...	...	500	400
260	600	400	350	325.00	...	...	600	400
270	600	400	350	338.00	...	...	600	450
280	600	400	350	338.00	...	...	600	450
290	700	500	350	363.00	...	...	600	450
300	700	500	400	375.00	...	...	600	450
320	750	600	400	400.00	...	...	...	500
340	900	600	450	425.00	...	...	...	600
360	1000	700	450	450.00	...	...	...	600
380	1250	750	500	475.00	...	...	...	600
400	1500	900	500	500.00	...	...	...	600
420	1750	1000	600	525.00				
440	2000	1250	600	550.00				
460	...	1250	600	575.00				
480	...	1500	600	600.00				
500	...	1500	...	625.00				

[a] For running protection of motors, notify values in columns 5 and 6, if nameplate-full-load current values different than those shown in the table. Reduce current values shown in columns 5 and 6 by 8 percent for all motors other than open-type motors marked to have a temperature rise of over 40° C. (72° F.)

[b] High-reactance, squirrel-cage motors are those designed to limit the starting current by means of deep-slot secondaries or double-wound secondaries and are generally started on full voltage.

TABLE 36

ALLOWABLE CURRENT-CARRYING CAPACITY IN AMPERES OF INSULATED CONDUCTORS (NOT MORE THAN THREE) IN RACEWAYS OR CABLES OR DIRECT BURIAL

(1) Size AWG MCM	(2) Rubber Type R Type RW Type RU Type RUW (14–2) / Type RH–RW / Thermoplastic Type T Type TW	(3) Rubber Type RH / Type RH–RW / Type RHW	(4) Paper / thermoplastic Asbestos Type TA / Var-Cam Type V / Asbestos Var-Cam Type AVB / MI Cable	(5) Asbestos Var-Cam Type AVA Type AVL	(6) Impregnated Asbestos Type AI (14–8) Type AIA	(7) Asbestos Type A (14–8) Type AA
14	15	15	25	30	30	30
12	20	20	30	35	40	40
10	30	30	40	45	50	55
8	40	45	50	60	65	70
6	55	65	70	80	85	95
4	70	85	90	105	115	120
3	80	100	105	120	130	145
2	95	115	120	135	145	165
1	110	130	140	160	170	190
0	125	150	155	190	200	225
00	145	175	185	215	230	250
000	165	200	210	245	265	285
0000	195	230	235	275	310	340
250	215	255	270	315	335	
300	240	285	300	345	380	
350	260	310	325	390	420	
400	280	335	360	420	450	
500	320	380	405	470	500	
600	355	420	455	525	545	
700	385	460	490	560	600	
750	400	475	500	580	620	
800	410	490	515	600	640	
900	435	520	555			
1,000	455	545	585	680	730	
1,250	495	590	645			
1,500	520	625	700	785		
1,750	545	650	735			
2,000	560	665	775	840		

These values are in accordance with the National Electrical Code.

TABLE 37

ALLOWABLE CURRENT-CARRYING CAPACITY IN AMPERES OF SINGLE, INSULATED CONDUCTOR IN FREE AIR

(1) Size AWG MCM	(2) Rubber Type R Type RW Type RU Type RUW (14-2) Type RH-RW Thermoplastic Type T Type TW	(3) Rubber Type RH Type RH-RW Type RHW	(4) Thermoplastic Asbestos Type TA Var-Cam Type V Asbestos Var-Cam Type AVB MI Cable	(5) Asbestos Var-Cam Type AVA Type AVL	(6) Impregnated Asbestos Type AI (14-8) Type AIA	(7) Asbestos Type A (14-8) Type AA	(8) Slow- burning Type SB Weather- proof Type WP Type SBW
14	20	20	30	40	40	45	30
12	25	25	40	50	50	55	40
10	40	40	55	65	70	75	55
8	55	65	70	85	90	100	70
6	80	95	100	120	125	135	100
4	105	125	135	160	170	180	130
3	120	145	155	180	195	210	150
2	140	170	180	210	225	240	175
1	165	195	210	245	265	280	205
0	195	230	245	285	305	325	235
00	225	265	285	330	355	370	275
000	260	310	330	385	410	430	320
0000	300	360	385	445	475	510	370
250	340	405	425	495	539	----------	410
300	375	445	480	555	590	----------	460
350	420	505	530	610	655	----------	510
400	455	545	575	665	710	----------	555
500	515	620	660	765	815	----------	630
600	575	690	740	855	910	----------	710
700	630	755	815	940	1005	----------	780
750	655	785	845	980	1045	----------	810
800	680	815	880	1020	1085	----------	845
900	730	870	940	----------	----------	----------	905
1000	780	935	1000	1165	1240	----------	965
1250	890	1065	1130	----------	----------	----------	----------
1500	980	1175	1260	1450	----------	----------	1215
1750	1070	1280	1370	----------	----------	----------	----------
2000	1155	1385	1470	1715	----------	----------	1405

These values are in accordance with the National Electrical Code.

TABLE 38

FULL-LOAD CURRENT FOR DIRECT-CURRENT MOTORS*

HP	120V	240V
¼	2.9	1.5
⅓	3.6	1.8
½	5.2	2.6
¾	7.4	3.7
1	9.4	4.7
1½	13.2	6.6
2	17	8.5
3	25	12.2
5	40	20
7½	58	29
10	76	38
15		55
20		72
25		89
30		106
40		140
50		173
60		206
75		255
100		341
125		425
150		506
200		675

*These values of full-load current are average for all speeds, and are in accordance with the National Electrical Code.

TABLE 39

FULL-LOAD CURRENT FOR SINGLE-PHASE AC MOTORS

HP	115V	230V
1/6	4.4	2.2
1/4	5.8	2.9
1/3	7.2	3.6
1/2	9.8	4.9
3/4	13.8	6.9
1	16	8
1½	20	10
2	24	12
3	34	17
5	56	28
7½	80	40
10	100	50

[a] These values of full-load current are in accordance with the National Electrical Code, and are for motors running at speeds usual for belted motors and motors with normal torque characteristics. Motors built for especially low speeds or high torques may require more running current, in which case the nameplate current rating should be used.

[b] For full-load currents of 208- and 200-volt motors, increase corresponding 230-volt motor full-load current by 10 and 15 percent, respectively.

TABLE 40

FULL-LOAD CURRENT FOR THREE-PHASE AC MOTORS

	Induction type squirrel-cage and wound motor Amperes				Synchronous type unity power factor† Amperes				
HP	110v	220v [b]	440v	550v	2300v	220v [b]	440v	550v	230v
½	4.0	2.0	1.0	0.8	----	----	----	----	----
¾	5.6	2.8	1.4	1.1	----	----	----	----	----
1	7.0	3.5	1.8	1.4	----	----	----	----	----
1½	10.0	5.0	2.5	2.0	----	----	----	----	----
2	13.0	6.5	3.3	2.6	----	----	----	----	----
3	----	9.0	4.5	4.0	----	----	----	----	----
5	----	15.0	7.5	6.0	----	----	----	----	----
7½	----	22.0	11.0	9.0	----	----	----	----	----
10	----	27.0	14.0	11.0	----	----	----	----	----
15	----	40.0	20.0	16.0	----	----	----	----	----
20	----	52.0	26.0	21.0	----	----	----	----	----
25	----	64.0	32.0	26.0	7.0	54.0	27.0	22.0	5.4
30	----	78.0	39.0	31.0	8.5	65.0	33.0	26.0	6.5
40	----	104.0	52.0	41.0	10.5	86.0	43.0	35.0	8.0
50	----	125.0	63.0	50.0	13.0	108.0	54.0	44.0	10.0
60	----	150.0	75.0	60.0	16.0	128.0	64.0	51.0	12.0
75	----	185.0	93.0	74.0	19.0	161.0	81.0	65.0	15.0
100	----	246.0	123.0	98.0	25.0	211.0	106.0	85.0	20.0
125	----	310.0	155.0	124.0	31.0	264.0	132.0	106.0	25.0
150	----	360.0	180.0	144.0	37.0	----	158.0	127.0	30.0
200	----	480.0	240.0	192.0	48.0	----	210.0	168.0	40.0

[a] These values of full-load current are in accordance with the National Electrical Code, and are motors running at speeds usual for belted motors and motors with normal torque characteristics. Motors built for especially low speeds or high torques may require more running current, in which case, the nameplate current rating should be used.

[b] For full-load currents of 208- and 200-volt motors, increase the corresponding 220-volt motor full-load current by 6 and 10 percent, respectively.

†For 90 and 80 percent power factor the above figures should be multiplied by 1.1 and 1.25, respectively.

TABLE 41

TYPICAL INSULATING PAINTS AND VARNISHES

Item	Thinner (By volume)		Method of application	Recoating time	Drying time to handle	Remarks
	Spray (percent)	Brush (percent)				
Paint, synthetic (glyptol)	Thinner, enamel, synthetic. 0–5	0	Brush or spray.	4 hr	12 hr	Identification of electrical circuits and devices. (Mica V-ring bands, armature bands).
Varnish, insulating	Thinner, enamel, synthetic, according to manufacturer's instructions.		Dip, spray, or brush.			Impregnating electrical components. Baking may be required.
Varnish, moisture and fungus resistant.	Thinner, enamel, synthetic. 15	5	Dip, spray, or brush.	12 hr	8 hr	For moisture-proofing and fungi-proofing electrical circuits.

Match colors to original finish.

TABLE 42

TYPICAL INSULATING COMPOUNDS AND CEMENTS

Item	Shrinkage	Melting point	Resistance		Character	Drying or setting time	Properties	Application
			Moisture	Oil				
Cement							Liquid cement.	Sealing and cementing.
Cement	None	None	Good	Good	Hard	24 hours	Good insulating, not strong.	Filling back of commutators.
Cement	5 per ct	235°C	Good	Good	Hard cement	Sets upon cooling.	Oilproof	Filling ground lead in bushings.
Cement	None	None	Good	Good	Hard	3 to 4 hours at 200°C.	High heat resistance.	Repairing arc deflectors and chutes.
Cement	None	None	Excellent.	Excellent	Hard cement	1 hour at at 200°C.	High heat resistance.	Filling cement.
Cement	None	None	Excellent.	Excellent	Hard cement	24 hours	Heat resisting.	Filling pits in commutators.
Cement	None	None	High	Excellent	Hard and tough.	12 to 14 hours at 115°C.	Tough and flexible.	Smoothing over windings or filling joints.
Cement	None	None	Good	Excellent	Hard	Bake 12 hours at 115°.	Hard packing.	Filling around coils.
Compound	None	None	Good	Good	Hard	Bake 12 hours at 110°C.	Good insulating and filling compound.	Treating field coils.

Appendix XII:

Component Color Code

COLOR	1ST DIGIT	2ND DIGIT	MULTIPLIER	TOLERANCE (percent)
Black	0	0	1	
Brown	1	1	10	
Red	2	2	100	
Orange	3	3	1,000	
Yellow	4	4	10,000	
Green	5	5	100,000	
Blue	6	6	1,000,000	
Violet	7	7	10,000,000	
White	9	9	1,000,000,000	
Gold			.1	5
Silver			.01	10
No color				20

(A)

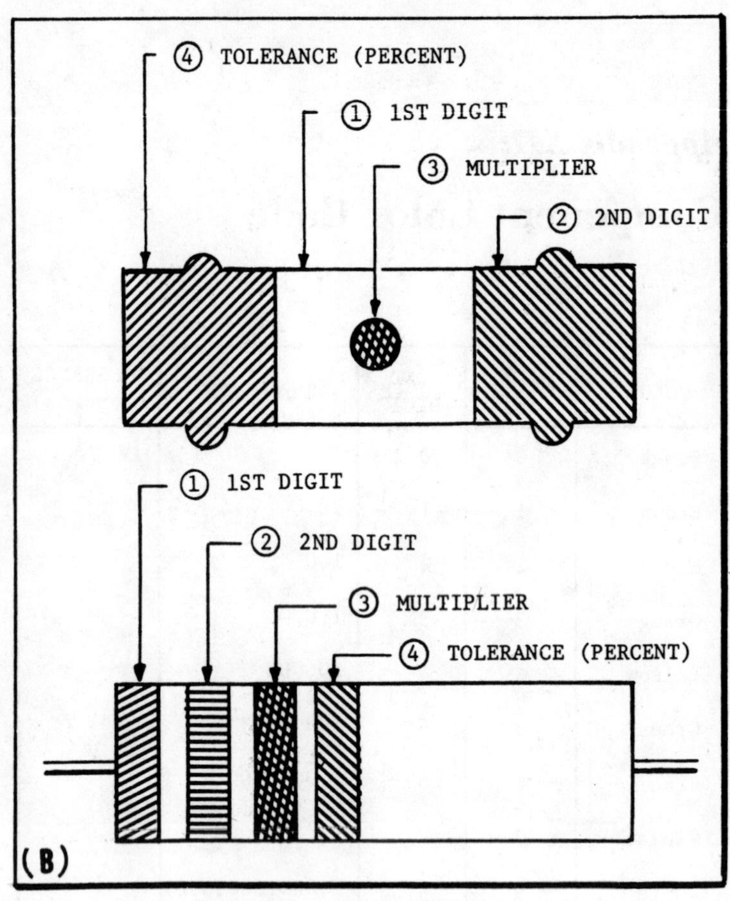

Resistor color code.

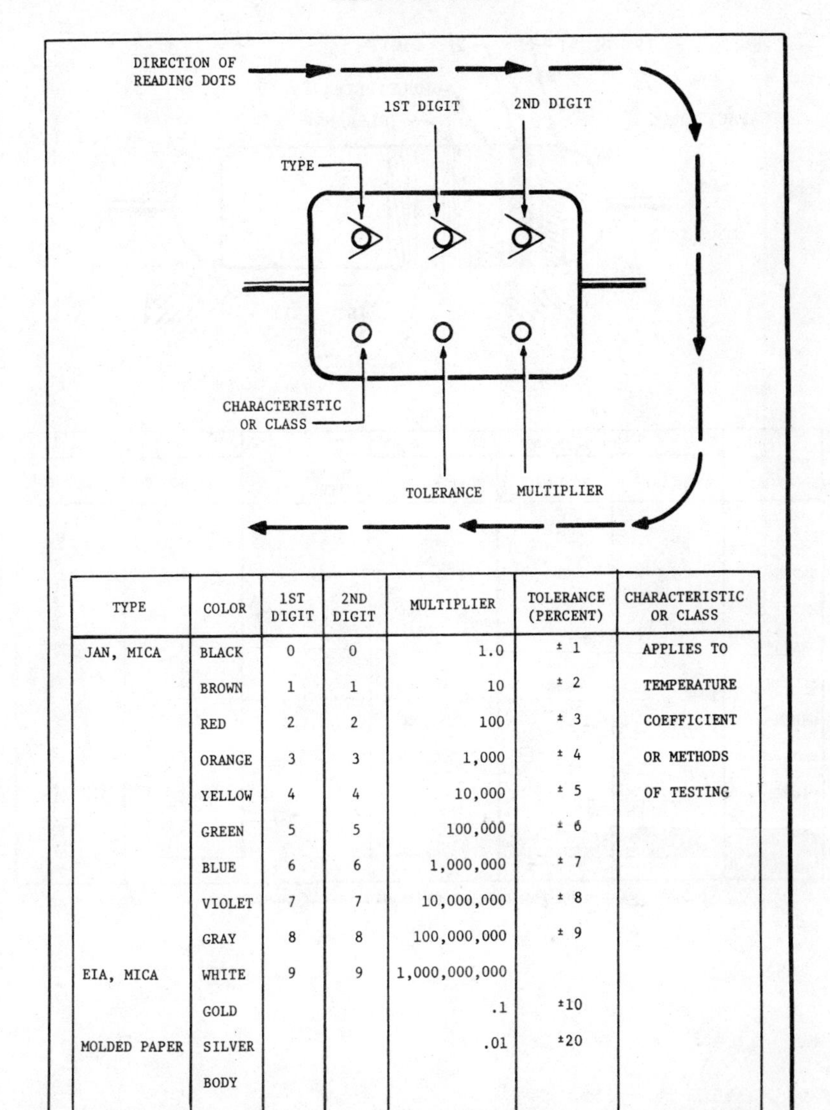

TYPE	COLOR	1ST DIGIT	2ND DIGIT	MULTIPLIER	TOLERANCE (PERCENT)	CHARACTERISTIC OR CLASS
JAN, MICA	BLACK	0	0	1.0	± 1	APPLIES TO
	BROWN	1	1	10	± 2	TEMPERATURE
	RED	2	2	100	± 3	COEFFICIENT
	ORANGE	3	3	1,000	± 4	OR METHODS
	YELLOW	4	4	10,000	± 5	OF TESTING
	GREEN	5	5	100,000	± 6	
	BLUE	6	6	1,000,000	± 7	
	VIOLET	7	7	10,000,000	± 8	
	GRAY	8	8	100,000,000	± 9	
EIA, MICA	WHITE	9	9	1,000,000,000		
	GOLD			.1	±10	
MOLDED PAPER	SILVER			.01	±20	
	BODY					

6-Dot color code for mica and molded paper capacitors.

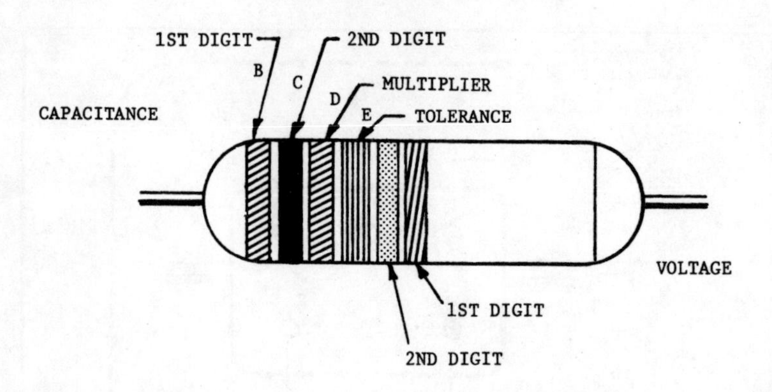

COLOR	CAPACITANCE			TOLERANCE (PERCENT)	VOLTAGE RATING	
	1ST DIGIT	2ND DIGIT	MULTIPLIER		1ST DIGIT	2ND DIGIT
BLACK	0	0	1	±20	0	0
BROWN	1	1	10		1	1
RED	2	2	100		2	2
ORANGE	3	3	1,000	±30	3	3
YELLOW	4	4	10,000	±40	4	4
GREEN	5	5	100,000	± 5	5	5
BLUE	6	6	1,000,000		6	6
VIOLET	7	7			7	7
GRAY	8	8			8	8
WHITE	9	9		±10	9	9

6-Band color code for tubular paper dielectric capacitors.

B - A - TEMPERATURE COEFFICIENT

B - 1ST DIGIT

C - 2ND DIGIT

D - MULTIPLIER

E - TOLERANCE

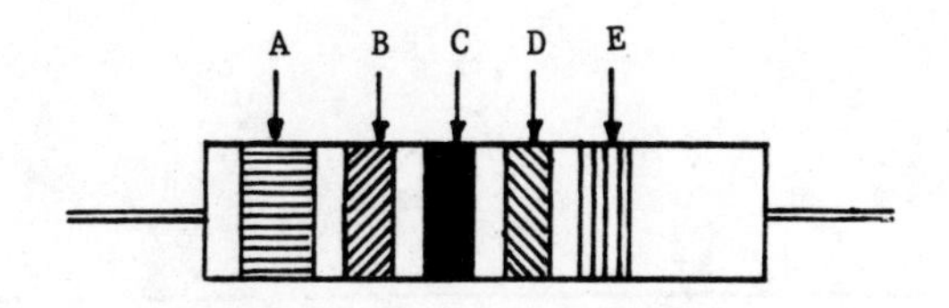

AXIAL LEAD CERAMIC

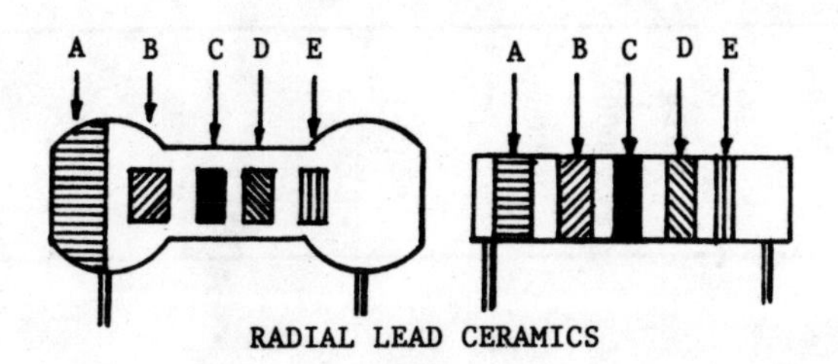

RADIAL LEAD CERAMICS

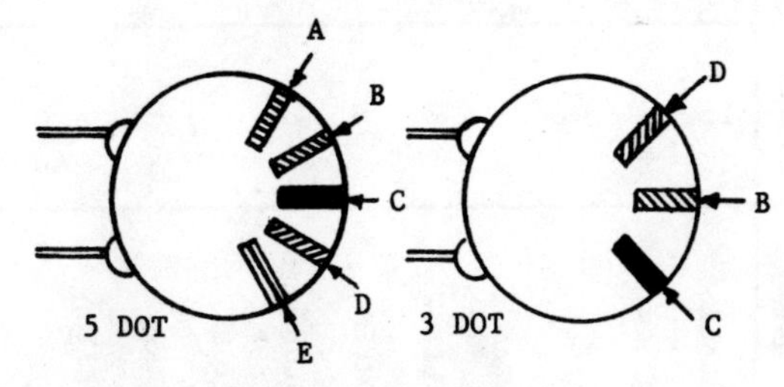

CERAMIC DISC CAPACITOR MARKING

COLOR	1ST DIGIT	2ND DIGIT	MULTIPLIER	TOLERANCE		TEMPERATURE COEFFICIENT*
				MORE THAN 10 pf (IN PERCENT)	LESS THAN 10 pf (IN pf)	
BLACK	0	0	1.0	±20	±2.0	0
BROWN	1	1	10	± 1		-30
RED	2	2	100	± 2		-80
ORANGE	3	3	1,000			-150
YELLOW	4	4	10,000			-220
GREEN	5	5		± 5	±0.5	-330
BLUE	6	6				-470
VIOLET	7	7				-750
GRAY	8	8	.01		±0.25	+30
WHITE	9	9	.1	±10	±1.0	+120 TO -750 (EIA)
						+500 TO -330 (JAN)
SILVER						+100 (JAN)
GOLD						BYPASS OR COUPLING (EIA)

*PARTS PER MILLION PER DEGREE CENTIGRADE.

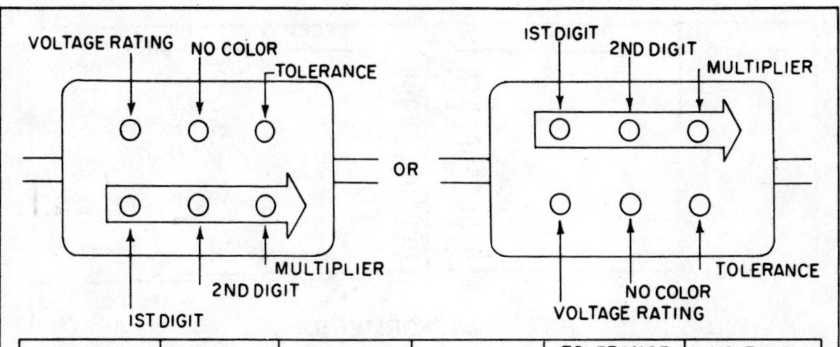

COLOR	1ST DIGIT	2ND DIGIT	MULTIPLIER	TOLERANCE (PERCENT)	VOLTAGE RATING
BLACK	0	0	1.0		
BROWN	1	1	10	±1	100
RED	2	2	100	±2	200
ORANGE	3	3	1,000	±3	300
YELLOW	4	4	10,000	±4	400
GREEN	5	5	100,000	±5	500
BLUE	6	6	1,000,000	±6	600
VIOLET	7	7	10,000,000	±7	700
GRAY	8	8	100,000,000	±8	800
WHITE	9	9	1,000,000,000	±9	900
GOLD			.1		1000
SILVER			.01	±10	2000
BODY				±20	*

* WHERE NO COLOR IS INDICATED, THE VOLTAGE RATING MAY BE AS LOW AS 300 VOLTS.

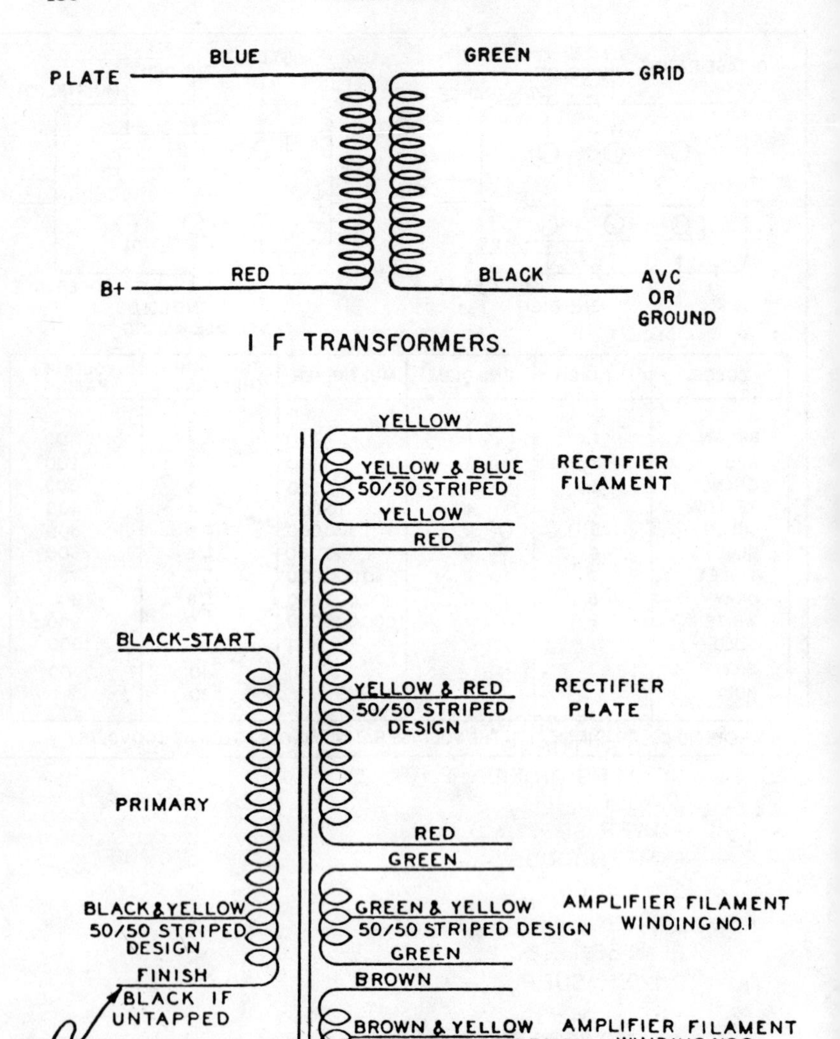

I F TRANSFORMERS.

POWER TRANSFORMERS.

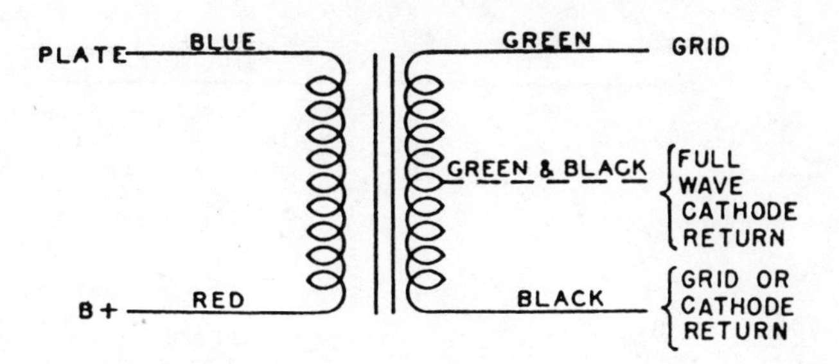

INTERSTAGE AUDIO TRANSFORMERS

STANDARD COLORS USED IN CHASSIS WIRING FOR THE PURPOSE OF CIRCUIT IDENTIFICATION OF THE EQUIPMENT ARE AS FOLLOWS:

CIRCUIT	COLOR
GROUNDS, GROUNDED ELEMENTS, AND RETURNS	BLACK.
HEATERS OR FILAMENTS, OFF GROUND	BROWN.
POWER SUPPLY B PLUS	RED.
SCREEN GRIDS	ORANGE.
CATHODES	YELLOW.
CONTROL GRIDS	GREEN.
PLATES	BLUE
POWER SUPPLY, MINUS	VIOLET (PURPLE).
A C POWER LINES	GRAY.
MISCELLANEOUS, ABOVE OR BELOW GROUND RETURNS, A V C, ETC	WHITE.

Index